Multiantenna Digital Radio Transmission

For a listing of recent titles in the *Artech House Mobile Communications Series*,
turn to the back of this book.

Multiantenna Digital Radio Transmission

Massimiliano "Max" Martone

Artech House
Boston • London
www.artechhouse.com

Library of Congress Cataloging-in-Publication Data
Martone, Massimiliano.
Multiantenna digital radio transmission / Massimiliano "Max" Martone.
p. cm. — (Artech House mobile communications series)
Includes bibliographical references and index.
ISBN 1-58053-318-3 (alk. paper)
1. Antenna arrays. 2. Radio—Transmitters and transmission. 3. Digital audio broadcasting. 4. Mobile communication systems. 5. Wireless communication systems. I. Title.

TK6565.A6 M37 2002
621.3845'6—dc21

2001053534

British Library Cataloguing in Publication Data
Martone, Massimiliano
Multiantenna digital radio transmission. — (Artech House mobile communications series)
1. Digital audio broadcasting 2. Antenna arrays
I. Title
621.3'845

ISBN 1-58053-318-3

Cover design by Igor Valdman

International Standard Book Number: 1-58053-318-3
Library of Congress Catalog Card Number: 2001053534

10 9 8 7 6 5 4 3 2 1

To my lovely wife Maria Luisa
and to my son Steven

Contents

Preface

A wireless communication system based on multiantenna technology can achieve impressively high data rates without impacting bandwidth usage, transmitted power, and the range of a system that employs traditional technology. Surprisingly, this is a recent discovery. Traditional wireless system engineering used diversity to improve reliability and performance of a wireless communication link, while the array processing community traditionally has been interested in target localization. The explosion of the worldwide demand for broadband data services has forced researchers to connect the two perspectives. At first *smart antennas* were considered advantageous in a cellular environment because they are able to suppress interference from users transmitting from different positions. Only several years after the realization of the advantages of smart antenna systems was it realized that the dimensionality of the wireless multiantenna channel allows digital communication at a much higher rate than predicted by conventional theory. The interest in this subject has grown to unexpected levels, and multiantenna (or spatio-temporal) transmission is now central to many technical events related to communication technology. Third-generation cellular systems will incorporate some forms of multiantenna technology. It was an obvious decision to start with a book that would treat the topic in a rather fundamental way. The first part of the book is really about the physical motivations that explain the capacity increase. This forces the study of the physics of wave propagation, which is usually neglected in communication theory books. From such a basic explanation the book covers several approaches to the demodulation of multiantenna signals and their experimental analysis, mostly the result of my research activity over the years. By no means does the book provide a comprehensive treatment. Thus a selective choice was made. As opposed to a superficial coverage of almost everything I have delved more deeply into physical layer issues particularly as they relate to baseband signal processing requirements. This leaves out an appropriate treatment of channel coding for the multiantenna channel (otherwise know as *space-time coding*) which would require by itself a separate book. To my fellow workers in the field who will find their favorite paper missing from the book, I would like to apologize.

Most of the material presented is intended for an engineering audience with a basic understanding of communication theory. The reader should have at least an undergraduate electrical engineering background. There are several excellent texts on statistical communication and information theory so I have not tried to provide all the fundamentals, but to produce a self-contained text: All of the significant results are derived either in the text or the appendixes to the chapter where they are first used.

Many have contributed to this book and it would be impractical to try to acknowledge all the help I have indirectly received through the publications of my colleagues. The energetic criticism of the many reviewers of my (over 100) technical papers still amazes me. WJ Communications, Inc. (formerly the Watkins-Johnson Company) deserves special mention for supporting multiantenna technology development. Many engineers at WJ were critical to my understanding of some concepts explained in this book. Relevant considerations from the many discussions with James Taylor have been included. John B. Anderson and Gary J. Saulnier were my mentors during my visiting period at Rensselaer Polytechnic Institute, New York, and I am indebted to them not only for supporting the early stages of my research, but also for being sources of inspiration. It is important to recognize Jack Winters as one of the first pioneers of multiantenna technology. Gerry Foschini, Reinaldo Valenzuela, and the entire BLAST group at Bell Labs have also greatly influenced many research directions in my work. From Stanford University the works by John Cioffi and coworkers, Arogyaswami Paulraj, and, in particular, Greg Raleigh, despite some animated technical discussions, have forced me to study and describe several practical implementation issues. In the statistical signal processing arena I have followed constantly Georgios Giannakis and William Gardner. It is also a pleasure to acknowledge the remarkable understanding I have gained being a student of Raymond Pickholtz and Branimir Vojcic during my stay at George Washington University. Atif Faheem contributed significantly to Chapter 5. I also extracted some descriptions from the excellent papers by Robert Watson and Charles Dexter. The Italian engineers from Alenia Spazio (now Alenia Aerospazio) have been important for my professional growth and ultimately for the content of this book. Finally, I would like to thank my family for their warm encouragement, patience, and continuing love throughout this project.

1

Introduction

Our eyes detect electromagnetic energy in the band 450×10^{12}–750×10^{12} Hz. The propagating radiation provides us with an immense amount of information allowing the most powerful computer, *our brain*, to react to the environment around us. Similarly to the human optical two-sensor array, a communication system, where radio frequency signals are simultaneously received and transmitted at different spatial positions, has a remarkable information transfer efficiency. This book, in fact, focuses on the description of multiantenna radio technology, an emerging area in communications that is attracting enormous interest. This chapter introduces the subject and discusses at a very high level radio equipment architectural implications. Moreover, it provides an outline of the book with some advice on how best to approach the study of the material.

1.1 Broadband Wireless Access

Broadband connectivity has been considered the future of the telecommunications industry for decades. It typically refers to those data services operating at data rates in excess of 1.544 Mbps (known as the T-1 rate) [1]. While the theoretical foundations of high data rate digital communications [2] are apparently well-known, there has not been any practical solution to implement the massive amount of signal processing required to operate reliably over the poor quality "last-mile" channel. The principal obstacle to mass deployment of broadband data services is without any doubt the trade-off between performance and cost. In particular the wireless terrestrial solution presents probably the most difficult technical challenges. Different users will have different needs: Terminals may be mobile (moving while communicating), or fixed (static while communicating); moving speeds can be as high as those of a train; more than one channel may be required for a single user; bandwidth can be statically or dynamically assigned; communication between terminals can be direct or through base stations; and asynchronous transfer mode may be required. As a consequence of these requirements, wireless access has become an extremely important sector of telecommunications. Today's second-generation (2G) mobile networks were designed and built primarily to handle voice traffic. Yet even as these networks are being built out, demand for wireless data services is driving third-generation (3G) air interface and network standards that can efficiently handle much higher-speed (384 Kbps to 2 Mbps) data to enable these services. Figure 1.1 summarizes the different generations of wireless communications, while Table 1.1 lists the spectrum, data rates, and ranges for indoor and outdoor wireless access systems. An attempt to characterize the main differences in mobility and data rates for currently envisioned wireless broadband access systems is reported in Figure 1.2. Even though many other cases could be listed, three major approaches are emerging: wireless local area networks (WLANs), mobile broadband cellular systems, and fixed wireless broadband systems (see Figure 1.3). Obviously the different requirements imposed by the various approaches have significant consequences on system design and development. The trade-off between flexibility and complexity and cost is always a difficult task to attack because it will impact not only the initial deployment, but also its future evolution. It is also extremely important to realize that not only market aspects are critical in the development and deployment of wireless broadband systems, but many technical challenges play an important role as well. As networks evolve, it is important for industry to provide platforms that allow service providers a cost-effective path to migrate toward 3G (and 4G) systems. Thus, new systems should

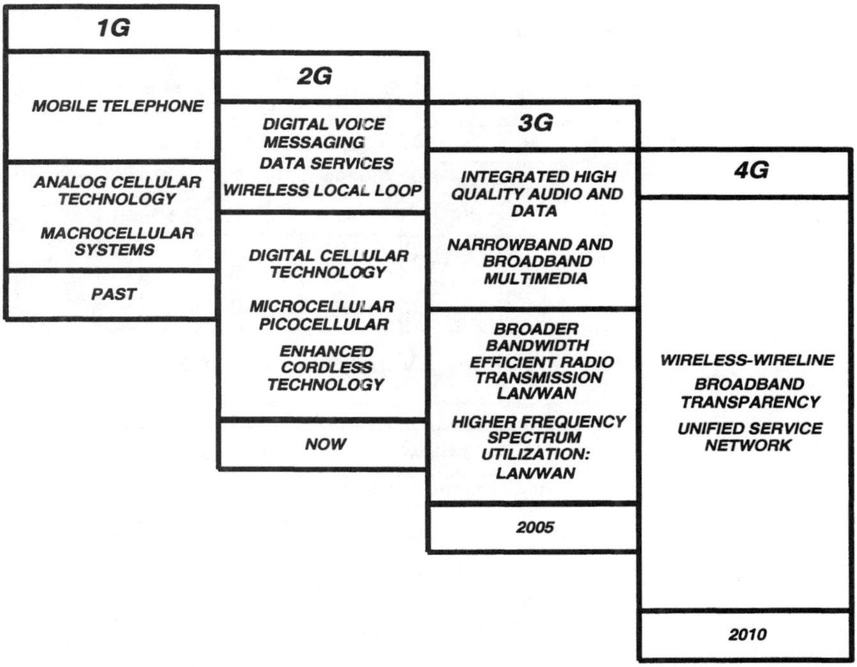

Figure 1.1 Wireless services evolution through different generations.

Table 1.1
Spectrum, Data Rates, and Range for Indoor and Outdoor Wireless Access

Parameter	***Indoor***	***Outdoor fixed***	***Outdoor mobile***
Spectrum	Licensed and unlicensed bands RF and infrared 5-150 MHz bandwidth	Licensed bands RF 5 MHz-1 GHz bandwidth	Licensed bands RF 5-30 MHz bandwidth
Data rate per user	1-2 bps/Hz 1-100 Mbps	0.5-2 bps/Hz 64 Kbps-2 Mbps	0.1-0.25 bps/Hz 10 Kbps-144 kbps
Number of users per cell	Tens	Hundreds	Tens
Cell range	60m	2 km	30 km

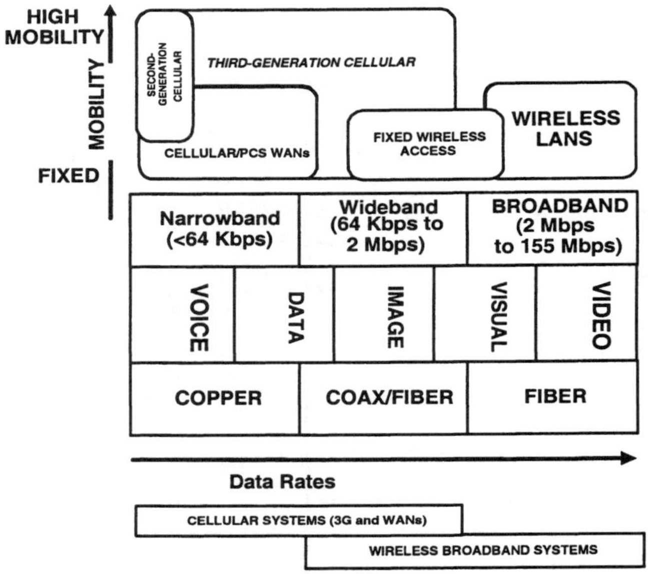

Figure 1.2 Comparison of mobility and data rates for several wireless communication systems.

be backward-compatible with currently deployed infrastructure equipment and should provide a gradual path to ever-increasing data rates based on market acceptance and demand. Further, air interfaces and network standards are still under consideration and can be expected to change several times before the systems are finally deployed. As a good example, the emergence of the fixed wireless market is driving requirements for radios to higher frequencies. The 2.4-GHz and 5.7-GHz bands are being used for point-to-multipoint data, while 3.5 GHz is becoming a de facto standard frequency for voice traffic, particularly internationally. The multipoint multichannel distribution service (MMDS) uses microwave transmission at a frequency approximately around 2.5 GHz. It is currently envisioned by the wireless communication industry for the MMDS system to broadcast video, voice, and data, allowing for interactive communications. MMDS offers a maximum of 33 analog video channels in about 500 MHz and a cell radius of approximately 25 to 30 miles. Given the evolution of digital systems, MMDS could provide two-way connectivity and transport to the

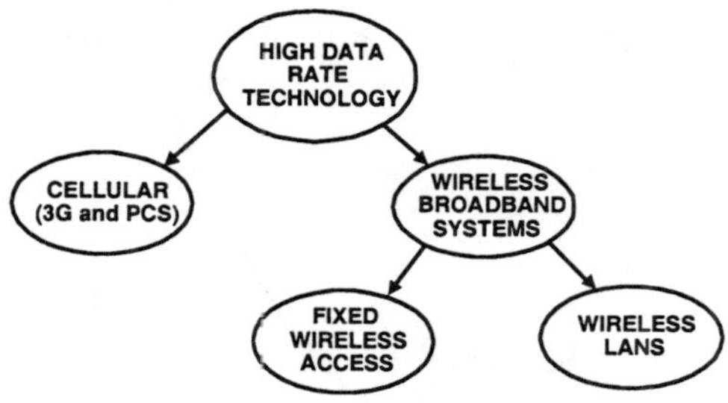

Figure 1.3 A general classification of the complex wireless access sector.

transmission nodes enabling ATM or SONET networks [1]. One of the drawbacks of the system is that MMDS relies on traditional high-capacity microwave technology, which requires line of sight (LOS) transmission and the development of high-performance radio frequency (RF) circuitry. This last factor has a dramatic impact on the cost of the customer premises equipment and appears to preclude the successful application of the idea to the residential market. In the United States there are several independent industry efforts that have attacked the problem of high data rate in non line of sight (NLOS) at MMDS frequencies. Members of the Broadband Wireless Internet Forum (BWIF) are committed to sponsoring the availability of interoperable solutions based on coded orthogonal frequency division multiplexing (OFDM) technology, inspired to the digital video broadcast (DVB) European standard [3]. Products are available now for point-to-point radio links where the data rate is up to about 25 Mbps in 6 MHz. The Wireless DSL Consortium is fundamentally competing for the same market. The companies that are part of the Wireless DSL Consortium share the goal of providing timely multivendor solutions for the 2 to 4 GHz point-to-multipoint broadband wireless access market. To accomplish that goal, the DSL consortium will pursue a standard based on a physical layer that is a multicarrier-based technology augmented by a system incorporating smart antennas to enable full NLOS operation, otherwise known as multiple-in multiple-out orthogonal frequency division multiplexing (MIMO-OFDM).

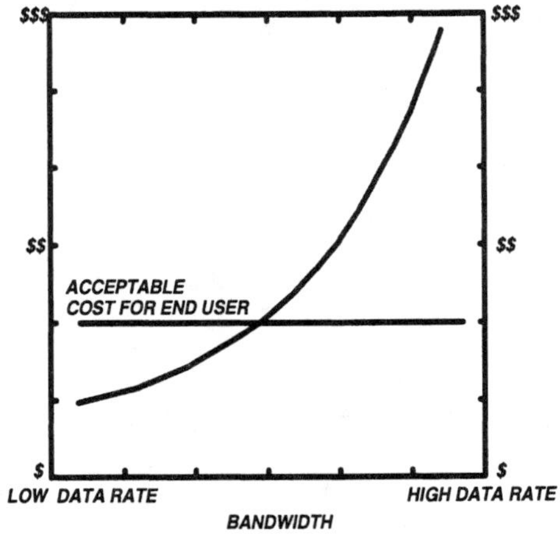

Figure 1.4 The current problem of the wireless communication industry. The data rates needed today intersect the cost of a wireless modem at a point that is significantly above the acceptable cost for the end user.

1.1.1 The Cost Factor

A wireless data modem for broadband data is inherently more expensive to manufacture than a "wired" modem with an equivalent data rate (see Figures 1.4 and 1.5). This is due to multiple reasons and very difficult to explain briefly. However, it is possible to identify a partial explanation of this for current modem technologies, namely, single-carrier and multicarrier modulation formats. Multicarrier systems perform a frequency domain decomposition of a channel characterized by frequency selective distortion in a multitude of subchannels that are affected by frequency-flat distortion. The distortion in each independent subchannel can then be compensated by simple gain and phase adjustments, somehow overcoming the need for one of the most complex components of the baseband section of a digital modem: the equalizer. This is a relatively old idea, which attracted a significant interest in recent years. However, an expensive implementation of the analog RF front end is extremely critical to the theoretical behavior of multicarrier transmission. Multicarrier signals, in fact, exhibit extreme vulnerability to distortions induced by RF electronics. So, while able to cope at baseband with

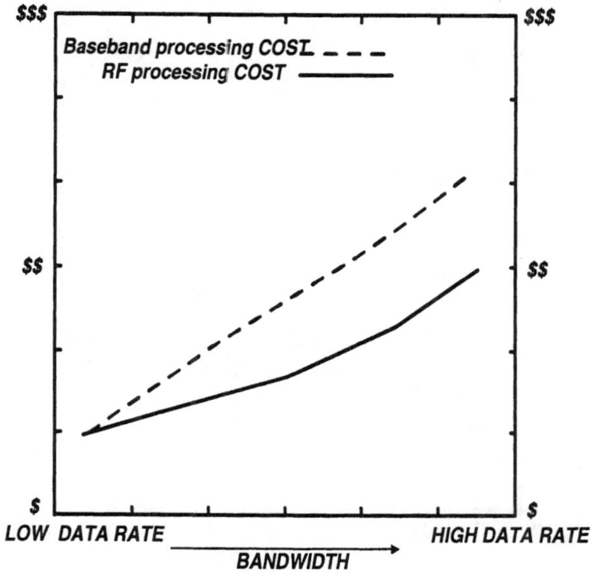

Figure 1.5 Breaking down the cost of a wireless modem in terms of baseband and RF processing.

channel time dispersions quite effectively (in terms of computational complexity), multicarrier techniques require an increased cost at the RF subsystem level. This is opposite of single-carrier modulations with traditional time-domain equalization. In other words, single-carrier schemes appear to be more demanding in terms of baseband processing resources but the RF front end requires significantly less stringent specifications (hence a reduced cost).

The basic concept we explore in this book is the following: Spatio-temporal modems can operate at rates well below their theoretical limit, while still providing the required throughput. A traditional modem is forced into a close-to-theoretical operating region, because only points close to the theoretical limit are able to satisfy the data rate requirement. This causes the intolerable cost increase of wireless modems based on traditional technologies. We can envision a day when industry will be able to deploy multiantenna (spatio-temporal) modems whose cost behaves as in Figure 1.6.

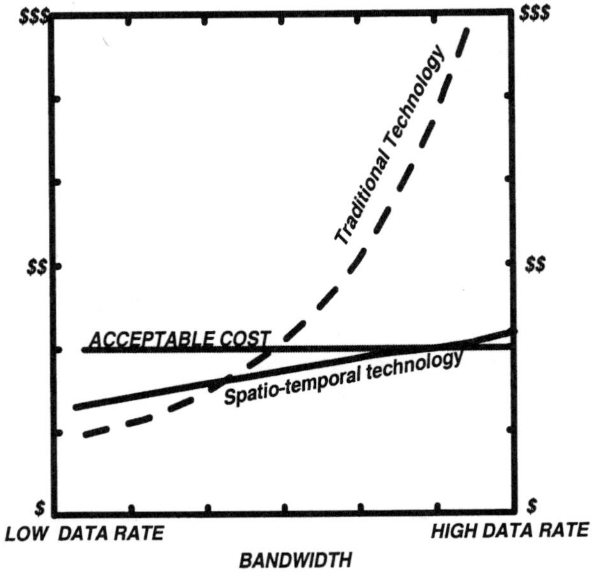

Figure 1.6 Multiantenna, spatio-temporal technology can solve the cost problem in the wireless communication industry.

1.2 Does Radio Frequency Scattering Help?

Traditional communication engineering states that an increase in data rate is achieved by means of an increase in bandwidth occupation and/or an increase in radiated power at the transmitter. A well-known upper bound on the maximum achievable data rate for the ideal band-limited additive white Gaussian noise (AWGN) channel is the Shannon-Nyquist [4, 5] criterion: Having an available channel bandwidth W (in Hertz) and signal-to-noise ratio over this bandwidth equal to SNR (in linear scale), it is impossible to transmit reliably data at a rate higher than $C = W\log_2(1 + \mathrm{SNR})$ (in bits per second). In other words, the maximum spectral efficiency is bounded at any signal-to-noise ratio by

$$\log_2(1 + \mathrm{SNR})$$

in bits per second per Hertz. The fundamental point in this famous formula is that the number of dimensions per unit-time that can be accommodated by an approximately W-band-limited channel is bounded by $2W$ at any SNR. Since in a sense (under the same bandwidth and power constraints) the AWGN is the most "benign" channel one can possibly expect in a real-life application, it is evident that the $\log_2(1 + \text{SNR})$ limit is in fact a hard limit for any communication system when signaling is limited to W Hz. Exploitation of the spatial wavefield associated with any type of signaling in presence of radio-frequency scattering, caused by multipath propagation, provides a dramatic increase of the achievable data throughput, well above the $\log_2(1 + \text{SNR})$ hard limit. Remarkably, without any increase in bandwidth occupation and signal power we can build digital radio systems able to transfer data with a spectral efficiency approaching $N\log_2(1+\text{SNR})$ bps/Hz, where N is an integer.

1.2.1 What Is a Wavefield?

A wavefield is a signal defined in time and space. Basic principles in physics [6] dictate that a field is always defined in a spatio-temporal domain. A formulation of the problem of RF communications in space-time (see Figure 1.7) leads to the idea that typical wireless channels can deliver data rates substantially higher than the rates achieved by currently deployed technology. A signal $s(t)$ in the wavefield modeling paradigm is defined as $s(t, \mathbf{r})$ [a wavefield[1]] where $\mathbf{r}$ is simply a vector of coordinates that identifies a spatial location. Evidently the impulse response of a wireless channel is also redefined as the response of the channel to an impulse generated at time t and spatial location $\mathbf{r}$, as measured at delay τ and spatial location $\mathbf{r}'$: $h(t, \tau, \mathbf{r}, \mathbf{r}')$. Define a spatial region around the transmitter as Γ_{tx} and a spatial region around the receiver as Γ_{rx}. All the possible impulse responses $h(t, \tau, \mathbf{r}, \mathbf{r}')$ generated by $\mathbf{r}$ spanning Γ_{rx} and $\mathbf{r}'$ spanning Γ_{tx} define a subspace, which we define the *wavefield space*. Properly sampling the wavefield space creates a multilink connection resource between the transmitter and the receiver. One of the possible implementations is the sampling of the wavefield space performed by means of antennas. At the transmitter, symbols are modulated and simultaneously transmitted using signals that occupy the same frequency portion of the spectrum, but are distinguishable because different is their position in the wavefield space. Evidently, the

1. Of course, we need to define wavefields obeying Maxwell's equations.

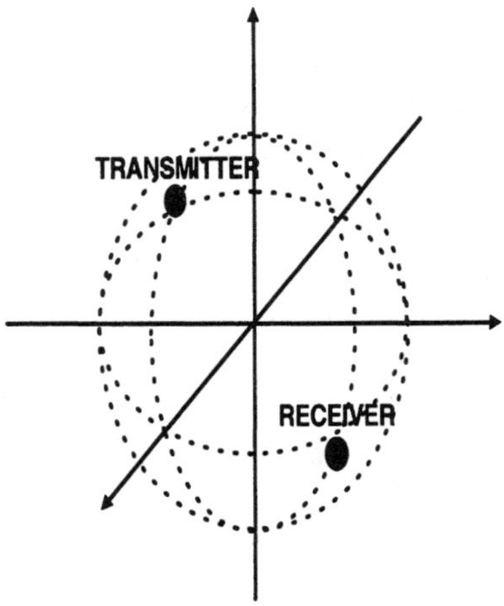

Figure 1.7 A spatio-temporal interpretation of the problem of wireless communication.

shape of the wavefield space depends on the particular propagation environment. The optimum demodulator is a device that estimates the *sampled* wavefield space parameters and performs optimum separation of the different signals to obtain the multiple streams. Evidently the throughput of such a communication system is an integer multiple of the throughput that is achievable at each different position in the wavefield space. The concept of wavefield modulation can be better understood by illustrating as in Figures 1.8 and 1.9 the transfer of digital information units from the baseband digital signal processing (DSP) domain to and from the radio frequency domain. Figure 1.8 depicts the conceptual operation performed by a traditional transceiver. The baseband DSP processors generate digital information units (packets) in a nonsynchronous fashion. The traditional transceiver is responsible for modulating such digital information units (packets) and allocate each unit (packet) in a preassigned position in the time-frequency two-dimensional space: One point in the time-frequency plane corresponds to a signal that exists in a particular time instant given by the projection of the point onto the time axis and with an instantaneous frequency given by the projection of the point onto the frequency axis. The traditional transceiver is also responsible for demodulating such

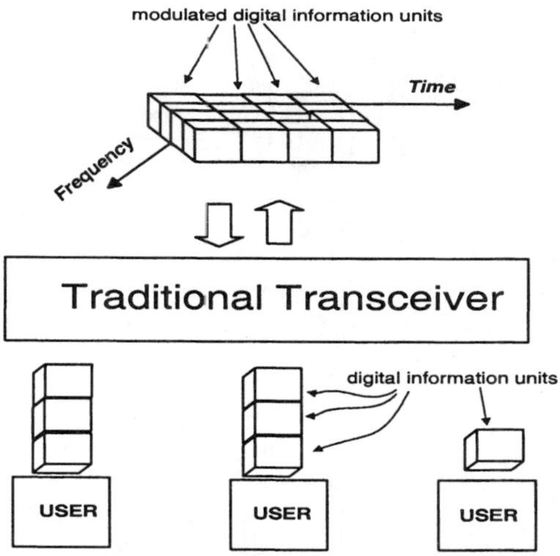

Figure 1.8 The conceptual operation of a traditional transceiver as a device that allocates and extracts digital information units in specific positions in the time-frequency plane.

digital information units (packets), extracting each unit (packet) from a preassigned position in the time-frequency two-dimensional space, and routing said units to the baseband DSP processors. It is evident that traditional transceivers are limited in throughput by the number of available slots in the time-frequency two-dimensional space: No two signals can occupy the same position in said space. The wavefield transceiver in Figure 1.9 is instead capable of much higher throughput given the same number of slots in the time-frequency two-dimensional space; in fact, two signals can occupy the same position in said space. The basic mechanism that allows this is the fact that the wavefield transceiver is capable of enabling the allocation of digital information units (packets) in a third dimension: the wavefield space. Basically the throughput is now limited by a three-dimensional volume. The wavefield transceiver enables the allocation of digital information units (packets) in the wavefield space by transmitting (and receiving) from different points in space. We will expand on this idea in Chapter 3 and Chapter 4, but it is possible to prove that transmission with an N_{tx}-sensor antenna array and reception with an N_{rx}-sensor antenna array is capable of achieving rates that increase linearly with $\min(N_{tx}, N_{rx})$. In other words, one can

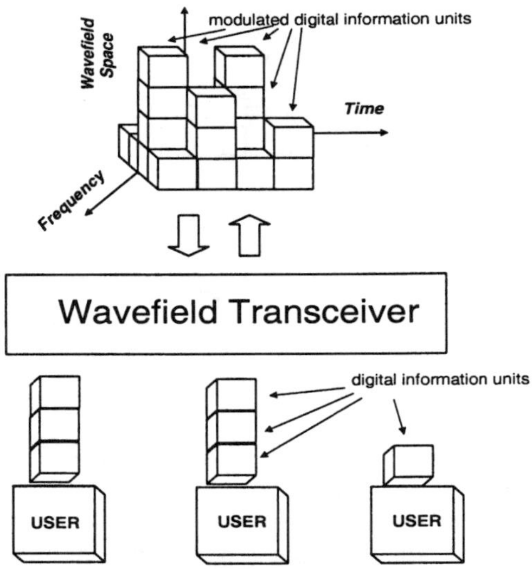

Figure 1.9 The conceptual operation of a wavefield transceiver as a device that allocates and extracts digital information units in specific positions in the time-frequency-wavefield three-dimensional space.

expect an increase in capacity directly proportional to the number of sensors at the antenna array without any penalty in power and bandwidth potentially achieving any point in the region highlighted in Figure 1.10. The subject has generated significant interest in the communication research community in recent years.

1.3 The Digital Radio

The viability of spatio-temporal technology is intimately related to the concept of a modular and reconfigurable digital radio design [7]. In this section we describe the architectural characteristics of a realistic digital radio for multiantenna communications to prove that industry is on the correct path to satisfy most of the requirements identified in the previous sections. Base stations and terminals must have the capability of offering post-deployment flexibility and long-term cost of ownership savings, to scale and support wideband transceivers for next-generation multichannel receivers. The basic idea is to interpret the digital radio as a collection of distributed

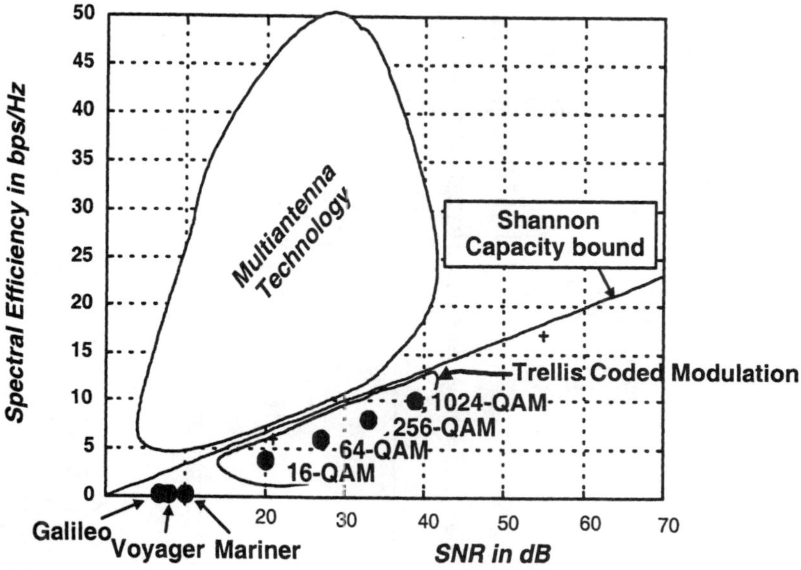

Figure 1.10 Spatio-temporal technology potential as compared to traditional modulations.

managed objects in a client-server framework. The immediate consequence of this approach is that hardware implementation details are decoupled from the controlling software. This, in turn, facilitates the rapid development of applications through a parallel, modular approach. The architecture identifies hardware modules that are partitioned along natural digital radio processing boundaries and interconnected via established, high-performance data distribution technologies. The most important features of such a new approach to radio development are:

- Support for user-downloadable signal processing reconfiguration;
- Management framework through common object request broker architecture (CORBA) and interface description language (IDL);
- High-speed ($> 10^9$ bps) serial digitized spectrum distribution via nonblocking switching with multicasting and combining capability;

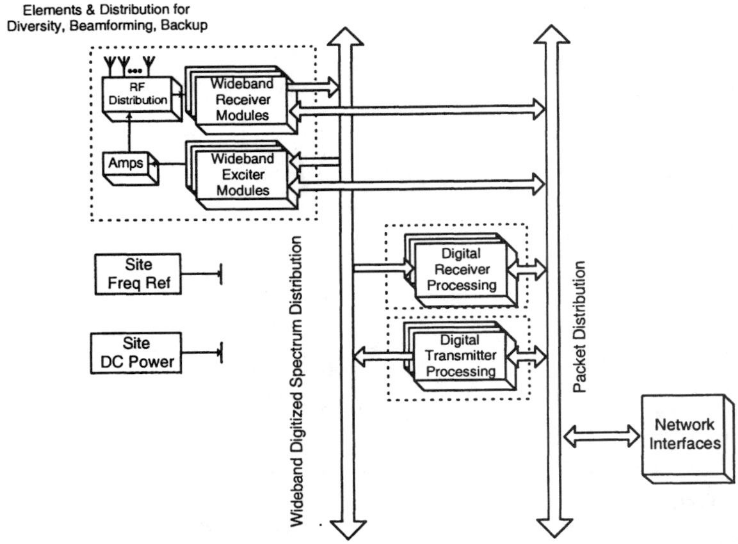

Figure 1.11 Architecture of a digital radio supporting multiantenna technology.

- User traffic and control distribution via fast Ethernet;
- Loosely coupled and hot swappable modules.

The dramatic commercial success of the wireless communication industry has started to skew the flexibility-cost compromise decision process. The wireless communication industry is currently characterized by exponential increases in customer usage (subscriber growth of 60% per year on average), increasing the demand for more capacity and more services. To meet this new demand, radio interfaces are increasing in complexity as more protocols and features are continually being introduced. A generic radio architecture satisfying the requirements identified in the previous paragraphs is presented in Figure 1.11. This architecture captures the fact that there are essentially two natural processing boundaries in digital radio systems: the digitized spectrum boundary and the user traffic boundary (the network side of the modem). This interpretation standardizes these two boundaries to facilitate maximum deployment flexibility and optimization, by allowing modules with different performance characteristics to seamlessly interwork through plug-

and-play data distribution networks. The RF section consists of one or more wideband receiver and wideband exciter modules along with the power amplifier and site RF distribution module. The wideband receiver and exciter modules interconnect with the digital receiver and transmitter modules through the wideband digitized spectrum distribution network. Both traffic and control are exchanged with the network interface module(s) via the packet distribution network, a fast Ethernet bus. The samples at the output of the analog-to-digital converter are routed to the wideband data distribution bus. This bus is a high-speed time division multiplexed resource that connects any RF module to any digital receiver module in a nonblocking fashion. All frequency and clock conversion oscillators are locked to the external site frequency reference ensuring coherent operation. In the wideband exciter module the samples from the wideband data distribution bus are routed to the digital-to-analog converter. The analog signal then is filtered, amplified, and mixed using the second local oscillator frequency. The RF signal is further filtered, amplified and upconverted using the first local oscillator frequency. After proper amplification and filtering, the signal goes to a multicarrier power amplifier that is responsible for providing enough RF power to the signal to reach the remote receiver after antenna radiation. Each module is controlled by a microcontroller, able also to provide interfaces with the data distribution resources (wideband digitized spectrum data distribution and user data distribution). Multiple modules can be replicated to meet traffic capacity and/or multiple antenna requirements.

1.4 Overview of the Book

Readers of this book need to have knowledge of multiple disciplines within electrical engineering, so we attempt to cover the fundamental concepts and the general ideas that were helpful to us. The book is analytical in nature, but mathematical derivations are only tools to better learn how to develop effective products. Much of the material covered represents the interpretation of hardware developments, and sometimes, lessons learned from hardware behavior. We attempt not to go into detailed circuit explanations. We simply describe the problem under analysis, ways to solve the problem, and intuitive motivations to move forward. Chapter 2 provides the reader with a review of communication theory necessary to approach the book. Chapter 3 explains the physics of wave propagation and the physical motivation of many commonly used wireless channel models with special attention to the multiantenna channel. Chapter 4 explores the dimensionality of the multiantenna spatio-temporal channel as the natural generalization of the traditional single-antenna system. In doing this we cover multicarrier and single-carrier methods to achieve capacity. Chapter 5 provides several practical algorithms for the detection of multiantenna signals using traditional approaches, while Chapters 6–8 describe more advanced solutions, mostly the result of the research activity of the author. Chapter 9 is dedicated to spread spectrum multiple access systems. The reader is assumed to have a background in random processes, Fourier analysis, and linear system theory. Some previous knowledge of basic communication and information theory is definitely helpful. The reader who is approaching the subject of multiantenna transmission for the first time is advised not to venture into Chapters 4 through 9, without having a complete understanding of the standard material covered in Chapter 2 and the numerous new perspectives on wireless propagation provided in Chapter 3. The topics covered in Chapter 2 should be reviewed with support from the references.

1.4.1 Mathematical Notation

In the book vectors and matrices are bold. For example, s is a scalar, $\mathbf{s}$ is a vector. $\mathbf{M}^T$, $\mathbf{v}^T$, $\mathbf{M}^H$, $\mathbf{v}^H$ designate transposition and Hermitian (complex conjugate transposed) for matrix $\mathbf{M}$ and vector $\mathbf{v}$, respectively. Complex conjugation for scalars, matrices, and vectors is indicated as u^*, $\mathbf{M}^*$, $\mathbf{v}^*$, respectively, while notations $[\mathbf{M}]_{l,m}$ and $[\mathbf{v}]_k$ stand for the l, m element of matrix $\mathbf{M}$ and the kth element of vector $\mathbf{v}$, respectively. We use the notation

$$||\mathbf{v}|| = \sqrt{\sum_{i=1}^{M} |v_i|^2}$$

for the 2-norm of the complex M-vector $\mathbf{v} = [v_1, ..., v_M]^T$, and $\mathbf{I}_M$ for the $M \times M$ identity matrix. A spatial position vector is indicated by $\mathbf{r}$ or $\vec{r}$; all vectorial fields are denoted as $\vec{E}$.

References

[1] J. C. Bellamy, *Digital Telephony*, New York: Wiley, 2000.
[2] M. R. Aaron, "Digital Communications–The Silent (R)evolution?" *IEEE Communications Magazine*, pp. 16–26, Jan. 1979.
[3] "Digital Video Broadcasting (DVB); Framing Structure, Channel Coding and Modulation for Digital Terrestrial Television (DVB-T)," *European Telecommunications Standard*, ETS-300-744, March 1997.
[4] J. G. Proakis, *Digital Communications*, New York: McGraw-Hill, 1989.
[5] C. E. Shannon, "A Mathematical Theory of Communication," *Bell Syst. Tech. J.*, Vol. 27, pp. 379–423, 623–656, July & Oct. 1948.
[6] P. M. Morse and H. Feshback, *Methods of Theoretical Physics, Part I*, New York: McGraw-Hill, 1953.
[7] J. Mitola, "Software Radios," *IEEE Communications Magazine*, May 1995.

2

Review

2.1 Introduction

In this chapter, we summarize those concepts of waveform analysis and communication theory that will be used in later developments. Moreover, the chapter should provide the necessary background for an understanding of the derivation of several algorithmic strategies for the detection of spatio-temporal signals.

2.2 Fundamentals of Waveform Analysis

A signal can be represented either as a function of time, $s(t)$ or as a function of frequency $S(f)$. The two representations are defined for finite-energy signals, and they are related through the Fourier transform

$$S(f) = \int_{-\infty}^{+\infty} s(t)e^{-j2\pi ft}dt$$

$$s(t) = \int_{-\infty}^{+\infty} S(f)e^{+j2\pi ft}df$$

The most relevant properties of the Fourier transform are

$$s^*(-t) \Leftrightarrow S^*(f),$$

$$s(t-\tau) \Leftrightarrow S(f)e^{-j2\pi f\tau},$$

$$s(t/a) \Leftrightarrow aS(af),$$

$$s(t) = \text{sinc}(t) = \frac{\sin\pi t}{\pi t} \Leftrightarrow S(f) = \text{rect}(f) = \begin{cases} 1 & |f| \leq 0.5 \\ 0 & |f| \geq 0.5 \end{cases},$$

$$r(t) = \int_{-\infty}^{+\infty} s(\tau)h(t-\tau)d\tau \Leftrightarrow R(f) = S(f)H(f),$$

$$r(t) = \int_{-\infty}^{+\infty} s(\tau)h^*(\tau-t)d\tau \Leftrightarrow R(f) = S(f)H^*(f),$$

$$\int_{-\infty}^{+\infty} s(\tau)h^*(\tau)d\tau \Leftrightarrow \int_{-\infty}^{+\infty} S(f)H^*(f)df.$$

2.2.1 Passband Systems

Assume we have a real valued signal $s(t)$ and its Fourier transform $S(f)$. Then we have the constraint $S(f) = S^*(-f)$. Define

$$H_+(f) = \begin{cases} 1, & f > 0 \\ 0.5 & f = 0 \\ 0 & f < 0 \end{cases}.$$

For a real $s(t)$ we define $\hat{s}(t)$ to be the analytic signal of $s(t)$ with Fourier transform

$$S(f) = \sqrt{2}S(f)H_+(f).$$

The relationship between $\hat{s}(t)$ and $s(t)$ is

$$s(t) = \sqrt{2}\text{Real}\{\hat{s}(t)\}.$$

The upconversion and downconversion of a signal can be described in a compact form using this notation [1, 2].

2.2.2 Modulation of a Real Signal

Let $s_e(t)$ be a baseband real signal of bandwidth W. The real part of the multiplication of the signal $s_e(t)$ by $\sqrt{2}e^{j2\pi f_0 t}$ is known as the process of upconversion

$$s(t) = Re\{s_e(t)\sqrt{2}e^{j2\pi f_0 t}\} = \sqrt{2}s_e(t)\cos(2\pi f_0 t)$$

where f_0 is the carrier frequency. Observe that the upconverted signal has a bandwidth of $2W$. Multiplying the analytic signal $\hat{s}(t)$ by $e^{j2\pi f_0 t}$ we perform a downconversion

$$\hat{s}(t)e^{-j2\pi f_0 t} = s_e(t).$$

2.2.3 Modulation of a Complex Signal

Let $s_I(t)$ and $s_Q(t)$ be two baseband real signals of bandwidth W

$$s_e(t) = s_I(t) + js_Q(t).$$

The real part of the multiplication of the signal $s_e(t)$ by $\sqrt{2}e^{j2\pi f_0 t}$ gives

$$s(t) = Re\{s_e(t)\sqrt{2}e^{j2\pi f_0 t}\} = \sqrt{2}s_I(t)cos(2\pi f_0 t) - \sqrt{2}s_Q(t)sin(2\pi f_0 t).$$

Observe that the upconverted signal has a bandwidth of $2W$, but it now contains two signals of bandwidth W. Multiplying the analytic signal $\hat{s}(t)$ by $e^{j2\pi f_0 t}$, we perform a downconversion

$$\hat{s}(t)e^{-j2\pi f_0 t} = s_e(t).$$

It is useful to be able to translate a signal into different frequency ranges because we can perform all signal processing tasks at a convenient frequency range and we can choose this frequency range independent of the actual transmission band. In this way only part of the transmitter/receiver architecture depends on the actual transmission frequency.

2.2.4 Passband White Noise

Assume that a wide sense stationary (WSS) process $Z(t)$ has power spectral density

$$S_Z(f) = \begin{cases} \frac{N_0}{2} & f_0 - W \le |f| \le f_0 + W \\ 0 & \text{elsewhere} \end{cases}.$$

The analytic signal of $Z(t)$ is $\hat{Z}(t)$ is also a WSS random process with power spectral density

$$S_{\hat{Z}}(f) = S_Z(f)|\sqrt{(2)}H^+(f)|^2 = \begin{cases} N_0 & f_0 - W \le f \le f_0 + W \\ 0 & \text{elsewhere} \end{cases}.$$

Now let $Z_e(t) = \hat{Z}(t)e^{-j2\pi f_0 t}$. The autocorrelation function of $Z_e(t)$ is

$$R_{Z_e}(t, \tau) = R_{\hat{Z}}(t - \tau)e^{-j2\pi f_0(t-\tau)}$$

and we can immediately see that

$$S_{Z_e}(f) = \begin{cases} \frac{N_0}{2} & -W \leq f \leq +W \\ 0 & \text{elsewhere} \end{cases}.$$

Moreover, if

$$Z_e(t) = Z_I(t) + jZ_Q(t)$$

it is possible to prove that $Z_I(t)$ and $Z_Q(t)$ are uncorrelated and that

$$S_{Z_I}(f) = S_{Z_Q}(f) = \begin{cases} \frac{N_0}{2} & -W \leq f \leq +W \\ 0 & \text{elsewhere} \end{cases}.$$

2.2.5 The Sampling Theorem

The "sinc" function is of fundamental conceptual importance in communication theory [2, 3]. It is defined as

$$\text{sinc}(t) \triangleq \frac{sin(\pi t)}{\pi t}$$

and its Fourier transform is

$$\text{rect}(f) = \begin{cases} 1 & |f| \leq 0.5/\tau \\ 0 & |f| \geq 0.5/\tau \end{cases}.$$

Many important properties of the sinc function are of interest:

$$s(t) = \text{sinc}(t/\tau) \Leftrightarrow S(f) = \begin{cases} \tau & |f| \leq 0.5/\tau \\ 0 & |f| \geq 0.5/\tau \end{cases}$$

$$s(t) = \text{sinc}(t/\tau - n) \Leftrightarrow S(f) = \text{rect}(f) = \begin{cases} \tau e^{-j2\pi n\tau f} & |f| \leq 0.5/\tau \\ 0 & |f| \geq 0.5/\tau \end{cases}$$

$$\int_{-\infty}^{+\infty} \text{sinc}(t/\tau - n)\text{sinc}(t/\tau - m)dtS(f) = \begin{cases} 0 & m \neq n \\ \tau & m = n \end{cases}.$$

In particular, the most important result relating to the sinc function is contained in the following theorem [3].

Sampling theorem. *Let $f(t)$ be a square integrable function that is low-pass limited to W. Then $f(t)$ is specified by its values at a sequence of points spaced $\tau = \frac{1}{2W}$ apart. In particular,*

$$f(t) = \sum_{n=-\infty}^{+\infty} f(n)\text{sinc}\left(\frac{t}{\tau} - n\right).$$

2.2.6 Nyquist Criterion

A by-product of the sampling theorem is the Nyquist criterion.

Nyquist criterion. *A function $\psi(t) \Leftrightarrow \Psi(f)$ and its time shifts $\psi(t - n\tau)$ form an orthonormal set if and only if*

$$\sum_{n} \left|\Psi\left(f + \frac{k}{\tau}\right)\right|^2 = \tau$$

for $-\frac{1}{2\tau} \leq f \leq +\frac{1}{2\tau}$.

2.2.7 Functions of Three-Dimensional Space

The Fourier transform of a function of a three-dimensional position (x, y, z)

$$s(x, y, z)$$

is actually a threefold transform over scalar coordinates

$$S(k_x, k_y, k_z) = \int_{-\infty}^{+\infty}\int_{-\infty}^{+\infty}\int_{-\infty}^{+\infty} s(x, y, z)e^{-j[k_x x + k_y y + k_z z]}dxdydz.$$

The unit vectors in the three spatial directions $\vec{i}_x$, $\vec{i}_y$, $\vec{i}_z$ allow the use of a shorthand notation

$$S(\vec{k}) = \int_{-\infty}^{+\infty} s(\vec{r})e^{-j\vec{k}\cdot\vec{r}}d\vec{r},$$

where

$$\vec{r} = x\vec{i}_x + y\vec{i}_y + z\vec{i}_z$$

$$\vec{k} = k_x\vec{i}_x + k_y\vec{i}_y + k_z\vec{i}_z$$

and $\vec{a}\cdot\vec{b}$ denotes the scalar product between the vectors $\vec{a}$ and $\vec{b}$. Also

$$s(\vec{r}) = \frac{1}{(2\pi)^3}\int_{-\infty}^{+\infty} S(\vec{k})e^{j\vec{k}\cdot\vec{r}}d\vec{k}.$$

2.3 Hypotheses Testing

Consider a set of hypotheses $H : \{0, 1, ..., (m-1)\}$ with prior probabilities $p_i = Prob\{H = i\}$. We observe Y, and the likelihood of observing y under the hypothesis i is $f_{Y|H}(y|i)$ [4]. The probability that i is the correct hypothesis given that y was observed, defined by $p_{H|Y}(i|y)$, is given by

$$p_{H|Y}(i|y) = f_{Y|H}(y|i)p_i / f_Y(y)$$

where $f_Y(y) = \sum_i f_{Y|H}(y|i)p_i$. The rule to maximize the probability of a correct decision is

$$\hat{H}(y) = arg \max_i p_{H|Y}(i|y).$$

This is the maximum a posteriori (MAP) decision rule. The MAP rule maximizes the probability of correct decision.

Example

The Gaussian case is extremely relevant in communication theory. Here we wish to decide for m given points a_i. Given hypothesis i, the observation Y is

$$Y = a_i + Z,$$

where $Z = (Z_1, ..., Z_n)$ is a Gaussian vector of independent zero mean random variables each of variance σ^2

$$f_{Z_i}(z) = \frac{e^{-\frac{z^2}{2\sigma^2}}}{\sqrt{2\pi\sigma^2}}.$$

So we have

$$f_{Y|H}(y|i) = \frac{1}{(2\pi\sigma^2)^{n/2}} e^{-\frac{||y-a_i||^2}{2\sigma^2}},$$

where $||x||$ is $\sqrt{\sum_n |x_n|^2}$. In this case we see that our decision can be taken on

$$||y - a_i^2||^2,$$

the *sufficient statistic.*

Consider a hypothesis testing case in which the hypotheses are parts of some set of signals $a_0(t), ..., a_{m-1}(t)$ and the observation is

$$Y(t) = a_i(t) + Z(t) \quad t \in [0, T]$$

where $Z(t)$ is a white Gaussian noise process with double-sided power spectral density $\frac{N_0}{2}$. If $\psi_1(t), ..., \psi_n(t)$ is an orthonormal set for the space spanned by $a_0(t), ..., a_{m_1}(t)$ and

$$a_{i,k} = \int a_i(t)\psi_k(t)dt,$$

$$Z_k = \int Z(t)\psi_k(t),$$

we can write

$$a_i(t) = \sum_{k=1}^{n} \psi_k(t)a_{i,k}(t)$$

and

$$Z(t) = \sum_{k=1}^{n} \psi_k(t)Z(t) + Z^+(t).$$

Due to the orthonormality of $\psi_k(t)$, $Z_1, Z_2, ..., Z_n$ is a vector of jointly Gaussian independent zero mean random variables with variance $\sigma^2 = N_0/2$, and we can express $Y(t)$ as

$$Y(t) = \sum_{k=1}^{n} \psi_k(t)a_{i,k}(t) + \sum_{k=1}^{n} \psi_k(t)Z(t) + Z^+(t).$$

$Z^+(t)$ is *irrelevant* for the hypothesis testing problem. We have then $Y = a_i + Z$, where the vector Y has the generic kth element as $Y_k = \int Y(t)\psi_k(t)dt$. The MAP rule is

$$\begin{aligned} \hat{H}(y) &= arg\max_i p_{H|Y}(i|y) = arg\max_i f_{Y|H}(y|i)p_i \\ &= arg\max_i ||y - a_i||^2/(2\sigma^2) - lnp_i \\ &= arg\max_i \frac{1}{\sigma^2}\sum_n y_n a_{i,n} - ||a_i||^2/(2\sigma^2) - lnp_i. \quad (2.1) \end{aligned}$$

If the prior probabilities p_i are equal for the m hypotheses, we obtain the maximum likelihood detector

$$arg\max_i \left[\frac{2}{N_0}\int_0^T a_i(t)Y(t)dt - \frac{1}{N_0}\int_0^T a_i(t)^2dt\right]. \quad (2.2)$$

Example

Transmission over the band-limited white Gaussian channel is an important example. Assume we use a Nyquist pulse $g(t)$ to signal as

$$s(t) = \sum_n a_n g(t - nT), \tag{2.3}$$

where a_n are the transmitted multilevel symbols. Such a signal is generally defined as pulse amplitude modulated (PAM) [2] signal.[1] Assume a $G(f)$ band limited to W with $W \geq \frac{1}{2T}$. We transmit $s(t)$ over a band-limited AWGN channel with bandwidth at least W. The optimal receiver structure is shown in Figure 2.1 where $n(t)$ is a process with autocorrelation $R_n(\tau) = N_0/2\delta(\tau)$. We can see that we

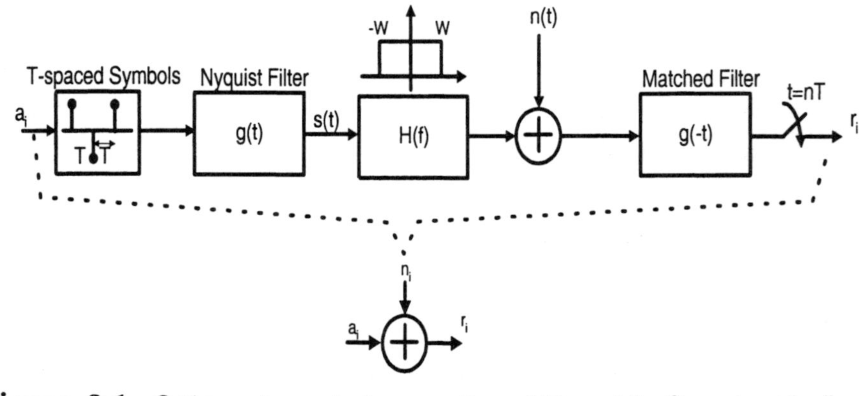

Figure 2.1 Optimum transmission over the additive white Gaussian ideally band-limited channel and its equivalence with a discrete additive white Gaussian channel.

can analyze and model the channel simply referring to the discrete-time channel

$$r_i = a_i + n_i,$$

where n_i is a white Gaussian noise sequence with variance $N_0/2$ and r_i is the output of a filter matched to the Nyquist filter $g(t)$ sampled exactly at $t = iT$.

1. The pulse $g(t)$ for a generic PAM signal need not be a Nyquist pulse.

2.4 Transmission over Linear Time-Invariant Channels

Let us consider the signal [1, 2]

$$x_e(t) = \sum_{n=0}^{N-1} x_n \psi(t - nT)$$

where x_n are complex symbols from a finite set $\mathcal{X}$. Assume that $h_e(t)$ is a baseband equivalent of the channel impulse response. The complex baseband equivalent received signal is expressed as

$$y_e(t) = \sum_{n=0}^{N-1} x_n g(t - nT) + Z(t) = s_e(\mathbf{x}, t) + Z(t),$$

where

$$g(t) = \int_{-\infty}^{+\infty} \psi(\tau) h_e(t - \tau) d\tau,$$

and we have $s_e(\mathbf{x}, t) = \sum_{n=0}^{N-1} x_n g(t - nT)$ to emphasize that $\sum_{n=0}^{N-1} x_n g(t - nT)$ depends on $\mathbf{x} = [x_0, x_1, ..., x_{N-1}]^T$, the sequence of N transmitted symbols. The pulse $g(t)$ will, in general, not satisfy the Nyquist criterion and have intersymbol-interference (ISI). We assume that the receiver knows the channel. Under the assumption of uniform prior probabilities, we now know that the optimal receiver is the receiver that maximizes the log-likelihood function expressed as

$$\tilde{L}(r|x) = \left\{ \frac{1}{N_0} \text{Real} \left\{ s_e^*(\mathbf{x}, t) y_e(t) \right\} - \frac{1}{2N_0} \int_{-\infty}^{+\infty} |s_e(\mathbf{x}, t)|^2 dt \right\}.$$

We see that all possible transmitted points are in a finite dimensional subspace. A basis for this subspace is canonically found by applying a Gram-Schmidt procedure to the set $g_e(t), ..., g_e(t - (N-1)T)$. The receiver can in practice [to compute $\tilde{L}(r|x)$] project the received signal $y_e(t)$ onto this subspace, call the result $\tilde{y}_e(t)$, and then find the most likely signal $\tilde{x}_e(t)$. For two (complex-valued) functions $a(t)$ and $b(t)$, define

$$\langle a(t), b(t) \rangle = \int_{-\infty}^{+\infty} a(t) b^*(t) dt.$$

Since

$$\int_{-\infty}^{+\infty} |\tilde{x}_e(t) - \tilde{y}_e(t)|^2 dt = \langle \tilde{x}_e(t), \tilde{x}_e(t) \rangle + \langle \tilde{y}_e(t), \tilde{y}_e(t) \rangle - 2\text{Real}\{\langle \tilde{x}_e(t), \tilde{y}_e(t) \rangle\}, \quad (2.4)$$

we can maximize the log-likelihood function

$$\begin{aligned} L(r|x) &= 2\text{Real}\{\langle \tilde{x}_e(t), \tilde{y}_e(t) \rangle\} - \langle \tilde{x}_e(t), \tilde{x}_e(t) \rangle \\ &= 2\text{Real}\left\{\sum_{n=0}^{N-1} \tilde{x}_n^* y_n\right\} - \sum_{n=0}^{N-1}\sum_{n=0}^{N-1} \tilde{x}_n^* \tilde{x}_m R_g(n-m), \end{aligned}$$

where

$$R_g(m) = \int g_e(t) g_e^*(t - mT) dt,$$

and

$$y_n = \int y_e(t) g_e^*(t - nT) dt.$$

Assuming that $g_e(t) = 0$ for $t \geq LT$, the problem of selecting the most likely transmitted sequence becomes

$$\max_{\tilde{x}_0, \tilde{x}_1, \ldots, \tilde{x}_{N-1}} \sum_{n=0}^{N-1} 2\text{Real}\left\{\tilde{x}_n^* \left(y_n - 0.5 R_g(0)\tilde{x}_n - \sum_{m=0}^{L-1} R_g(m)\tilde{x}_{n-m}\right)\right\}.$$

or

$$\max_{\tilde{x}_0, \tilde{x}_1, \ldots, \tilde{x}_{N-1}} \sum_{n=0}^{N-1} m(y_n, \tilde{x}_n, \sigma_n),$$

where $m(y_n, \tilde{x}_n, \sigma_n)$ is the *metric* and $\sigma_n = (\tilde{x}_{n-L+1}, \ldots, \tilde{x}_{n-1})$ is the state at time n (basically the last L symbols in the sequence). The maximization can be implemented by means of the Viterbi algorithm [5] further explained in Chapter 5. It is, however, important to recognize that the *discrete-time* equivalent channel

$$y_n = \sum_k R_g(k) x_{n-k} + z_n$$

where $z_n = \int Z(t) g_e^*(t - nT) dt$ is a Gaussian noise sample, fully characterizes our detection problem. The only remarkable issue is that the process z_n is correlated as

$$E\left\{z_n z_{n-k}^*\right\} = N_0 R_g(k),$$

where N_0 is the power spectral density of the process $Z(t)$. When a wide sense stationary process a_n with correlation $R_a(n) = E\left\{a_n a_{n-k}^*\right\}$ is filtered by a finite impulse response filter h_n, we have that the z-transform of the output of the filter

$$w_n = \sum_m h_m a_{n-m}$$

is related to the z-transform of the input as

$$S_w(z) = \sum_k R_w(k) z^{-k} = H(z) H^*(1/z^*) S_a(z)$$

where $S_a(z) = \sum_k R_a(k) z^{-k}$ and $H(z) = \sum_k h_k z^{-k}$. Since $R_a(n)$ is symmetric, $S_a(z)$ can be always factored as

$$S_a(z) = F(z) F^*(1/z^*)$$

where $F(z)$ is a polynomial in z^{-1}. Assume now that we factor the z-transform of $R_z(k)$ defined $S_z(z)$ as

$$S_z(z) = F(z) F^*(1/z^*)$$

and we pass y_n through $1/F^*(1/z^*)$. The result is

$$r_n = \sum_m f_m x_{n-m} + w_n,$$

and we have that w_n is a white Gaussian sequence: The filter $1/F^*(1/z^*)$ is a whitening filter. We also have that f_m is simply the inverse z-transform of $F(z)/N_0$. Moreover, we can design $F(z)$ in a way that it is stable and causal.

2.5 Equalization

The output of the matched filter (sampled at symbol period) gives the following representation

$$y_n = \sum_k R_g(k) x_{n-k} + z_n,$$

where $R_z(k) = N_0 R_g(k)$. A suboptimum [1, 2], but effective, way to compensate for intersymbol interference is to employ the non-linear device of Figure 2.2. The basic idea to use past decisions to

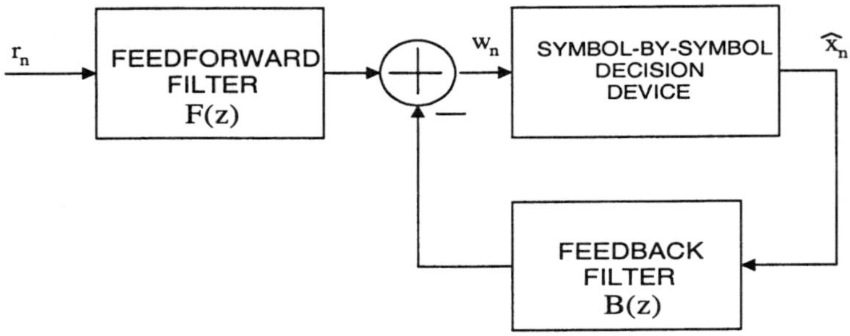

Figure 2.2 A decision feedback equalizer.

cancel part of the intersymbol interference. In order for the backward filter to be realizable, we assume that it is strictly causal. It will be convenient to define the monic and causal filter $B_0(z) = 1 + B(z)$. Using the z-transform, the input to the decision device, $W(z)$, is expressed as

$$W(z) = F(z)Y(z) - B(z)\hat{X}(z).$$

The filters $F(z)$ and $B(z)$ can be selected to minimize the variance of "noise" at the input to the decision device. A standard derivation shows that the $F(z)$ that minimizes the variance of "noise" (more properly, the MSE) at the input to the decision device is given by

$$F_{MMSE}(z) = \frac{S_{yy}(z)}{S_{yx}(z)} = \frac{B_0(z)}{S_g(z) + N_0}.$$

We note that we expressed the optimal forward filter as a function of the backward filter. The optimal setting of the feedback filter $B_0(z)$ is

$$B_0(z) = \frac{S_u^+(z)}{K_u}$$

where $S_u^+(z)$ is the causal part of $S_u(z) = S_u^+(z)S_u^-(z)$. $S_u(z)$ is the z-transform of $R_u(k)$, the autocorrelation function of the process $U(z)$ defined as

$$U(z) = \frac{Y(z)}{S_g(z) - N_0} - X(z).$$

The constant K_u should be ideally selected as

$$K_u = \sqrt{\int_{-0.5}^{0.5} \log_e \left[S_u(e^{j2\pi f})\right] df}.$$

The noise variance (the MSE) at the input to the decision device is

$$\sigma_{MMSE,DFE} = \exp \int_{-0.5}^{0.5} \log_e \left[\frac{N_0}{N_0 + S_g(e^{j2\pi f})} \right] df.$$

A linear equalizer is obtained avoiding the use of $B(z)$, and imposing $B_0(z) = 1$. The noise variance (the MSE) at the input to the decision device for a linear equalizer is

$$\sigma^2_{MMSE,LE} = \int_{-0.5}^{0.5} \left[\frac{N_0}{S_g(e^{j2\pi f})} \right] df.$$

Evidently

$$\sigma^2_{MMSE,LE} \geq \sigma_{MMSE,DFE},$$

which basically states that DFE is a more effective equalizer [1].

2.6 Shannon's Framework

Information theory [6–8] establishes the limits of communications. Coding theory [9] attempts to determine low-complexity schemes that approach these limits [8]. The general problem of communications is to convey a source reliably to a receiver over a given channel. The source can be a picture, a sound pattern or some text. A channel can be a radio link, a phone line, or a fiber-optic cable. The receiver simply attempts to reconstruct the transmitted information given the channel output with high reliability. In his paper from 1948, Shannon showed that the problem can be divided down into two separate problems. First, we should represent the source as accurately as possible given a desired upper bound on the distortion. This is called source coding. In the second stage an appropriate amount of redundancy is incorporated to the bits to protect them against the errors in the channel. This is called channel coding. Shannon's channel coding theorem states that there is a maximum rate at which information can be transmitted reliably, with zero probability of error, over a given channel. This maximum transmission rate is called the *capacity* of the channel.

2.6.1 Channel Capacity

Consider a generic scalar channel with input x and output y that is fully specified by the conditional probability function $f_{Y/X}(y/x)$. The designer can define the input distribution $f_X(x)$ as to transfer the maximum amount of information through the channel. The channel capacity [6, 7, 10] for such a channel is defined as

$$C = \max_{f_X(x)} I(X;Y) \quad \text{bits/channel use} \tag{2.5}$$

where $I(X;Y)$ is the (average) mutual information defined as

$$I(X;Y) = H(Y) - H(Y|X)$$

and the entropies are

$$H(Y) = -\int f_Y(y)\log f_Y(y)dy$$

and

$$H(Y|X) = -\int\int f_{X,Y}(x,y)\log f_{Y|X}(y|x)dxdy.$$

When the logarithms used in the definition of entropy are in base 2 we obtain capacity in bits, while when the logarithms are natural logarithms we obtain capacity in *nats*. An even more obvious way of expressing capacity is

$$C = \max_{f_X(x)} \int\int f_X(x) f_{Y|X}(y|x)\log_2\left[\frac{f_{Y|X}(y|x)}{\int f_X(z) f_{Y|Z}(y|z)dz}\right] dydx.$$

2.6.2 The Scalar AWGN Channel

The case of interest in communications is evidently the AWGN channel shown in Figure 2.3,

$$Y = X + Z$$

where Z is normally distributed with variance σ^2

$$f_Z(z) = \frac{e^{\frac{-z^2}{2\sigma^2}}}{\sqrt{2\pi\sigma^2}}$$

and there is a power constraint expressed as

$$E_X\{|X|^2\} \le P.$$

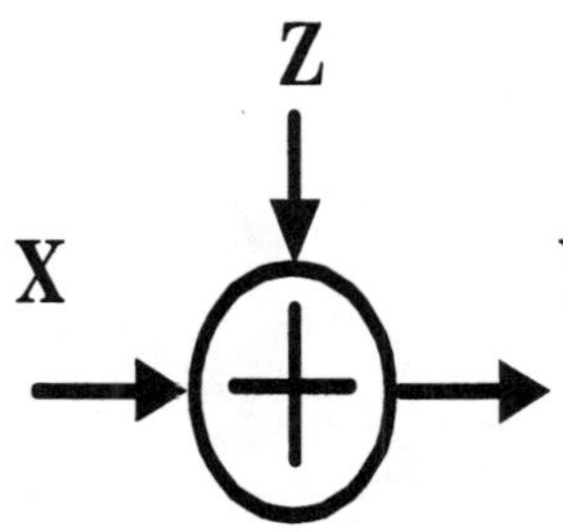

Figure 2.3 The scalar AWGN channel.

Since it is easy to show that

$$H(Z) = \log_e\left(\sqrt{2\pi e\sigma^2}\right),$$

a standard derivation shows that capacity is

$$C = \frac{1}{2}\log_e(1 + P/\sigma^2) \quad \text{nats/channel use.}$$

2.6.3 The Vector AWGN Channel

It is possible to generalize the scalar case to a plurality of scalar channels

$$Y_i = X_i + Z_i \quad i = 1, ..., N,$$

where we have a constraint on the power transmitted (see Figure 2.4) over the N channels

$$\sum_{i=1}^{N} P_i = \sum_{i=1}^{N} P_i E_{X_i}\{|X_i|^2\} \leq P.$$

In this case capacity is obtained from the so-called water-filling power assignment

$$C = \frac{1}{2}\sum_{i=1} \log_e(1 + \tilde{P}_i/\sigma_i^2) \text{ nats/channel use,}$$

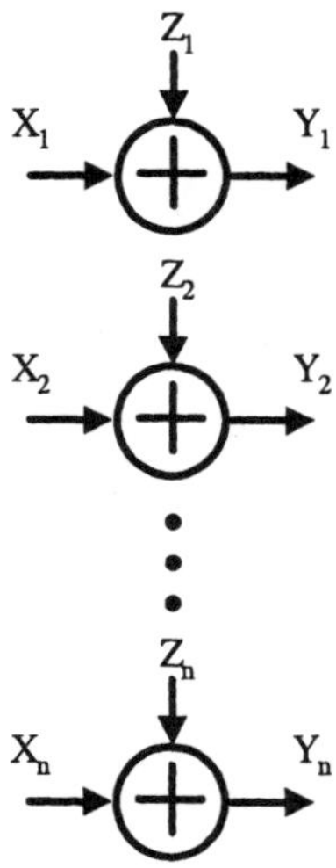

Figure 2.4 The vector AWGN channel.

where the optimal powers $\tilde{P}_i$ are obtained by

$$P_i + \sigma_i^2 = K \;\; \text{if } K \geq \sigma_i^2$$

$$P_i = 0 \;\; \text{if } K < \sigma_i^2.$$

where K is adjusted to maintain the power constraint.

2.6.4 The Waveform AWGN Channel

The waveform channel is characterized by a continuous time linear time-invariant $h(t)$ and an additive continuous time white Gaussian noise process $N(t)$ with double-sided power spectral density equal to $N_0/2$,

$$Y(t) = X(t) * h(t) + N(t) = \tilde{X}(t) + N(t), \tag{2.6}$$

where $*$ simply denotes convolution. The first step in the computation of the capacity for (2.6) is to assume that the time interval is limited to T seconds and obtain an orthonormal representation for $\tilde{X}(t)$. The second step is to find the capacity of this time-limited channel C_T.

The final step is to increase T ideally to an infinite interval to obtain the capacity per unit time

$$C = \lim_{T \to \infty} \frac{1}{T} C_T.$$

Orthonormal Basis Expansions

An orthonormal series expansion for a deterministic waveform $x(t)$ over some arbitrary time interval T with $T_i \leq t \leq T_f$ is given by

$$x(t) = \sum_n x_n \phi_n(t),$$

where x_n are coefficients of the expansion and $\phi_n(t)$ are the basis waveforms required to satisfy

$$\int_{T_i}^{T_f} \phi_n(t)\phi_m^*(t)dt = \begin{cases} 1 & n = m \\ 0 & n \neq m \end{cases} .$$

We also have

$$x_n = \int_{T_i}^{T_f} x(t)\phi_n(t)dt.$$

Observe that in general it is very difficult (and sometimes impossible) to find orthonormal expansions with *finite size.* If we denote $x_N(t)$ as the expansion of $x(t)$ with N terms, we typically want that

$$\lim_{N \to \infty} \int_{T_i}^{T_f} |x(t) - x_N(t)|^2 dt = 0,$$

which makes the expansion *complete.* The problem of finding the expansion is equivalent to solving an eigenvalue problem in fact the expansion coefficients and expansion functions can be found from the solutions of the integral equation

$$\lambda_n \phi_n(t) = \int_{T_i}^{T_f} R_x(t - s)\phi_n^*(s)ds,$$

where $R_x(\tau)$ is the (deterministic for now) correlation function of the waveform $x(t)$. The scalar quantities λ_n are *eigenvalues* and the functions $\phi_n(t)$ are *eigenfunctions* of the kernel $R_x(\tau)$. Eigenfunctions and eigenvalues allow the expansion of $x(t)$ as

$$x_N(t) = \sum_{n=0}^{N-1} \lambda_n \phi_n(t).$$

Observe that if N is selected large enough as to make $\int_{T_i}^{T_f} |x(t) - x_N(t)|^2 dt \simeq 0$, we can think of N as the essential degrees of freedom or *dimensionality* of $x(t)$.

Orthonormal Expansions of Random Processes

When $x(t)$ is a random process (in general, nonstationary) and

$$R_x(t,s) = E\{x(t)x^*(s)\},$$

the integral equation to solve is

$$\lambda_n \phi_n(t) = \int_{T_i}^{T_f} R_x(t,s)\phi_n^*(s)ds \qquad t \in [T_i, T_f]. \tag{2.7}$$

The expansion coefficients are uncorrelated

$$E\{x_n x_m^*\} = \begin{cases} \lambda_n & n = m \\ 0 & n \neq m \end{cases}$$

and the expansion is known as Kharunen-Loeve expansion of a random process.

Example

The signal expansion for a band-limited process is a good and important application of the Karhunen-Loeve expansion. If $X(t)$ is a stationary, zero mean perfectly band-limited to W real process with two-sided power spectral density $N_0/2$, its autocorrelation function is

$$R_X(t,s) = R_X(t-s) = N_0 W \frac{\sin[\pi W(t-s)]}{\pi W(t-s)}.$$

The solutions of the integral eigenequation (2.7) with $T_i = -T/2$ and $T_f = T/2$ are known as *prolate spheroidal functions.* These functions are infinite duration waveforms orthonormal not only over the infinite real line but over the interval $[-T/2, T/2]$. The eigenvalue λ_n is the energy of $\phi_n(t)$. The most remarkable point [11] is that even if in principle there are an infinite number of such funtions, the eigenvalues λ_n drop sharply for

$$n > \lceil 2WT + 1 \rceil,$$

which means that only about $2WT + 1$ eigenfunctions $\phi_n(t)$ have significant energy. A band-limited process has approximately

$$N = 2WT + 1$$

degrees of freedom. So it is appropriate to say that all band-limited processes lie in a subspace with $2WT + 1$ *dimensions*. It is known that this argument "hardens" as WT increases. In fact for large WT the eigenfunctions $\phi_n(t)$ approach sinusoids and the Fourier basis expansion emerges.

The problem of finding the capacity of the band-limited waveform channel is intimately related to the expansion of band-limited random processes. Assume that $h(t)$ is strictly band-limited to W and that

$$E\left\{|\tilde{X}(t)|^2\right\} < P,$$

which means that the power available over the time interval T is PT. Using the prolate spheroidal functions basis the waveform channel of (2.6) can be easily seen equivalent to the vector Gaussian channel

$$Y_i = \tilde{X}_i + N_i \quad i = 1, 2, ..., 2WT + 1,$$

where

$$\sigma_{\tilde{X}_i} = \lambda_i$$

and

$$\sigma_{N_i} = \frac{N_0}{2}.$$

We know that the capacity of such channel is

$$C_T = \frac{1}{2} \sum_{i=1}^{2WT+1} \log\left[1 + \frac{\lambda_i}{N_0/2}\right].$$

We can then write

$$C = \lim_{T\to\infty} \frac{1}{T} C_T = \lim_{T\to\infty} \frac{2WT+1}{2T} \log\left[1 + \frac{2PT}{(2WT+1)N_0}\right],$$

where we have used the fact that for large WT the eigenvalues are approximately equal to $PT/(2WT+1)$. The above expression is easily rewritten as

$$C = W \log\left(1 + \frac{P}{N_0 W}\right) = W \log\left(1 + \text{SNR}\right).$$

2.7 Elements of QAM Modulation

A quadrature amplitude modulation (QAM) system has a baseband equivalent, that is essentially a pulse amplitude modulation (PAM) signal, except that the signal is complex rather than real. The data sequence has a rate of R_b bps and is converted, b bits at a time, into a sequence of symbols a_k chosen from a signal set (alphabet, constellation) $\mathcal{A}$ of size $M = 2^b$. The symbol (or baud) rate is $R_s = R_b/b$ symbols per second, and the symbol period is $T = 1/R_s$ seconds. The transmitted symbols a_k are complex numbers, rather than real numbers like in conventional PAM. The square $M \times M$-QAM signal set is the Cartesian product of two M-PAM sets

$$\mathcal{A}_{PAM} = \{-d(M-1)/2, ... - d/2, d/2, ..., d(M-1)/2\}$$

The signal set $\mathcal{A}_{QAM}$ consists of a square array of M^2 points as illustrated in Figure 2.5 for $M^2 = 64$ for $d = 2$. The energy per

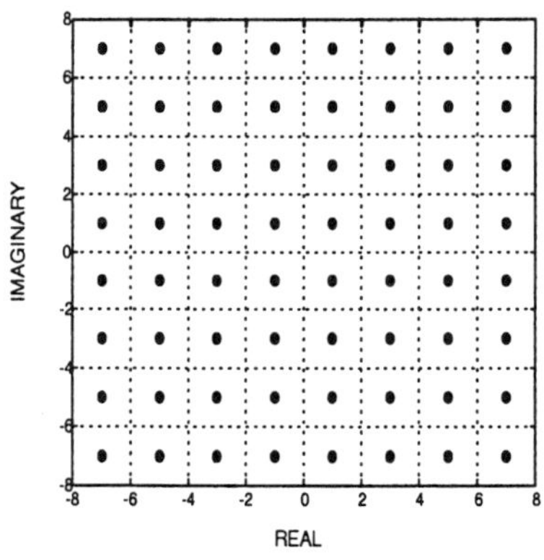

Figure 2.5 64-QAM constellation.

symbol is on average

$$E_s = \frac{d^2(M^2-1)}{6}.$$

The symbol train

$$s_a(t) = \sum_k a_k \delta(t - kT)$$

where $\delta(t)$ is the delta function, is input to the *pulse shaping filter* $p_T(t)$ and translated to the carrier frequency $e^{j2\pi f_o t}$ to obtain the QAM-modulated signal

$$s_{QAM}(t) = \sum_k a_k p_T(t - kT) e^{j2\pi f_o t}.$$

Removal of the carrier at the receiver and the effect of the linear AWGN channel result in the received signal

$$r(t) = \int \sum_k a_k p_T(\tau - kT) f(t - \tau) d\tau + n(t),$$

where $f(t)$ is a complex filter that represents the linear effect of the channel and $n(t)$ is AWGN with double-sided power spectral density equal to $N_0/2$. The typical receiver attempts a matched filtering operation followed by symbol-rate sampling

$$r_k = \int r(\tau) p_R(t - \tau) d\tau \Big|_{t=kT} = \int r(\tau) p_R(kT - \tau) d\tau,$$

where $p_R(t)$ is a filter that should be ideally matched to

$$\int p_T(t - \tau) f(\tau) d\tau = p_T(t) * f(t).$$

When the cascaded filtering operations $p_T(t) * f(t) * p_R(t) = q(t)$ result in an impulse response function that satisfies the Nyquist criterion,

$$q(t)|_{t=kT} = \begin{cases} 1 & k = 0 \\ 0 & k \neq 0 \end{cases},$$

we obtain

$$r_k = a_k + \int n(\tau) p_R(kT - \tau) d\tau = a_k + n_k,$$

where n_k is a white Gaussian complex sequence with zero-mean and variance $N_0/2$. The optimum receiver is the *slicer*, a device that selects the symbol $\hat{a}_k$ *closest* (in the Euclidean sense) to r_k

$$\hat{a}_k = \texttt{slice}\{r_k\} = arg \min_{a_k \in \mathcal{A}_{QAM}} |r_k - a_k|^2.$$

The probability of making an error is easily found from the probability of making an error with an M-level PAM detector.

2.8 Key Points of This Chapter

This chapter briefly touched on several fundamental points of communication theory. We have reviewed some basic concepts associated with waveform analysis with an emphasis on those principles particularly important in communications, including modulation, the sampling theorem, the Nyquist criterion, and passband representations. Next, we summarized the basic hypothesis testing problem both discrete and continuous time, and moved on to apply the theory to the optimum transmission scheme over a time-invariant linear channel. A look at the equalization problem led the discussion to the Shannon's interpretation of the communication problem. The important point concluding the chapter is that the dimensionality of a linear channel is directly proportional to the capacity of such a channel, because it is possible to establish a discrete-time-AWGN channel across each dimension by means of simple mathematical manipulations. The review in this chapter is by no means comprehensive. Readers will find solid and excellent treatments of information and communication theory cited next in the references.

References

[1] J. G. Proakis, *Digital Communications*, New York: McGraw-Hill, 1989.

[2] S. Benedetto, E. Biglieri and V. Castellani, *Digital Transmission Theory*, Englewood Cliffs, NJ: Prentice-Hall, 1988.

[3] J. Wozencraft and I. M. Jacobs, *Principles of Communication Engineering*, New York: Wiley, 1967.

[4] L. Van Trees, *Detection, Estimation, and Modulation Theory, Part I*, New York: Wiley, 1968.

[5] G. D. Forney, "Maximum Likelihood Sequence Estimation of Digital Sequences in the Presence of Intersymbol Interference," *IEEE Trans. Inform. Theory*, Vol. IT-18, pp. 363–378, May 1972.

[6] T. Cover and J. Thomas, *Elements of Information Theory*, New York: Wiley 1991.

[7] R. Gallager, *Information Theory and Reliable Communication*, New York: Wiley, 1968.

[8] A. J. Viterbi and J. K. Omura, *Principles of Digital Communications and Coding*, New York: McGraw-Hill, 1980.

[9] R. Blahut, *Theory and Practice of Error Control Codes*, Reading, MA: Addison-Wesley, 1983.

[10] C. E. Shannon, "A Mathematical Theory of Communication," *Bell Syst.*

Tech. J., Vol. 27, pp. 379–423, 623–656, July & Oct. 1948.

[11] H. J. Landau and H. O. Pollak, "Prolate Spheroidal Wave Functions, Fourier Analysis and Uncertainty-III: The Dimensions of the Space of Essentially Time and Band-Limited Signals," *Bell Systems Tech. J.*, Vol. 41, pp. 1295–1336, July 1962.

3

Signals in Space and Time

3.1 Introduction

In this chapter we will discuss the importance of a spatial interpretation of the wireless channel. In engineering applications we are concerned with waveforms of a single variable, which typically represents time. The fact that a waveform may be also a function of space, that is, of a position in space, requires some fundamental understanding of the physics of wave propagation. Beginning with Maxwell's equations, we describe in this chapter the wireless spatio-temporal channel and explain the assumptions at the basis of commonly used channel models.

3.2 Wave Propagation

Many physical phenomena are either a result of waves propagating through a medium or exhibit a wave-like physical manifestation. Electromagnetic phenomena are produced by the displacement of a fundamental constituent called *electric charge*. The description of electromagnetic interactions in terms of *fields* simplifies the treatment of many engineering problems, and the empirical existence of electromagnetic waves, carrying energy and momentum, is important evidence for the physical reality of the fields themselves.

While the concept of a scalar field is intuitively clear, the concept of a vectorial field appears in a sense abstract. The typical example that helps the physical understanding of vectorial fields is the following. Consider an empty room with an open window. The temperature inside the room is clearly a function of the spatial position x, y, z and time t so it is possible to define a *scalar* field $T(x, y, z, t)$ that maps temperature to space and time. Velocity of the air inside the room is also clearly a function of the spatial position x, y, z and time t but it is not possible to fully characterize velocity by a single number, a scalar. Velocity is necessarily a vector $\vec{v}$ with components in the coordinate system $\vec{v} = \vec{i}_x v_x + \vec{i}_y v_y + \vec{i}_z v_z$. We need to define a *vector* field $\vec{V}(x, y, z, t)$ which maps velocity, a vectorial quantity, of air in the room to space and time. Electric fields are generated by electric charges, while magnetic fields are generated by charges in motions (i.e., currents). In addition, the two fields are coupled. The electrical field $\vec{E}$ (or $\mathbf{E}$) and the magnetic field $\vec{H}$ (or $\mathbf{H}$) can be expressed in terms of integrals over the spatial region containing their sources. The starting point for an accurate description of wave propagation is in fact Maxwell's equations. From these we can derive the wave equation and the Helmholtz equation.

3.2.1 Coordinate Systems

A spatio-temporal signal is written $s(t, x, y, z)$ with x, y, z being the three spatial variables in a right-handed coordinate system. The unit vectors in the three spatial directions are $\vec{i}_x$, $\vec{i}_y$, $\vec{i}_z$. Evidently we can represent

$$s(t, x, y, z) = s(t, \vec{r}) = s(t, x, y, z) = s(t, \mathbf{r})$$

where $\vec{r}$ or $\mathbf{r}$ denotes the triple of the spatial variables (x, y, z). For some problems the spherical coordinates represent space more appropriately (see Figure 3.1). The spherical coordinates r, ϕ, θ are related to Cartesian coordinates as follows

$$x = r\text{sin}\phi\text{cos}\theta, y = r\text{sin}\phi\text{sin}\theta, z = r\text{cos}\theta,$$

$$r = \sqrt{x^2 + y^2 + z^2},$$

$$\theta = \cos^{-1}\left(\frac{x}{x^2 + y^2}\right),$$

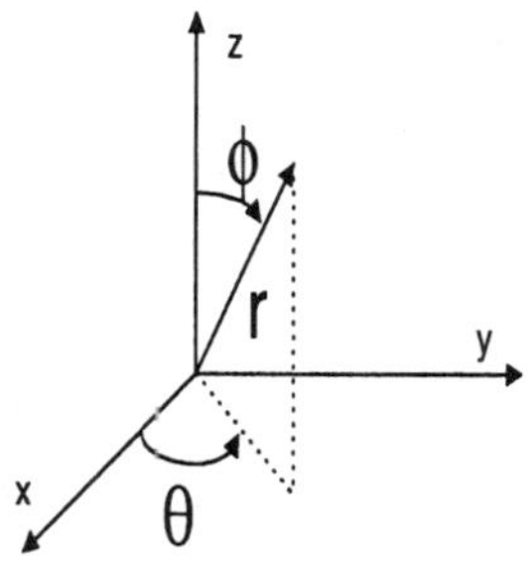

Figure 3.1 An orthogonal coordinate system.

$$\phi = \cos^{-1}\left(\frac{z}{x^2 + y^2 + z^2}\right).$$

3.2.2 Maxwell's Equations

Maxwell's equations [1, 2] in differential form are

- Faraday's law:

$$\nabla \times \vec{E} = -\frac{\partial}{\partial t}\vec{B}$$

- Ampere's law:

$$\nabla \times \vec{H} = \frac{\partial}{\partial t}\vec{D} + \vec{J}$$

- Gauss's laws:

$$\nabla \cdot \vec{D} = \rho$$

$$\nabla \cdot \vec{B} = 0$$

where $\vec{E}$ is the electric field, $\vec{H}$ is the magnetic field, $\vec{D}$ is the displacement or electric flux density, $\vec{B}$ is the magnetic flux density, $\vec{J}$ is the electric current density, and ρ is the electric charge density. The flux densities are related to the fields through the constitutive relations:

$$\vec{D} = \epsilon_0 + \vec{P} = \epsilon\vec{P}$$

$$\vec{B} = \mu_0\vec{H} + \mu_0\vec{M} = \mu\vec{H}$$

where $\vec{P}$ is the electric polarization, $\vec{M}$ is the magnetization, ϵ is the dielectric permittivity, and μ is the magnetic permeability; ϵ_0 is the permittivity in free space, and μ_0 is the permeability in free space. The operator ∇ is defined in Cartesian coordinates as

$$\nabla = \frac{\partial}{\partial x}\vec{i}_x + \frac{\partial}{\partial y}\vec{i}_y + \frac{\partial}{\partial z}\vec{i}_z$$

and in spherical coordinates

$$\nabla = \frac{\partial}{\partial r}\vec{i}_r + \frac{\partial}{\partial \theta}\vec{i}_\theta + \frac{\partial}{\partial \phi}\vec{i}_\phi.$$

The *curl* operator acts on a field $\vec{A}$ as

$$\nabla \times \vec{A} = \begin{vmatrix} \vec{i}_x & \vec{i}_y & \vec{i}_z \\ \frac{\partial}{\partial x} & \frac{\partial}{\partial y} & \frac{\partial}{\partial z} \\ E_x & E_y & E_z \end{vmatrix}$$

or in spherical coordinates

$$\nabla \times \vec{A} = \begin{vmatrix} \frac{\vec{i}_r}{r^2 \sin\theta} & \frac{\vec{i}_\theta}{r \sin\theta} & \frac{\vec{i}_\phi}{r} \\ \frac{\partial}{\partial r} & \frac{\partial}{\partial \theta} & \frac{\partial}{\partial \phi} \\ E_r & rE_\theta & r\sin\theta E_\phi \end{vmatrix}.$$

The *divergence* operator acts on a field $\vec{A}$ as

$$\nabla \cdot \vec{A} = \frac{\partial}{\partial x}A_x + \frac{\partial}{\partial y}A_y + \frac{\partial}{\partial z}A_z$$

or in spherical coordinates

$$\nabla \cdot \vec{A} = \frac{1}{r^2}\frac{\partial}{\partial r}r^2 A_r + \frac{1}{r\sin\theta}\frac{\partial}{\partial \theta}\sin\theta A_\theta + \frac{1}{r\sin\theta}\frac{\partial}{\partial \phi}A_\phi.$$

Taking the curl on both sides of Faraday's law and reversing the order of derivation for time and space, we get

$$\nabla \times (\nabla \times \vec{E}) = -\mu\frac{\partial}{\partial t}\left(\nabla \times \vec{H}\right).$$

We now use Ampere's law assuming zero current density, and the constitutive relation for the electric displacement to find an equation for the electric field, assuming that the permittivity is time-invariant

$$\nabla \times (\nabla \times \vec{E}) = -\mu\frac{\partial}{\partial t}\left(\frac{\partial}{\partial t}\vec{D}\right) = -\mu\epsilon\frac{\partial}{\partial t}\left(\frac{\partial}{\partial t}\vec{E}\right).$$

The following identity

$$\nabla \times (\nabla \times \vec{A}) = \nabla \left(\nabla \cdot \vec{A}\right) - \nabla^2 \vec{A}$$

can be applied to obtain

$$\nabla \left(\nabla \cdot \vec{E}\right) - \nabla^2 \vec{E} = -\mu\epsilon \frac{\partial}{\partial t}\left(\frac{\partial}{\partial t}\vec{E}\right).$$

If the divergence of the electric field is zero, we can write the homogeneous wave equation for the electric field

$$\nabla^2 \vec{E} - \mu\epsilon \frac{\partial^2}{\partial t^2}\vec{E} = 0. \tag{3.1}$$

The same derivation gives for the magnetic field

$$\nabla^2 \vec{H} - \mu\epsilon \frac{\partial^2}{\partial t^2}\vec{H} = 0. \tag{3.2}$$

The wave equation is of crucial importance because it describes wave propagation at any point in space and time for any type of wave and, in particular, electromagnetic waves.

3.2.3 Plane Waves

The simplest solution of the wave equation (3.1) is

$$\vec{E}(t, \vec{r}) = \vec{E}(t, x, y, z) = \vec{E}_0 f(\omega t - k_x x - k_y y - k_z z) \tag{3.3}$$

where $\vec{E}_0$ is a constant vector, $f(.)$ is an arbitrary function and $\omega \geq 0$, k_x, k_y, k_z are real constants. Substitution of this form into the wave equation gives

$$\left(-k_x^2 - k_y^2 - k_z^2 + \omega^2\mu\epsilon\right) f'' \vec{E}_0 = 0,$$

where $f''(.)$ is the second-order derivative of $f(.)$, and a nontrivial solution requires

$$k_x^2 + k_y^2 + k_z^2 = \omega^2 \mu\epsilon.$$

A *monochromatic* wave in the z direction is

$$\vec{E} = \vec{i}_x E_0 \cos(\omega t - kz).$$

This particular wave appears to be harmonic in space and time and it propagates in free space with a velocity

$$v = \frac{dz}{dt} = \frac{\omega}{k} = \frac{1}{\epsilon_0 \mu_0} = c,$$

the familiar propagation speed of light $c = 2.998 \cdot 10^8\ m/s$. Since Maxwell's equations are first-order differential equations in space and time, they are also linear provided that the constitutive relations are linear. In this case, super-positions of solutions to Maxwell's equations are themselves solutions, and monochromatic plane waves can be added to form solutions of arbitrary time waveforms

$$\vec{E}(\vec{r}, t) = \vec{i}_x \frac{1}{2\pi} \int_0^\infty E_x(\vec{r}) \cos(\omega t - kz) d\omega. \tag{3.4}$$

The linearity of Maxwell's equations allows the use of phasor notation. The plane wave solution can be written in the following form

$$\vec{E}(\vec{r}, t) = \text{Real}\left\{E_x(\vec{r}) e^{j(\omega t - \phi_x(\vec{r}))}\right\} = \text{Real}\left\{\tilde{E}_x(\vec{r}) e^{j\omega t}\right\}$$

and the phase factor can be included in the expression for the field amplitude component, which then becomes a complex quantity, a phasor. In phasor notation we drop the explicit taking of the real part such that the plane wave solution is written

$$\vec{E}(\vec{r}, t) = \tilde{E}_x(\vec{r}) e^{j\omega t}.$$

Combining the three vector-components of the electric field in a single phasor, we obtain

$$\vec{E}(\vec{r}, t) = \vec{E}(\vec{r}) e^{j\omega t}$$

where the phasor $\vec{E}(\vec{r})$ has six components (three vector components each with amplitude and phase). In general,

$$\vec{E}(\vec{r}, t) = \vec{E}_0 e^{j(\omega t - \vec{k} \cdot \vec{r})}$$

and $\vec{k}$ is called the *wave number vector*. After one temporal period $T = 2\pi/\omega$, the wave has propagated a distance equal to λ, the *wavelength*

$$\lambda = \frac{2\pi c}{\omega}.$$

The magnitude of the wave number vector expresses the number of cycles in radians per meter length that the monochromatic wave exhibits in the direction of propagation

$$|\vec{k}| = \frac{2\pi}{\lambda}.$$

The vector $\vec{k}$ is therefore considered as a *spatial frequency variable.*

3.2.4 The Helmholtz Equation

The Maxwell's equations in phasor form are easily written:

- Faraday's law:

$$\nabla \times \vec{E} = -j\omega\vec{B};$$

- Ampere's law:

$$\nabla \times \vec{H} = j\omega\vec{D} + \vec{J};$$

- Gauss's laws:

$$\nabla \cdot \vec{D} = \rho,$$

$$\nabla \cdot \vec{B} = 0.$$

The wave equation is now expressed as

$$\left(\nabla^2 + \omega^2\mu_0\epsilon_0\right)\vec{E} = 0, \tag{3.5}$$

known as the Helmholtz equation.

3.2.5 Characterization of Electromagnetic Waves

Consider the wave propagating in the z-direction

$$\vec{E} = \vec{E}_0 e^{j\omega t - kz}$$

with $\omega/k = c$ and $\vec{E} = \vec{i}_x E_x - \vec{i}_y E_y + \vec{i}_z E_z$ with associated magnetic field

$$\vec{H} = \vec{H}_0 e^{j\omega t - kz}$$

where $\vec{H} = \vec{i}_x H_x + \vec{i}_y H_y + \vec{i}_z H_z$. From the Maxwell's equations we recognize that

$$E_{0,z} = H_{0,z} = 0,$$

which means that the electric and magnetic fields must necessarily lie in the plane *perpendicular* to the direction of travel. That is why electromagnetic waves are called transverse and can be polarized. Moreover, we also get

$$E_{0,x} = cH_{0,y},$$

$$E_{0,y} = -cH_{0,x}.$$

Polarization is that property of an electromagnetic wave describing the time-varying direction and relative magnitude of the electric-field vector. In other words, polarization is the figure traced as a function of time by the extremity of the vector at a fixed location in space, and the sense in which it is traced, as observed along the direction of propagation. Polarization may be linear, circular, or elliptical. If the vector that describes the electric field at a point in space as a function of time is always directed along a line, the field is linearly polarized. In general the figure that the vector traces is an ellipse. When the trace becomes a circle, the wave is circularly polarized. The wave *impedance* in free space is

$$Z_0 = \sqrt{\frac{\mu_0}{\epsilon_0}} \simeq 377 \text{ ohms}.$$

3.2.6 Refraction

Radio wave propagation is characterized in a real-world communication scenario by refraction and diffraction. Consider a monochromatic plane wave of the form

$$\vec{E}_i = E_{0,i}\cos\left(\omega_i t - \vec{k}_i \cdot \vec{r}\right)$$

incident on a dielectric interface described by the surface normal $\vec{S}$. The corresponding reflected and transmitted fields are

$$\vec{E}_r = E_{0,r}\cos\left(\omega_r t - \vec{k}_r \cdot \vec{r} + \phi_r\right)$$

$$\vec{E}_t = E_{0,t}\cos\left(\omega_t t - \vec{k}_t \cdot \vec{r} + \phi_t\right)$$

where ϕ_r and ϕ_t are constant phase terms. Imposing the continuity of the tangential of the electric field

$$\vec{S} \times \vec{E}_t + \vec{S} \times \vec{E}_r = \vec{S} \times \vec{E}_i,$$

we obtain an equation valid at any point in space and time. So we can write that at the dielectric surface it must hold

$$\omega_i t - \vec{k}_i \cdot \vec{r} = \omega_t t - \vec{k}_t \cdot \vec{r} + \phi_t = \omega_r t - \vec{k}_r \cdot \vec{r} + \phi_r.$$

Since the three waves undergo forced vibrations at the frequency of the incident wave we also have $\omega_r = \omega_i = \omega_t$. We then have at the interface

$$-\vec{k}_i \cdot \vec{r} = -\vec{k}_t \cdot \vec{r} + \phi_t = -\vec{k}_r \cdot \vec{r} + \phi_r.$$

Looking at the first two equations, we can state that because the position vector is in the plane of the interface, this equation can only hold if the difference of the k-vectors is perpendicular to the plane. We can therefore conclude that $\vec{k}_r$ is in the incident plane (the plane defined by the incident wave vector and the surface normal), and the fact that the magnitude of the k-vectors is the same (because they are in the same medium and have the same frequency), allow us to further obtain that

$$\sin\theta_i = \sin\theta_r$$

and

$$|\vec{k}_i|\sin\theta_i = |\vec{k}_t|\sin\theta_t.$$

More in general, we have to define different polarizations. Horizontal polarization is defined as the electric field of the incident wave being parallel to the surface off of which it is reflecting. Thus, the magnetic field is perpendicular to the surface and lies in a plane known as the plane of incidence (Figure 3.2). Vertical polarization is defined as the electric field being perpendicular to the surface and the magnetic field being parallel (Figure 3.3).

In the case of horizontal polarization we have

$$\frac{E_{0,t}}{E_{0,i}} = \frac{2n_i/\mu_i\cos\theta_i}{n_i/\mu_i\cos\theta_i + n_t/\mu_t\cos\theta_t},$$

$$\frac{E_{0,r}}{E_{0,i}} = \frac{n_i/\mu_i\cos\theta_i - n_t/\mu_t\cos\theta_t}{n_i/\mu_i\cos\theta_i + n_t/\mu_t\cos\theta_t},$$

while for vertical polarization

$$\frac{E_{0,t}}{E_{0,i}} = \frac{2n_i/\mu_i\cos\theta_i}{n_i/\mu_i\cos\theta_t + n_t/\mu_t\cos\theta_i},$$

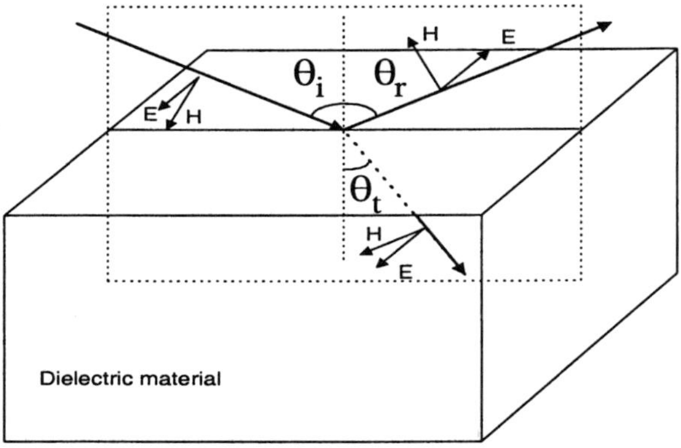

Figure 3.2 Horizontal polarization of an electromagnetic field as it is transmitted and reflected off a dielectric surface.

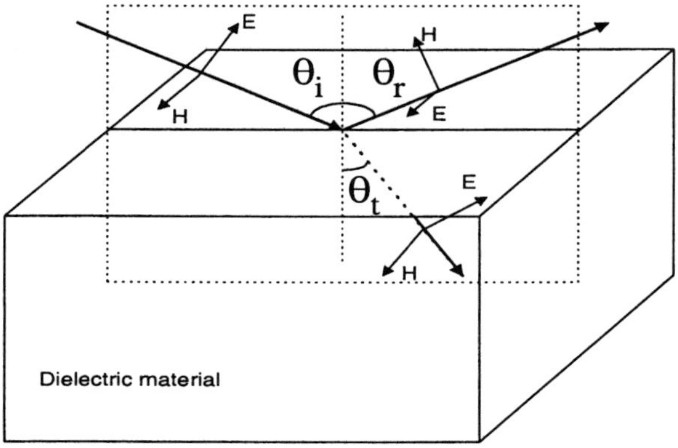

Figure 3.3 Vertical polarization of an electromagnetic field as it is transmitted and reflected off a dielectric surface.

$$\frac{E_{0,r}}{E_{0,i}} = \frac{n_i/\mu_i \cos\theta_t + n_t/\mu_t \cos\theta_i}{n_i/\mu_i \cos\theta_t + n_t/\mu_t \cos\theta_i},$$

where n_i and n_t are indices of refraction, defined as the ratios of the speed of light over the speed of propagation for the dielectrics. We can define the intrinsic impedance, $\eta = \vec{E}/\vec{H}$ of a medium. The relative permeability can be taken as unity for nonmagnetic materials, thus $\mu = \mu_0 = 4\pi 10^{-7}$. Assuming this to be the case, we can define n' as a function of the two media as

$$(n')^2 = \frac{\eta_i^2}{\eta_t^2}$$

where the incident and reflected fields exist within medium I, and the refracted component within medium II. The reflected field can then be defined in terms of the incident field for both vertical and horizontal polarizations. For horizontal polarizations we have

$$\frac{E_{0,t}}{E_{0,i}} = \frac{2\sin\psi}{\sin\psi + \sqrt{n'^2 - \cos^2\psi}}$$

$$\frac{E_{0,r}}{E_{0,i}} = \frac{\sin\psi - \sqrt{n'^2 - \cos^2\psi}}{\sin\psi + \sqrt{n'^2 - \cos^2\psi}}$$

and for vertical polarizations we have

$$\frac{E_{0,t}}{E_{0,i}} = \frac{2n'\sin\psi}{n'\sin\psi + \sqrt{n'^2 - \cos^2\psi}}$$

$$\frac{E_{0,r}}{E_{0,i}} = \frac{n'\sin\psi - \sqrt{n'^2 - \cos^2\psi}}{n'\sin\psi + \sqrt{n'^2 - \cos^2\psi}}.$$

The angle ψ is such that

$$\cos\psi = \sin\theta_i = n'\sin\theta_t.$$

3.2.7 Diffraction

Diffraction is caused by discontinuities in a surface where an electromagnetic wave impinges on that surface. The mechanism results in the discontinuity acting as a radiating point or edge for a fraction of the electromagnetic wave. Thus, an electromagnetic wave appears to propagate around a corner. The diffraction caused by an

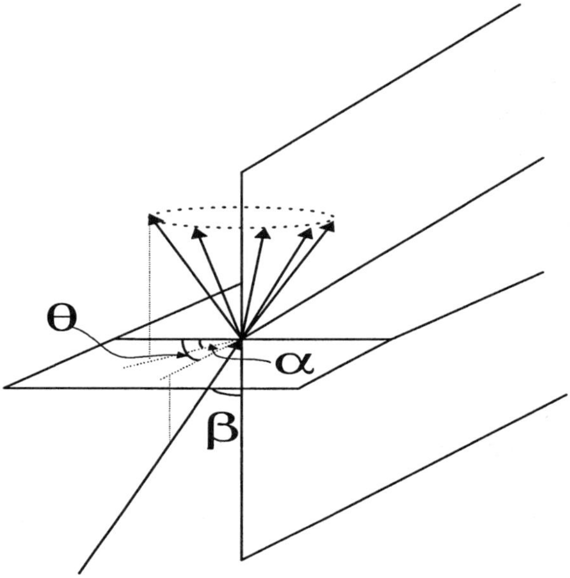

Figure 3.4 Diffraction by a screen.

infinite edge of a perfectly conducting plane is the simplest case of diffraction. A wave incident on an edge produces a cone of scattered components as we show in Figure 3.4.

For the angles α, θ, and β, as defined in Figure 3.4, the diffraction coefficient

$$\mathbf{D} = -\frac{e^{j\pi/4}}{2\sqrt{2\pi k}\sin\beta}\left[\sec(1/2\theta - 1/2\alpha) + \operatorname{cosec}(1/2\theta - 1/2\alpha)\right]$$

defines the diffracted field as

$$\vec{E}_d = \mathbf{D}\vec{E}_i r^{1/2} e^{jkr}$$

at a distance r from the diffraction point. This formula is obtained using the geometrical theory of diffraction (GTD). The removal of the discontinuities at the boundaries of reflection and shadowing for the perfectly conducting wedge is the unified theory of diffraction (UTD).

3.2.8 Scattering

Surfaces in an environment are always, to some degree, rough surfaces that affect the reflection of electromagnetic waves. It is possible to modify the Fresnel reflection coefficients by a scattering loss factor, ρ_s, to account for the lower energy caused by rough surface scattering. The loss factor is defined by

$$\rho_s = \exp\left[\left(\frac{\pi}{\lambda}\sigma_h \cos\theta_i\right)^2\right] J_0\left[8\left(\frac{\pi}{\lambda}\sigma_h \cos\theta_i\right)^2\right]$$

where σ_h is the is the standard deviation of the surface height about its mean value, and $J_0(.)$ is the modified Bessel function of order zero.

3.3 The Antenna

In essence, an antenna in transmit mode is a device that maps a vector field to a scalar voltage quantity, while an antenna in receive mode is a device that maps a scalar voltage to a vector field [3].

3.3.1 The Antenna as an Aperture

An antenna in the receive mode is used to capture electromagnetic waves and to extract a voltage (signal) from them, while an an antenna in the transmit mode is used to radiate electromagnetic waves from a voltage (signal) provided to it. The electric field produced by radiation from the antenna is a function of the amplitude and phase of the current distribution across the aperture of the antenna. Adding vectorially all the contributions from the current elements constituting the aperture one can obtain the field intensity at any point in space and time. This straightforward way of proceeding is mathematically involved. Approximations to this solution can be given dividing the area surrounding the antenna aperture into three regions.

- The region in the immediate vicinity of the aperture is the *near field.*
- In the *Fresnel* region rays from the aperture to the observation point do not appear parallel and the antenna radiation pattern is not constant with distance.

- The farthest region from the antenna aperture is the *far-field* region.

The boundary between the Fresnel and far-field regions is usually

$$R_F = \frac{2D^2}{\lambda}$$

where D is the size of the aperture and λ is the wavelength. In communications we are mostly interested in the far-field region.

Example

Let us consider the rectangular aperture (see Figure 3.5) with a current at distance z equal to $I(z)$ assumed to flow in the x direction. The electrical field ϕ component (relative radiation intensity[1]) is

$$E_\phi = \int_{-a/2}^{a/2} I(z)\exp\left\{j2\pi\frac{z\sin\phi}{\lambda}\right\} dz.$$

A receive antenna in response to the field E_ϕ would then generate the current distribution

$$I(z) = \int_{-\infty}^{+\infty} E_\phi\exp\left\{-j2\pi\frac{z\sin\phi}{\lambda}\right\} d(\sin\phi).$$

The simplest distribution to conceive is the uniform $I(z) = I_0$ for $-a/2 \leq z \leq a/2$, which gives the field

$$E_\phi = I_0\frac{\sin\left[\pi(a/\lambda)\sin\phi\right]}{(\pi/\lambda)\sin\phi}.$$

Example

Another good example is the circular aperture with current distribution $I(r,\theta)$ where r,θ are polar coordinates. Assuming that the radius of the antenna is r_0 we have

$$E_r = \int_0^{2\pi} d\theta \int_0^{r_0} I(s,\theta)\exp\left\{j2\pi\frac{R}{\lambda}\right\} sds.$$

1. In this extremely simplified model the field-intensity pattern is $|E_\phi|$, while $|E_\phi|^2$ is the power radiation pattern.

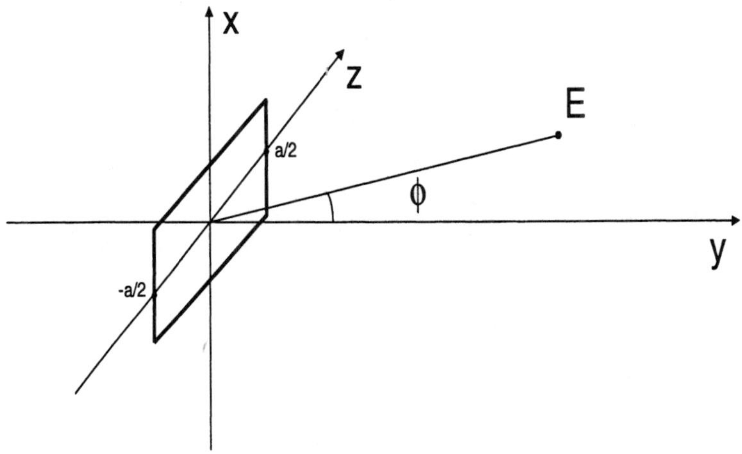

Figure 3.5 The rectangular aperture.

3.3.2 The Rigorous Approach

It should be clear that the way to attack the problem of antenna modeling at the transmitter is to do the following.

- Specify the source-current densities $\vec{J}$ and $\vec{M}$;
- Calculate the fields $\vec{E}$ and $\vec{H}$ by integration.

By reciprocity the problem at the receiver is solved proceeding in the opposite direction. In practice, antenna theory suggests the use of vector potentials to solve radiation synthesis problems. As far as we are concerned this is a technicality that we just need to be aware of. For aperture antennas (which are the most commonly used antennas at microwave frequencies) the total $\vec{E}$ and $\vec{H}$ fields are found in spherical coordinates r, θ, ϕ by determining the fields outside a closed surface $\mathcal{S}$ that encloses the original sources $\vec{J}$ and $\vec{M}$ (see Figure 3.6)

$$E_r \simeq H_r \simeq 0$$

$$E_\theta \simeq -\frac{jke^{-jkr}}{4\pi r}(L_\phi + \eta N_\theta)$$

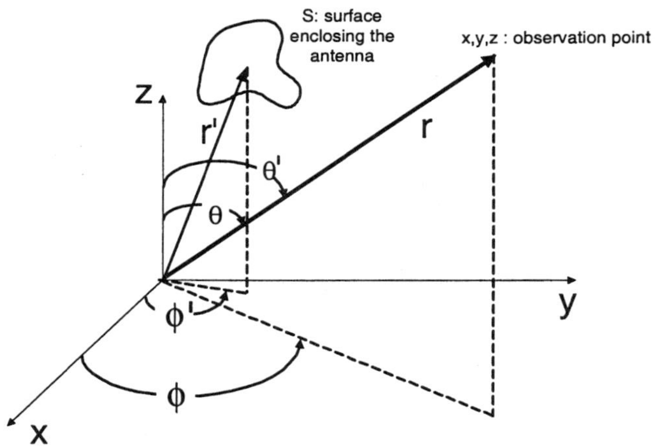

Figure 3.6 Coordinate system for an antenna with generic aperture.

$$E_\phi \simeq +\frac{jke^{-jkr}}{4\pi r}(L_\eta - \eta N_\phi)$$

$$H_\theta \simeq \frac{jke^{-jkr}}{4\pi r}(N_\phi - L_\theta/\eta)$$

$$H_\phi \simeq -\frac{jke^{-jkr}}{4\pi r}(N_\theta - L_\phi/\eta)$$

where the primed coordinates indicate the space occupied by the aperture over which we have to perform integration, and the vectors $\vec{L}$ and $\vec{N}$ can be found from

$$\vec{N} = \int_S \left(\vec{i}_x J_x + \vec{i}_y J_y + \vec{i}_z J_z\right) e^{jkr'\cos\psi} ds',$$

$$\vec{L} = \int_S \left(\vec{i}_x M_x + \vec{i}_y M_y + \vec{i}_z M_z\right) e^{jkr'\cos\psi} ds'.$$

Example

Consider an infinitesimal linear wire of length $l << \lambda$ with uniform current distribution

$$\vec{I}(z) = \vec{i}_z I_0$$

where I_0 is a constant. Using the outlined procedure in spherical coordinates it is possible to obtain in the far-field region

$$E_r \simeq \eta I_0 l \frac{e^{-jkr}}{2\pi r^2} \cos\theta,$$

$$E_\theta \simeq j\eta I_0 kl \frac{e^{-jkr}}{4\pi r} \sin\theta,$$

$$E_\phi \simeq H_r \simeq H_\theta \simeq 0,$$

$$H_\phi \simeq jI_0 kl \frac{e^{-jkr}}{4\pi r} \sin\theta,$$

and

$$\vec{E} = \vec{i}_r E_r + \vec{i}_\theta E_\theta,$$

$$|\vec{E}| = \sqrt{|E_r|^2 + |E_\theta|^2}.$$

3.4 From the Antenna to the Radio

In general, the effects of the antenna gain, phase change, and polarization mismatch for a propagating wave may be accounted for by a single *polarization vector* $\vec{a}$. If the propagation of a single time-harmonic electromagnetic wave is described by an electric field vector, $\vec{E}(\vec{r})$, the baseband voltage, $e(\vec{r})$ at the output of an antenna can be expressed as

$$e(\vec{r}) = \vec{a} \cdot \vec{E}(\vec{r}).$$

The dot product of the electric field and the polarization vector of the antenna produces the baseband voltage at the terminals of the receiver antenna. The polarization vector, $\vec{a}$, has units of distance (meters) to be dimensionally correct. The magnitude of the polarization vector is proportional to the gain of the antenna. Since the polarization vector is complex, it is capable of modeling the phase change of the incoming signal. The orientation of the vector is capable of modeling polarization mismatch, and it changes as the incident angle of the impinging wave changes. If multiple waves from different directions impinge upon the antenna, then a polarization vector must be calculated for each wave

$$e(\vec{r}) = \sum_n \vec{a}_n \cdot \vec{E}_n(\vec{r}).$$

This operation basically makes the voltage at the output of the antenna a scalar field.

3.4.1 A Solution Basis

It should be clear from the previous section that the electromagnetic field of an arbitrarily shaped antenna can be terribly complicated and probably impossible to fully describe analytically. In fact, the antenna pattern synthesis and analysis are fields of active research. In order to describe complicated propagation in a bounded region of linear free space, it is useful to break down the complicated received voltage levels as a function of space into a solution basis. We may postulate that a solution basis for *any baseband received voltage* is the set of all complex sinusoidal waves

$$e(\vec{r}) = \sum_n A_n \exp\left\{-j\vec{k}_n \cdot \vec{r} - j\phi_n\right\}, \tag{3.6}$$

where A_n are real amplitudes, ϕ_n are real phases, and $\vec{k}_n$ are real vectors. According to basic Fourier analysis, any function of space can be written in this form, moreover, since every term of (3.6) solves Maxwell's equations, this is an analytically satisfactory way of handling complicated propagation phenomena. However, if we really wanted to represent in this form arbitrary wavefields, there is an important point for which we have to account. The wave vector $\vec{k}$ of (3.6) may be complex valued. A complex $\vec{k}$ still solves Maxwell's time-harmonic equations and is needed to describe every possible instance of the received voltage in free space. To have a complete representation of the signal at the input to a radio receiver, it is best to classify the terms of the plane wave basis of (3.6) into *homogeneous* plane waves with strictly real-valued wave vectors and *inhomogeneous* plane waves with complex-valued wave vectors.

3.4.2 Homogeneous Plane Waves

Homogeneous plane waves are also called uniform plane waves because their envelope is a constant value that does not depend on position in space. Since the wave vector of a homogeneous plane wave is real-valued, the plane wave may be written as

$$e(\vec{r}) = A_0 \exp\left\{-j\mathcal{K}\vec{k}_0 \cdot \vec{r} - j\phi_0\right\}$$

where $\vec{k}_0$ is a unit vector that points in the direction of propagation and $\mathcal{K}$ is the free space wave number. In a time-harmonic free space analysis,

$$\mathcal{K} = 2\pi f/\lambda.$$

Regardless of the direction of travel, all homogeneous plane waves propagate at the same speed with the same wave number. This is not true of inhomogeneous plane waves.

3.4.3 Inhomogeneous Plane Waves

The complex wave vector

$$\vec{k} = \vec{\beta}\beta - j\vec{\alpha}\alpha$$

where $\vec{\beta}$ is a unit vector pointing in the direction of the real part of $\vec{k}$ and $\vec{\alpha}$ is a unit vector pointing in the direction of the imaginary part of $\vec{k}$. Since $\vec{\beta}$ and $\vec{\alpha}$ must be orthogonal and $\vec{k} \cdot \vec{k} = \mathcal{K}^2$, we can write

$$e(\vec{r}) = A_0 \exp\{\vec{\alpha} \cdot \vec{r}\} \exp\left\{-\beta\vec{\beta} \cdot \vec{r} - \phi_0\right\}.$$

The value of β is the actual wave number of the inhomogeneous plane wave, determining the rate of phase progression through space in the direction $\vec{\beta}$. The value of α represents the rate of amplitude decay and the amplitude of the plane wave attenuates in the direction of $\vec{\alpha}$. The direction of amplitude attenuation must always be orthogonal to the direction of propagation.

The wave number of the inhomogeneous plane wave is always greater than the free space wave number, $\mathcal{K}$. An effective wavelength, λ_{eff}, for inhomogeneous plane waves, can be defined as

$$\lambda_{eff} = \frac{2\pi}{\beta}.$$

Equiphase fronts of the inhomogeneous plane wave are more densely packed than the homogeneous plane wave. The second key difference between homogeneous and inhomogeneous plane waves is the exponential decay of amplitudes perpendicular to the direction of propagation. If the decay rate is small, then the wave number approaches the homogeneous free space wave number. Inhomogeneous plane waves are caused by scatterers or sources in the propagation medium outside the bounded free space region.

The larger wave numbers of inhomogeneous plane waves help meet the boundary conditions of electromagnetic fields close to material surfaces with fine spatial structure. Boundary conditions near large scatterers typically introduce inhomogeneous plane waves into a region of bounded free space. The direction of decay for the wave tends to point away from the scatterer. The direction of propagation for homogeneous plane waves is typically away from the scatterer.

3.5 Transmitting Information Using Waves

In wireless channel modeling it is usual to approximate the channel in a bounded region of free space where only homogeneous waves are present. Of course, the size of the free space region that satisfies these approximations varies from case to case and is largely a function of scatterer proximity. We know that inhomogeneous plane waves are caused by scatterers in free space and that their contribution to propagation diminishes in regions away from the significant scatterer. Thus, the proximity to significantly large scatterers usually determines the acceptable size of the region of space where we model the channel.

3.5.1 Antennas in Free Space

The first important case to study by means of the basis solution (3.6) is the ideal wireless channel. This channel results from a single uniform plane wave propagating through free space. The plane wave impinges upon a receiver antenna and excites a constant-amplitude voltage at the receiver that does not change in magnitude as a function of position. Indeed, this channel is representative of two antennas when their distance is large enough to make their size negligible. Assuming that the transmitter is located at the origin of a coordinate system the voltage measured at the output of the receive antenna located at position $\vec{r}$ and distance R, using the representation we have previously discussed is

$$e(\vec{r}) = A_0(R)\exp\left\{j\phi_0 - j\mathcal{K}\vec{k}_0 \cdot \vec{r}\right\}$$

and unveiling the carrier frequency f_c gives

$$e(\vec{r}) = A_0(R)\exp\left\{j\phi_0 - j\frac{2\pi f_c}{c}\vec{k}_0 \cdot \vec{r}\right\}.$$

For small frequency displacements about the carrier frequency, we can write

$$e(f,\vec{r}) = A_0(R)\exp\left\{j\phi_0 - j\frac{2\pi(f+f_c)}{c}\vec{k}_0\cdot\vec{r}\right\}, \tag{3.7}$$

and if a band-limited transmitted signal $s(t) = \int_{-\infty}^{+\infty} S(f)\exp\{j2\pi ft\}df$ is exciting the transmit antenna

$$\begin{aligned} e(t,\vec{r}) &= A_0(R)\int_{-\infty}^{+\infty} A(f)\exp\left\{j\phi(f) - j\frac{2\pi(f+f_c)}{c}\vec{k}_0\cdot\vec{r} + j\phi_0\right\} \\ &\quad \exp\{j2\pi ft\}df, \end{aligned} \tag{3.8}$$

where $S(f) = A(f)e^{j\phi(f)}$. If the distance between the two antennas R is such that the bandwidth of $S(f)$ (say, W) satisfies

$$\vec{k}_0\cdot\vec{r}/c << \frac{1}{W},$$

we can simplify the expression as

$$e(t,\vec{r}) = A_0(R)s(t)\exp\left\{-j\frac{2\pi f_c}{c}\vec{k}_0\cdot\vec{r} + j\phi_0\right\}. \tag{3.9}$$

For example, this model represents exactly two lossless Hertzian dipoles located at distance R.

Example

We know from basic electromagnetic theory that for the receiver at location $\vec{r} = (R,\theta,0)$ expressed in spherical coordinates the field components for a time-harmonic current with amplitude I_0 are

$$\vec{E} = E_\theta = j\mathcal{K}I_0 l\frac{e^{-j\mathcal{K}R}}{4\pi R}\sin\theta\vec{i}_\theta$$

where l is the length of the dipole. The induced current at the receiving dipole is simply $E_\theta l$ so that for a signal $s(t)$ with narrow bandwidth, (3.9) becomes

$$e(t,\vec{r}) = jl\mathcal{K}I_0 l^2 s(t)\frac{e^{-j\mathcal{K}R}}{4\pi R}\sin\theta.$$

In practice for free space communications we can show that (3.9) is fairly accurate for arbitrarily shaped antennas once we specify properly the power radiated and captured by the antennas.

Friis Transmission Equation

The *directivity* of the antenna can be found from the field intensity using the following formula (see the Poynting theorem in [1])

$$U(\theta,\phi) = \frac{1}{2\eta}\left(|E_\theta|^2 + |E_\phi|^2\right).$$

The directive gain of an antenna is

$$D_g(\theta,\phi) = \frac{U(\theta,\phi)}{U_0(\theta,\phi)}$$

where $U_0(\theta,\phi)$ is the radiation intensity of an isotropic source. An isotropic source is a point source with ideal isotropic radiating properties in all directions. The antenna efficiency $0 \leq e \leq 1$ is a dimensionless parameter that takes into account losses at the input/output terminals and within the structure itself of the antennas. Polarization loss ρ accounts for the mismatch in polarization between a wave generated by a transmit antenna and the polarization of the receive antenna. If the input power at the terminals transmit antenna is P_t, the received power is [4]

$$P_r = e_t e_r \frac{\lambda^2 D_{g,t}(\theta,\phi) D_{g,r}(\theta,\phi) P_t}{(4\pi R^2)} \rho_{t,r} \tag{3.10}$$

where

- e_t and e_r are the antenna efficiency of transmit and receive anetnna, respectively.
- $D_{g,t}(\theta,\phi)$ and $D_{g,r}(\theta,\phi)$ are the antenna directive gain of transmit and receive antenna, respectively.
- $\rho_{t,r}$ is the polarization loss between the polarization of the transmit antenna and the polarization of the receive antenna.
- λ is the wavelength of the carrier.

See also Figure 3.7 for the geometrical orientation of the two antennas. If the antennas are aligned for maximum directional gain,

$$P_r = \frac{\lambda^2 G_{0,t} G_{0,r} P_t}{(4\pi R^2)} \tag{3.11}$$

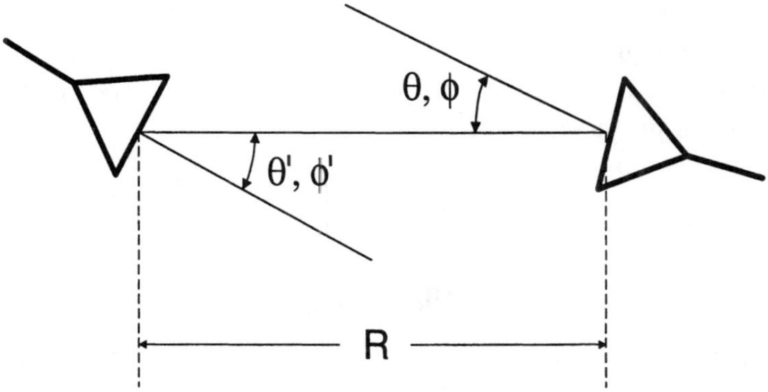

Figure 3.7 Geometry used in the Friis equation.

where $G_{0,t}$ and $G_{0,r}$ are the maximum gains of transmit and receive antennas, respectively. We can define a system gain as $|\Gamma_s(R)|^2 = P_r/P_t$ and express the signal received as

$$e(t, \vec{r}) = \Gamma_s(R)s(t)\exp\left\{-j\frac{2\pi f_c}{c}\vec{k}_0 \cdot \vec{r} + j\phi_0\right\}. \qquad (3.12)$$

More in general, if the transmitter is at position $\vec{s}$, we have

$$e(t, \vec{r}, \vec{s}) = \Gamma_s(R)s(t)\exp\left\{-j\frac{2\pi f_c}{c}\vec{k}_0 \cdot (\vec{r} - \vec{s}) + j\phi_0\right\}.$$

Example

Consider N_{tx} transmit antennas located at spatial coordinates $\vec{r}_1, \vec{r}_2, ..., \vec{r}_{N_{tx}}$ and N_{rx} receive antennas located at spatial coordinates $\vec{s}_1, \vec{s}_2, ..., \vec{s}_{N_{rx}}$. Denoting $s_i(t)$ the signal transmitted at the ith transmit antenna, we can express the noiseless signal received at kth receive antenna as

$$r_k(t) = \sum_{i=1}^{N_{tx}} \Gamma_s(R_{i,k})s_i(t)\exp\left\{-j\frac{2\pi f_c}{c}\vec{k}_{i,k} \cdot (\vec{r}_k - \vec{s}_i)\right\} \qquad (3.13)$$

where $R_{i,k}$, and $\vec{k}_{i,k}$ are, respectively, distance and unity vector of the plane wave between ith transmit antenna and kth receive antenna. In a more simplified expression we can write

$$\mathbf{r}(t) = \mathbf{H}\mathbf{s}(t)$$

where $\mathbf{r}(t)$ is an N_{rx} vector with components equal to the signals received at the N_{rx} receive antennas, $\mathbf{s}(t)$ is an N_{tx} vector with components equal to the signals transmitted at the N_{tx} transmit antennas, and $\mathbf{H}$ is an $N_{rx} \times N_{tx}$ matrix with the (i,k) component equal to the propagation factor

$$[\mathbf{H}]_{i,k} = \Gamma_s(R_{i,k})\exp\left\{-j\frac{2\pi}{\lambda}R_{i,k}\right\}.$$

3.5.2 Antennas in Multipath Propagation

Assuming that the transmitter is located at the origin of a coordinate system the voltage measured at the output of the receive antenna located at position $\vec{r}$ and distance R, using the representation we have previously discussed

$$e(\vec{r}) = \sum_{n=1}^{N} A_n \exp\left\{j\phi_n - j\mathcal{K}\vec{k}_n \cdot \vec{r}\right\}$$

where $\mathcal{K} = \omega/c = 2\pi f/c$, $\vec{k}_n$ are unit vectors in the direction of propagation, $A_n e^{j\phi_n}$ are the coefficients of the plane wave expansion which characterizes well the scalar field as measured at position $\vec{r}$. The transmitted signal $s(t)$ is now included to obtain

$$\begin{aligned} e(t,\vec{r}) &= \sum_{n=1}^{N} A_n e^{j\phi_n} \int_{-\infty}^{+\infty} A(f)\exp\left\{j\phi(f) - j\frac{2\pi(f+f_c)}{c}\vec{k}_n \cdot \vec{r}\right\} \\ &\quad \exp\{j2\pi ft\}df. \end{aligned} \tag{3.14}$$

To connect the model with a physical situation, consider Figure 3.8. There are three significant scattering objects and many waves are being reflected so that the prediction of the electrical field at the receive antenna becomes a formidable challenge. However, it is possible to observe that as the distance among the scattering objects increases, a single plane wave can be used to approximate propagation between each of the objects. In other words, the expansion in (3.6) assumes also a physical meaning in addition to being mathematically consistent. For example, two main waves could approximate the extremely complicated propagation scenario of Figure 3.8. Observe that now multiple waves are being scattered,

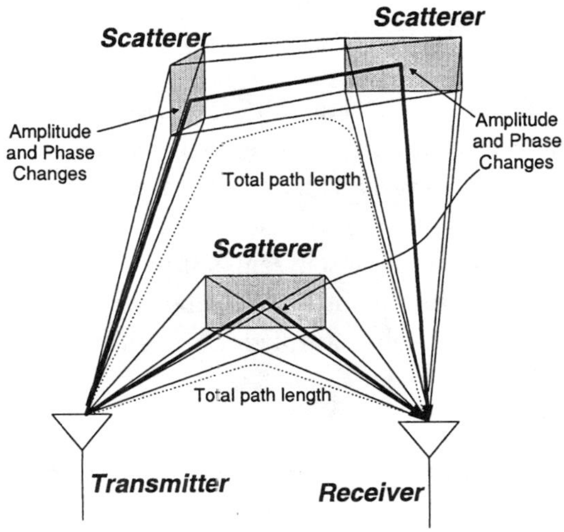

Figure 3.8 Time-harmonic propagation approximates a complicated propagation scenario.

and some of them may have a cumulative time of travel whose inverse is significant with respect the bandwidth of the signal $s(t)$.

$$e(t,\vec{r}) = \sum_{n=1}^{N_2} A_n e^{j\phi_n} s\left(t - \frac{\vec{k}_n \cdot \vec{r}}{c}\right) \exp\left\{j\phi_n - jK\vec{k}_n \cdot \vec{r}\right\}. \quad (3.15)$$

Evidently a prediction of the power received and scattered is in this case a difficult problem to attack. Ray-tracing theory attempts such a prediction but it inevitably gives results of limited applicability. It is important, however, to observe that the above model clearly shows the connection with the well-known tapped-delay line model for multipath characterization [5]

$$e(t,\vec{r}) = \sum_l \beta_l e^{j\phi_l} s(t-\tau_l) = \int_{-\infty}^{t} h(\tau,\vec{r}) s(t-\tau) d\tau \quad (3.16)$$

where

$$h(\tau,\vec{r}) = \sum_l \beta_l(\vec{r}) e^{j\psi_l(\vec{r})} \delta(\tau - \tau_l(\vec{r})) \quad (3.17)$$

$$\beta_n(\vec{r})e^{j\psi_n(\vec{r})} = A_n e^{j\phi_n} \exp\left\{j\phi_n - j\mathcal{K}\vec{k}_n \cdot \vec{r}\right\}, \tag{3.18}$$

$$\tau_n(\vec{r}) = \frac{\vec{k}_n \cdot \vec{r}}{c}. \tag{3.19}$$

Now, we remove the restriction that the transmitter is at the origin of the coordinate system. Due to the reciprocity principle of antenna theory, that basically states that the field-response of the transmit antenna to a current source is also the same exact field in which reception generates the same current source, we can think that the channel seen from the transmitter can be modeled similarly to the receiver so the scalar field sampled at position $\vec{r}$ as generated at position $\vec{s}$ is

$$e(\vec{r},\vec{s}) = \sum_{m=1}^{M} A'_m \exp\left\{j\phi'_m - j\mathcal{K}\vec{k}'_m \cdot \vec{s}\right\} \sum_{n=1}^{N} A_n \exp\left\{j\phi_n - j\mathcal{K}\vec{k}_n \cdot \vec{r}\right\}.$$

Following the same exact procedure

$$e(t,\vec{r},\vec{s}) = \sum_{m=1}^{M}\sum_{n=1}^{N} A_n e^{j\phi_n} A'_m e^{j\phi'_m} s\left(t - \frac{\vec{k}_n \cdot \vec{r}}{c} - \frac{\vec{k}'_m \cdot \vec{s}}{c}\right)$$
$$\exp\left\{j(\phi_n + \phi'_m) - j\mathcal{K}\vec{k}_n \cdot \vec{r} - j\mathcal{K}\vec{k}'_m \cdot \vec{s}\right\}. \tag{3.20}$$

Again, the model clearly shows the connection with the well-known tapped-delay line model for multipath characterization [5]

$$e(t,\vec{r},\vec{s}) = \sum_{n,m} \beta_{n,m}(\vec{r},\vec{s}) e^{j\phi_{n,m}(\vec{r},\vec{s})} s\left(t - \tau_{n,m}(\vec{r},\vec{s})\right)$$
$$= \int_{-\infty}^{t} h(\tau,\vec{r},\vec{s}) s(t-\tau) d\tau \tag{3.21}$$

where

$$h(\tau,\vec{r},\vec{s}) = \sum_{n,m} \beta_{n,m}(\vec{r},\vec{s}) e^{j\psi_{n,m}(\vec{r},\vec{s})} \delta(\tau - \tau_{n,m}(\vec{r},\vec{s}))$$

$$\beta_{n,m}(\vec{r},\vec{s}) e^{j\psi_{n,m}(\vec{r},\vec{s})} = A_n A'_m e^{j(\phi_n + \phi'_m)}$$
$$\exp\left\{j(\phi_n + \phi'_m) - j\mathcal{K}\vec{k}_n \cdot \vec{r} - j\mathcal{K}\vec{k}'_m \cdot \vec{s}\right\},$$

$$\tau_{n,m}(\vec{r}) = \frac{\vec{k}_n \cdot \vec{r}}{c} + \frac{\vec{k}'_m \cdot \vec{s}}{c}. \tag{3.22}$$

Example

Consider N_{tx} transmit antennas located at spatial coordinates $\vec{r}_1, \vec{r}_2, ..., \vec{r}_{N_{tx}}$ and N_{rx} receive antennas located at spatial coordinates $\vec{s}_1, \vec{s}_2, ..., \vec{s}_{N_{rx}}$. Denoting $s_i(t)$ as the signal transmitted at the ith transmit antenna, we can express the noiseless signal received at the kth receive antenna as

$$r_k(t) = \sum_{i=1}^{N_{tx}} \int_{-\infty}^{t} h(\tau, \vec{r}_k, \vec{s}_i) s_i(t-\tau) d\tau \tag{3.23}$$

where $h(\tau, \vec{r}, \vec{s})$ is defined in (3.22). In a more simplified expression we can write

$$\mathbf{r}(t) = \int_{-\infty}^{t} \mathbf{H}(\tau)\mathbf{s}(t-\tau)d\tau$$

where $\mathbf{r}(t)$ is an N_{rx}-vector with components equal to the signals received at the N_{rx} receive antennas, $\mathbf{s}(t)$ is an N_{tx}-vector with components equal to the signals transmitted at the N_{tx} transmit antennas, $\mathbf{H}(t)$ is an $N_{rx} \times N_{tx}$ matrix with the (i, k) component equal to the propagation factor

$$[\mathbf{H}(t)]_{i,k} = h(\tau, \vec{r}_k, \vec{s}_i).$$

3.5.3 Mobility

Mobility of the transmitter and the receiver will require that the position vectors are changing in time

$$\vec{r}(t) = [r_r(t), r_\theta(t), r_\phi(t)]^T = [r_x(t), r_y(t), r_z(t)]^T,$$

$$\vec{s}(t) = [s_r(t), s_\theta(t), s_\phi(t)]^T = [s_x(t), s_y(t), s_z(t)]^T.$$

The simplest model accounts for constant velocity of point $\vec{r}$ or $\vec{s}$. For radio communications this is appropriate. For example, 900 MHz with a maximal mobile speed of 200 km/hr gives a maximum Doppler spread of 333 Hz and coherence is on the order of 3 ms. This time scale is then extremely fast with respect to mobile dynamics so we can use the simplified position evolution

$$\vec{r}(t) = \vec{r}(t_0) + \vec{v}_r(t - t_0)$$

$$\vec{s}(t) = \vec{s}(t_0) + \vec{v}_s(t - t_0)$$

where $\vec{v}_r(t)$ and $\vec{v}_s(t)$ are the velocity vectors for receiver and transmitter, respectively, and t_0 is some arbitrary reference time. The space-time channel model is then

$$h(t, \tau, \vec{r}, \vec{s}) = \sum_{n,m} \beta_{n,m}\left(\vec{r}(t), \vec{s}(t)\right) e^{j\psi_{n,m}(\vec{r}(t),\vec{s}(t))} \delta(\tau - \tau_{n,m}(\vec{r}(t), \vec{s}(t)). \tag{3.24}$$

where the parameters of the model $\beta_{n,m}(\vec{r}(t), \vec{s}(t))$, $\beta_{n,m}(\vec{r}(t), \vec{s}(t))$, $\tau_{n,m}(\vec{r}(t), \vec{s}(t))$ are obtained from (3.22) with $\vec{r}(t)$ and $\vec{s}(t)$ instead of $\vec{r}$ and $\vec{s}$.

3.6 The Channel Coherence Concept

Wireless channels are functions of time, frequency, and space. A wireless channel is in general affected by some type of selectivity. A channel is selective if it changes as a function of time, frequency, or space. A channel has coherence if it does not change as a function of time, frequency, or space.

3.6.1 Temporal Coherence

Mathematically, we express temporal coherence in terms of a narrowband (no frequency dependence), fixed (no spatial dependence) channel,

$$\tilde{h}(t) \simeq \text{constant for } |t_0 - t| \leq T_c$$

where T_c is the size of the time window of interest, and t_0 is an arbitrary observation instant. The largest value of T_c is called the *coherence time* and is the time interval over which the channel appears static. In RF communications the most common source of temporal incoherence is motion by the transmitter, the receiver, or the scatterers in the propagation environment. If the transmitted data rate is comparable to the temporal coherence, it becomes challenging for the receiver to demodulate the transmitted signal reliably. One simple method for reliable communications in a time-varying channel is to transmit data using symbols that are much longer in duration than the channel coherence time. When the envelope of the carrier wave fluctuates at a rate faster than the transmitted symbol rate, the channel is said to be flat fading.

3.6.2 Frequency Coherence

A wireless channel has frequency coherence if the magnitude of the carrier does not change over a frequency interval of interest. The interval is usually the bandwidth of the transmitted signal

$$H(f) \simeq \text{constant for } |f_c - f| \leq B_c$$

where B_c is the span of the frequency interval, and f_c is the center carrier frequency. The largest value of B_c is called the coherence bandwidth and is the approximate range of frequencies over which the channel appears static. The loss of frequency coherence in a wireless communications system is caused by the dispersion of multipath propagation. Since each received multipath wave has traveled a different path from the transmitter, the same transmitted signal will arrive at the receiver dispersed in time. In the time domain, a dispersive channel introduces *intersymbol interference.* This behavior in the frequency domain gives two classes of fading in wireless communications. A wireless channel with a coherence bandwidth that is less than the bandwidth of a transmitted signal is said to have frequency-selective fading. A channel with a coherence bandwidth that is greater than the transmitted signal bandwidth is said to have frequency-flat fading.

3.6.3 Spatial Coherence

A wireless channel has spatial coherence if the magnitude of the carrier does not change over a spatial displacement of the receiver. Spatial coherence is expressed in terms of the static narrowband channel, $\tilde{h}(\mathbf{r})$, which is a function of three-dimensional position vector, $\mathbf{r}$

$$\tilde{h}(\mathbf{r}) \simeq \text{constant for } |\mathbf{r}_0 - \mathbf{r}| \leq \mathbf{V}_c$$

where $\mathbf{V}_c$ is the volume of the position displacement, and $\mathbf{r}_0$ is an arbitrary position in space. The largest value of $\mathbf{V}_c$ is called the coherence distance and is the approximate distance that a wireless receiver can move with the channel appearing to be static. Spatial incoherence is caused by multiple paths arriving from many different directions in space. These waves create constructive and destructive interference so that the received signal power does not appear to be constant. This type of channel has spatial selectivity.

3.6.4 Mobility

Spatial incoherence in a channel leads directly to temporal incoherence. As a receiver moves through space, the fading may treated as a function of time instead of space by relating the position $\mathbf{r}$ and the time t by a simple equation of motion

$$\mathbf{r}(t) = \mathbf{r}(t_0) + \mathbf{v}t$$

where $\mathbf{v}$ is the velocity vector of the receiver. Small-scale fading causes temporal incoherence for wireless systems with mobile receivers.

3.6.5 Large-Scale Fading

Large-scale fading, refers to fluctuations in spatially averaged received power due to shadowing and scattering of objects in the propagation environment. Typically, small-scale fluctuations occur when a receiver moves a distance comparable to the size of the electromagnetic wavelength of the carrier. Large-scale fading occurs when the receiver moves over many wavelengths.

3.7 Temporal Statistical Characterizations

Consider now

$$h(t, \tau) \triangleq h(t, \tau, \vec{r}, \vec{s})$$

The generally accepted assumption is that $h(t, \tau)$ are wide-sense stationary (WSS) processes [5, 6]. Recalling the definition of the WSS process, this means that the autocorrelation of the process can be expressed as

$$\frac{1}{2} E\left\{h(t, \tau_1)^* h(t + \Delta t, \tau_2)\right\} = \phi(\tau_1, \tau_2, \Delta t).$$

Making the additional assumption that any path attenuation and phase shift is uncorrelated with delay and phase shift of any other path, we can write

$$\frac{1}{2} E\left\{h(t, \tau_1)^* h(t + \Delta t, \tau_2)\right\} = \phi^c(\tau_1, \Delta t)\delta(\tau_1 - \tau_2).$$

The function $\phi^c(\tau, 0) = \phi^c(\tau)$ is called the *multipath intensity profile.* The range of values of τ over which $\phi^c(\tau)$ is nonzero is called the *multipath spread* of the channel. Characterizing the channel in the frequency domain and defining the Fourier transform with respect to τ, we obtain

$$\tilde{H}(f,t) = \int_{-\infty}^{+\infty} h(\tau,t)e^{-j2\pi f\tau}d\tau.$$

It is important to define the correlation function

$$\Phi^c(f_1,f_2,\Delta t) = \frac{1}{2}E\left\{\tilde{H}^*(f_1,t)\tilde{H}(f_2,t+\Delta t)\right\} = \Phi^c(\Delta f,\Delta t)$$

where we have used again the uncorrelated scattering assumption. Observe that

$$\Phi^c(\Delta f,\Delta t) = \mathcal{F}\left\{\phi^c(\tau,\Delta t)\right\}.$$

Taking again the Fourier transform of $\Phi^c(\Delta f,\Delta t)$ with respect to Δt

$$S^c(\Delta f,\lambda) = \int_{-\infty}^{+\infty} \Phi^c(\Delta f,\Delta t)e^{-j2\pi\lambda\Delta t}d\Delta t$$

and setting $\Delta f = 0$, we obtain

$$S^c(0,\lambda) = S^c(\lambda)$$

the *Doppler power spectrum*. The range of values of λ over which $S_c(\lambda)$ is nonzero is called the *Doppler spread* of the channel.

A function that totally characterizes the channel and that is obtained as

$$\mathcal{S}^c(t,\lambda) = \int_{-\infty}^{+\infty} \phi^c(\tau,\Delta t)e^{-j2\pi\lambda\Delta t}d\Delta t$$

or

$$\mathcal{S}^c(t,\lambda) = \int_{-\infty}^{+\infty} S^c(\Delta f,\lambda)e^{j2\pi\Delta f\tau}d\Delta f,$$

is called the *scattering function* of the channel.

3.8 Spatial Statistical Characterizations

Parameters characterizing the spatial dispersion of the channel can be defined similarly to what we have described in the temporal domain. For simplicity, in this section we will neglect the time dependence of the channel. In other words,

$$\int_{-\infty}^{+\infty} h(t,\tau,\vec{r},\vec{r}')e^{-j\omega\tau}d\tau = H(t,\omega,\mathbf{r},\mathbf{r}') \triangleq H(\mathbf{r},\mathbf{r}').$$

Moreover, we assume that a point $\mathbf{r}$ is defined by D spatial coordinates:

$$\mathbf{r} = [r_1, r_2, ..., r_D].$$

First, one makes the assumptions

$$\frac{1}{2}E\left\{H(\mathbf{r}, \mathbf{r}_1)^* H(\mathbf{r} + \Delta\mathbf{r}, \mathbf{r}_2)\right\} = \phi_S(\mathbf{r}_1, \mathbf{r}_2, \Delta\mathbf{r})$$

and

$$\frac{1}{2}E\left\{H(\mathbf{r}, \mathbf{r}_1)^* H(\mathbf{r} + \Delta\mathbf{r}, \mathbf{r}_2)\right\} = \phi^S(\Delta\mathbf{r}, \mathbf{r}_1)\delta(\mathbf{r}_1 - \mathbf{r}_2).$$

The function $\phi^S(\mathbf{r}, 0) = \phi^S(\mathbf{r})$ can be defined as the *spatial multipath intensity profile.* The spatial range of values of $\mathbf{r}$ over which $\phi^S(\mathbf{r})$ is nonzero is called the *spatial multipath spread* of the channel. Characterizing the channel in the D-dimensional spatial frequency domain $\mathbf{f} = [f_1, f_2, ..., f_D]$, and defining the Fourier transform with respect to $\mathbf{r} = [r_1, r_2, ..., r_D]$, we obtain

$$\mathcal{H}(\mathbf{f}, \mathbf{r}) = \int_{s_1=-\infty}^{+\infty} \cdots \int_{s_D=-\infty}^{+\infty} H(\mathbf{r}, \mathbf{s}) \prod_{i=1}^{D} e^{-j2\pi f_i s_i} ds_i.$$

It is important to define the correlation function

$$\Phi^S(\mathbf{f}_1, \mathbf{f}_2, \Delta\mathbf{r}) = \frac{1}{2}E\left\{\mathcal{H}^*(\mathbf{f}_1, \mathbf{r})\mathcal{H}(\mathbf{f}_2, \mathbf{r} + \Delta\mathbf{r})\right\} = \Phi^S(\Delta\mathbf{f}, \Delta\mathbf{r})$$

where we have used again a kind of spatial uncorrelated scattering assumption. Observe that

$$\Phi^S(\Delta\mathbf{f}, \Delta\mathbf{r}) = \mathcal{F}\left\{\phi^S(\mathbf{r}, \Delta\mathbf{r})\right\}.$$

Taking again the Fourier transform of $\Phi^S(\Delta\mathbf{f}, \Delta\mathbf{r})$ with respect to $\Delta\mathbf{r}$, we obtain

$$\begin{aligned} S^S(\Delta\mathbf{f}, \lambda_1, ..., \lambda_D) &= \int_{\Delta r_1=-\infty}^{+\infty} \cdots \int_{\Delta r_D=-\infty}^{+\infty} \Phi^S(\Delta\mathbf{f}, \Delta\mathbf{r}) \\ &\quad \cdot \prod_{i=1}^{D} e^{-j2\pi\lambda_i \Delta r_i} d\Delta r_i \end{aligned}$$

and setting $\Delta\mathbf{f} = \mathbf{0}$, we obtain

$$S^S(\mathbf{0}, \lambda_1, ..., \lambda_D) = S^S(\lambda_1, ..., \lambda_D)$$

the *spatial Doppler power spectrum.* The range of values of λ over which $S_c(\lambda_1, ..., \lambda_D)$ is nonzero is called the *spatial Doppler spread* of the channel.

3.9 Communication Modes of a Wave

We end the chapter with a result from applied physics that also should illuminate some of the concepts we have attempted to explain. Chapter 2 showed that the band-limited waveform channel cannot have more than $2W$ dimensions (or communication modes) per unit time. We observe that an ideally band-limited wave with bandwidth W available for transmission can be proven to lie in a subspace with more than $2W$ dimensions per unit time. The underlying assumption of the traditional framework is the one of a *point source* approximation of the "transmitter" and the "receiver." This interpretation is conceptually satisfactory but not sufficiently justified from a physical point of view. An extremely important and common class of situations arises in the context of *optics.* Suppose we are looking at an object through an imaging lens and we wish to know the number of *resolvable* spots. A *resolved* spot is effectively a distinct channel from the object to the image, and it could be used to establish a separate information channel. The *diffraction* approach was formalized by di Francia in 1955 [7] and extensively documented by Gabor in 1961 [8]. The diffraction approach says that *the smallest spot we can use is the one from which the diffracted light approximately fills the aperture of an imaging lens or surface.* Using a smaller spot will make the diffraction angle larger, and we will establish a weak connection because significant portions of the light will be diffracted, instead of being collected by the image surface. Observe that a spot (or spatial channel) is one dimension at a particular frequency (since we are assuming monochromatic light). We know from Chapter 2 that the transfer of information over each dimension is further limited by the presence of disturbances (noise). The *diffraction* approach predicts

$$N \simeq \Omega A/\lambda^2$$

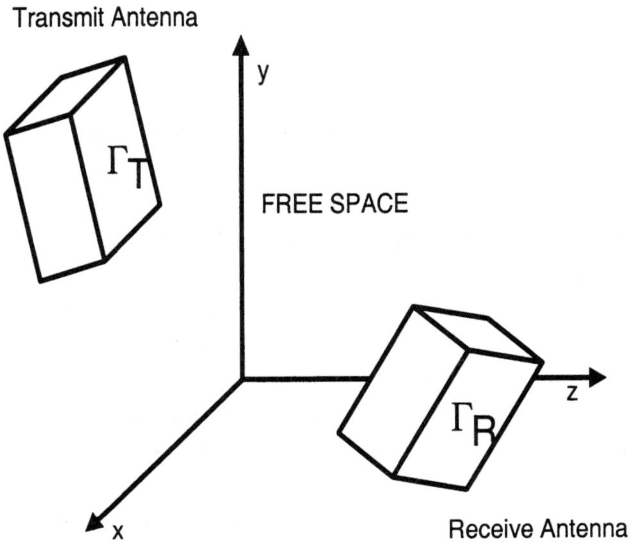

Figure 3.9 The problem of communicating in free space with two antennas of non-negligible size.

dimensions over which communication can take place. In this expression Ω is the solid angle projected by the image surface at the object surface, A is the object surface area (assuming that the object is of planar shape), λ is the wavelength. So optics already gives us a first indication that spatial dimensions must have an impact on the ability to transfer information.

3.9.1 Formal Approach

In the following we sketch the work of Miller [9]. The reader interested in the mathematical details of the rigorous derivation should refer to the original article [9]. Consider the situation depicted in Figure 3.9. A transmit antenna with spatial volume Γ_T generates electromagnetic waves $\psi(t, \mathbf{r})$ which are received by the receive antenna with spatial volume Γ_R as $\phi(t, \mathbf{r})$. The *scalar* wave equation for this case is slightly different from the homogeneous wave equation previously derived, and it is expressed as

$$\left(\nabla^2 - \frac{1}{c^2}\frac{\partial^2}{\partial t^2}\right)\phi(t,\mathbf{r}) = -\psi(t,\mathbf{r}). \tag{3.25}$$

Assuming that the wavefields have a Fourier representation $\psi(t,\mathbf{r}) = \int d\omega \Psi(\omega,\mathbf{r})e^{j\omega t}$, and $\phi(t,\mathbf{r}) = \int d\omega \Phi(\omega,\mathbf{r})e^{j\omega t}$, each Fourier component satisfies the inhomogeneous Helmholtz equation [1]

$$\left(\nabla^2 + \mathcal{K}^2\right)\Phi(\omega,\mathbf{r}) = -\Psi(\omega,\mathbf{r}). \tag{3.26}$$

At any particular ω the Green's function $G(\mathbf{r},\mathbf{r}')$ describes the possible waves $\psi(t,\mathbf{r})$, $\phi(t,\mathbf{r})$ obeying the Helmoltz equation

$$\Phi(\mathbf{r}) = \int_{\Gamma_T} G(\mathbf{r},\mathbf{r}')\Psi(\mathbf{r}')d\mathbf{r}',$$

where for simplicity we have neglected the explicit notation for ω. The Green's function for *outgoing waves* in free space is

$$G(\mathbf{r},\mathbf{r}') = e^{-j\mathcal{K}|\mathbf{r}-\mathbf{r}'|} / \left(4\pi|\mathbf{r}-\mathbf{r}'|\right).$$

Define the two eigenvalue problems[2]

$$|\xi|^2\Psi(\mathbf{r}) = \int_{\Gamma_T} K(\mathbf{r},\mathbf{r}')\Psi(\mathbf{r}')d\mathbf{r}' \tag{3.27}$$

$$|\xi|^2\Phi(\mathbf{r}) = \int_{\Gamma_R} J(\mathbf{r},\mathbf{r}')\Phi(\mathbf{r}')d\mathbf{r}' \tag{3.28}$$

where

$$K(\mathbf{r},\mathbf{r}'') = \int_{\Gamma_R} G^*(\mathbf{r}',\mathbf{r})G(\mathbf{r}',\mathbf{r}'')d\mathbf{r}'$$

$$J(\mathbf{r},\mathbf{r}') = \int_{\Gamma_T} G^*(\mathbf{r},\mathbf{r}'')G(\mathbf{r}',\mathbf{r}'')d\mathbf{r}''.$$

Due to the fact that $J(\mathbf{r},\mathbf{r}')$ and $K(\mathbf{r},\mathbf{r}')$ are Hermitian kernels, a complete set of eigenfunctions with real eigenvalues $|\xi|^2$ exists. The eigenvalues and eigenfunctions solutions of (3.27) and (3.28) are

$$|\xi_1|^2,\ |\xi_2|^2, \ldots,$$

2. Remember from Chapter 2 that when we try to determine dimensionality of a signal we typically have to solve eigenvalue problems.

$$\Phi_1(\mathbf{r}),\ \Phi_2(\mathbf{r}),\ldots$$

$$\Psi_1(\mathbf{r}),\ \Psi_2(\mathbf{r}),\ldots$$

Each solution is a spatial dimension of the system across which we can establish $2W$ dimensions per second. In practice the number of dimensions is finite because finite is the volume occupied by the transmit antenna and the receive antenna. Consider the case when the two prisms of Figure 3.9 are oriented along the same axis and have size $2\Delta x_T, 2\Delta y_T, 2\Delta z_T$ in the x, y, z directions for the transmitter and $2\Delta x_R, 2\Delta y_R, 2\Delta z_R$ in the x, y, z directions for the receiver. The distance between the two prisms (center to center) is r. The volumes are far apart with respect to their sizes. The first eigenvalue problem becomes

$$|\xi|^2\Psi(\mathbf{r}_T) = \int_{\Gamma_T}\int_{\Gamma_R} \frac{e^{j\mathcal{K}|\mathbf{r}_R-\mathbf{r}_T|}}{4\pi|\mathbf{r}_R-\mathbf{r}_T|}\frac{e^{-j\mathcal{K}|\mathbf{r}_R-\mathbf{r}'_T|}}{4\pi|\mathbf{r}_R-\mathbf{r}'_T|}\Psi(\mathbf{r}'_T)d\mathbf{r}_R d\mathbf{r}'_T.$$

This problem can be solved analytically as shown by Miller in [9] by

$$\Psi(\mathbf{r}) = F_T(\mathbf{r})\beta_T(\mathbf{r})$$

where

$$F_T(\mathbf{r}) = e^{-j\mathcal{K}\left(z-\frac{1}{2r}(x^2+y^2)\right)}$$

and

$$\beta_T(\mathbf{r}) = \alpha_{T,x}(x)\alpha_{T,y}(y)C_z$$

where C_z accounts for the fact that the eigenproblem is solved by a constant in the z direction. The generic solution has been shown by Miller to be

$$\Psi_{n_x,n_y}(\mathbf{r}) = F_T(\mathbf{r})\alpha_{n_x}(x)\alpha_{n_y}(y)$$

where $\alpha_{n_x}(x)$ is the n_xth prolate spheroidal function with bandwidth

$$\Omega_{T,x} = \frac{\mathcal{K}\Delta x_R}{r},$$

scale Δx_T and eigenvalue $\nu_{T,x,n}$. The same applies to the y direction. Following similar procedures for the second eigenvalue problem, it was found that

$$\Phi(\mathbf{r}) = F_R(\mathbf{r})\beta_R(\mathbf{r})$$

where

$$F_R(\mathbf{r}) = \epsilon^{-j\mathcal{K}\left(z-\frac{1}{2r}(x^2+y^2)\right)}$$

and

$$\beta_R(\mathbf{r}) = \alpha_{R,x}(x)\alpha_{R,y}(y)C_z$$

where C_z accounts for the fact that the eigenproblem is solved by a constant in the z direction. The generic solution was shown to be

$$\Phi_{n_x,n_y}(\mathbf{r}) = F_R(\mathbf{r})\alpha_{n_x}(x_R)\alpha_{n_y}(y_R)$$

where $\alpha_{n_x}(x)$ is the n_xth prolate spheroidal function with bandwidth $\Omega_{R,x} = \frac{\mathcal{K}\Delta x_T}{r}$, scale Δx_R and eigenvalue $\nu_{R,x,n}$. The same applies to the y direction. The eigenvalues $\nu_{T,x,n}$, $\nu_{T,y,n}$, $\nu_{R,x,n}$, $\nu_{R,y,n}$ are known to decrease toward 0 in value as n increases and $\nu_n > \nu_{n-1}$ for any n. Moreover, there is a single critical value for n, $n_{x,\text{max}} = \frac{2\Delta x_T \Delta x_R}{\lambda r}$ beyond which they fall off extremely rapidly in value. This critical value defines the spatial dimensionality of the system in each direction. The essential number of dimensions per second available for communications are then

$$N = 2W n_{x,\text{max}} n_{y,\text{max}} = 2W\frac{16\Delta x_T \Delta x_R \Delta y_T \Delta y_R}{\lambda^2 r^2}.$$

A small antenna is characterized by

$$\Delta_{x,T}\Delta_{x,R}/(\lambda r) << 1$$

where $\Delta_{x,T}$ and $\Delta_{x,R}$ are maximum sizes of the antennas in the x direction. In the case of small antennas only one solution to the eigenequations exists [9]

$$\Psi(\mathbf{r}) = \frac{1}{\sqrt{\Gamma_T}}e^{-j\mathcal{K}z},$$

and

$$\Phi(\mathbf{r}) = \frac{1}{\sqrt{\Gamma_R}}e^{-j\mathcal{K}z}.$$

This implies that for small antennas there is only one mode for communication and the optimum choice is a plane wave that travels with direction from the transmit antenna to the receive antenna.

3.10 Key Points of This Chapter

This chapter has prepared the necessary ground for the algorithmic developments of the following chapters. The physics of wave propagation, fundamentally explained by Maxwell's equations, allowed us to understand the important spatial properties of the wireless channel. We have also reviewed many important concepts related to antenna modeling and their connections with well-established communication theory assumptions. We focused in particular on the spatial characterization of a signal transmitted or received by an antenna both in free space and in a multipath scenario, and emphasized the point that plane waves have the capability of characterizing complicated scattering environments, just like sinusoids in Fourier analysis can characterize arbitrary signals. We have also introduced the reader to the characterization of a signal transmitted or received by an array of sensors (or by multiple antennas). The most important point of this chapter is that the dimensionality of a band-limited electromagnetic wave collected at different spatial positions exceeds $2W$ per unit time where W is the bandwidth occupied by the wave.

References

[1] P. M. Morse and H. Feshback, *Methods of Theoretical Physics, Part I,* New York: McGraw-Hill, 1953.

[2] J. D. Kraus and K. R. Carver, *Electromagnetics*, New York: McGraw-Hill, 1973.

[3] J. D. Kraus, *Antennas*, New York: McGraw-Hill, 1950.

[4] S. K. Schelkunoff and H. T. Friis, *Antennas: Theory and Practice,* New York: Wiley, 1952.

[5] J. G. Proakis, *Digital Communications,* New York: McGraw-Hill, 1989.

[6] P. A. Bello, "Characterization of Randomly Time Variant Linear Channels," *IEEE Trans. Comm. Syst. Tech.*, Vol. 11, pp. 360–393, Dec. 1963.

[7] G. Toraldo di Francia, "Resolving Power and Information," *Opt. Soc. Am.*, Vol. 5, pp. 497–501, 1955.

[8] D. Gabor, "Light and Information," *Progress in Optics*, Vol. 1, E. Wolf, Ed., Amsterdam, The Netherlands: North-Holland, pp. 109–153, 1961.

[9] D. A. B. Miller, "Communicating with Waves Between Volumes: Evaluating Spatial Orthogonal Channels and Limits on Coupling Strengths," *Appl. Opt.*, Vol. 39, pp. 1681–1699, April 2000.

4

Dimensionality of the Wireless Channel

4.1 Introduction

The operation of the simplest wireless communication system is characterized by the following:

- A transmitted signal, $s(t)$;
- A received signal, $r(t)$;
- A channel, $h(t)$, [or $h(t, \tau)$ in the mobile channel case];
- An additive noise random process, $n(t)$.

If the channel is linear and time-invariant, then it is possible to relate these quantities as

$$r(t) = \int_{-\infty}^{+\infty} h(\tau)s(t-\tau)d\tau + n(t).$$

When the channel is linear but time-varying, the input/output relationship is

$$y(t) = \int_{-\infty}^{+\infty} h(t,\tau)s(\tau)d\tau + n(t). \tag{4.1}$$

This simple model captures the fundamental properties of the single-antenna communication system shown in Figure 4.1. From Chapter 3 we know how to relate the impulse response of the channel $h(t)$ or $h(t, \tau)$ to a particular propagation environment. We know in fact that if $\vec{r}$ is the position of the receive antenna and $\vec{s}$ is the position of the antenna at the receiver,

$$h(t) \triangleq h(t, \vec{r}, \vec{s})$$

or in the time-varying case

$$h(t, \tau) \triangleq h(t, \tau, \vec{r}, \vec{s})$$

where $h(t, \vec{r}, \vec{s})$ is expressed in (3.21) [and $h(t, \tau, \vec{r}, \vec{s})$ is expressed in (3.25)] as the sum of a plurality of waves with different directions of travel. The waves impress different delays, attenuations, and phase shifts to different frequency components of the original signal. In this chapter we analyze the dimensionality of such a channel and quickly move on to the multiantenna channel as a generalization of the typical single-antenna channel.

4.1.1 Single Antenna Considerations

As we have already seen in Chapter 2, for some canonically simple cases the ultimate throughput of a linear (wireless) communication channel is characterized by the modes (or dimensions) of the channel. The most generic type of *linear* channel can be represented by the complex baseband equivalent input/output expression (4.1) where $h(t, \tau)$ is the response of the channel in Figure 4.1 to an impulse at time $t - \tau$ as measured at time t, $n(t)$ is additive Gaussian noise such that

$$\sigma^2 \delta(t-s) = E\{n(t)n^*(s)\},$$

and $s(t)$ is the transmitted signal. We are interested in a finite signaling time T. Define

$$R_h(t, \tau) = \int_T h(\theta, t) h^*(\theta, \tau) d\theta \tag{4.2}$$

and consider the integral eigenequation

$$\int_T R_h(t, u)\phi(u) du = \lambda \phi(t). \tag{4.3}$$

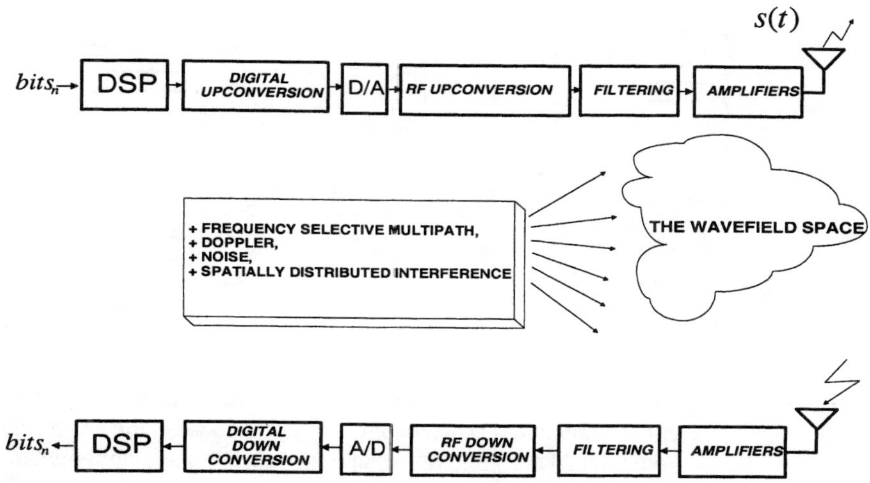

Figure 4.1 The parameters characterizing multipath propagation in a single-antenna system can be easily lumped into the effect of a linear operator $h(t, \tau)$.

The $\mathcal{R}$ solutions of (4.3) describe the $\mathcal{R}$ "modes" of the channel in turn defined by the orthonormal basis

$$\phi_1(t),\ \varphi_2(t),\ ...,\ \phi_{\mathcal{R}}(t),$$

and the eigenvalues

$$\lambda_1,\ \lambda_2,\ ...,\ \lambda_{\mathcal{R}}.$$

Signaling as

$$s(t) = \sum_{n=1}^{\mathcal{R}} a_n \phi_n(t), \qquad (4.4)$$

and processing the received signal as in Figure 4.2 can achieve capacity with Gaussian symbols a_n [1] because it creates $\mathcal{R}$ parallel channels

$$r_i = \sqrt{\lambda_i} a_i + \eta_i, \quad i = 1, 2, ..., \mathcal{R},$$

with SNR equal to $\frac{\sigma_{a,n}^2 \lambda_n}{\sigma^2}$, where $\sigma_{a,n}^2 = \mathrm{Var}\{a_n\}$.

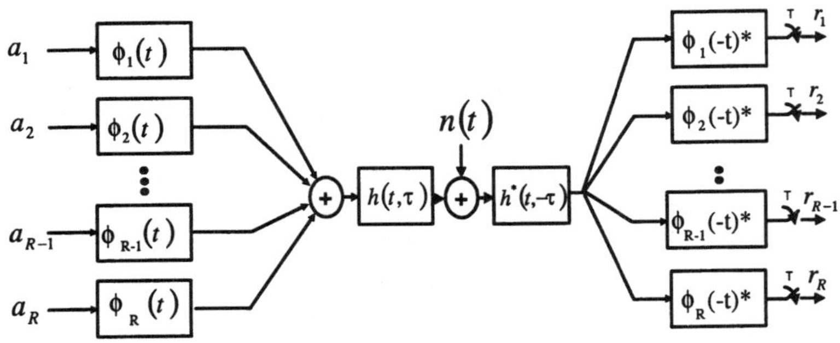

Figure 4.2 Capacity-achieving transmitter/receiver realization for a time-varying channel. The waveforms $\phi_i(t)$ diagonalize the correlation kernel for $h(t,\tau)$ and are solutions of an integral eigenequation.

Selection of the transmitted powers $\sigma^2_{a,n}$ according to the water-filling argument [1] amounts to reach the capacity of the channel as

$$C = \sum_{i=1}^{\mathcal{R}} \log_2 \left[1 + \frac{\left(\xi + \frac{\sigma^2}{\lambda_i}\right)^+}{\frac{\sigma^2}{\lambda_i}} \right] \quad \text{bits}/T \text{ sec} \tag{4.5}$$

where ξ is such that

$$\sum_{i=1}^{\mathcal{R}} \left(\xi - \frac{\sigma^2}{\lambda_i}\right)^+ = P,$$

P is the total transmission power and $(x)^+ = \max(x, 0)$. If the channel is time-invariant [which means $h(t,\tau) = h(t-\tau)$], $R_h(t,\tau) = R_h(t-\tau)$ becomes simply

$$R_h(t) = \int_T h(\tau)h^*(\tau - t)d\tau.$$

In [2] it is proved that the ideal real band-limited (lowpass, time-invariant) channel has optimal eigenfunctions related to the prolate spheroidal wave functions. For $T \to [-\infty; +\infty]$, evidently $\mathcal{R} \to \infty$ and the density of the eigenvalues tends to $|H(f)|^2$ where $H(f) = \int_{-\infty}^{+\infty} h(\tau)e^{-j2\pi f\tau}d\tau$ is the frequency domain representation of the time-invariant channel. The eigenfuctions in this case reduce to complex sinusoids. A standard derivation [1] gives that the capacity of a static frequency selective multipath channel $h(\tau)$ as

$$C = \int_B \log_2 \left[1 + P_s^{(o)}(f)\frac{|H(f)|^2}{N(f)}\right] df \text{ bit/sec} \tag{4.6}$$

where $P_s^{(o)}(f)$ [power spectral density of the transmitted signal $s(t)$] satisfies

$$P_s^{(o)}(f) = \begin{cases} K - \frac{N(f)}{|H(f)|^2} & f \in B \\ 0 & \text{elsewhere} \end{cases},$$

$B = \{f : P_s^{(o)}(f) > 0\}$ is called the *capacity achieving band*, K is a constant selected to satisfy a power constraint $\int |s(t)|^2 dt \leq P$, and $N(f)$ is the power spectral density of the additive channel Gaussian noise.

4.1.2 Approaching Capacity with Multicarrier Transmission

The derivation of (4.6) suggests multicarrier modulation. Multicarrier techniques transmit data by dividing the stream into several parallel bit streams. Each of the subchannels has a much lower bit rate and is modulated onto a different carrier. OFDM [3] is a special case of multicarrier modulation [4] with equally spaced subcarriers [5] and overlapping spectra [6]. The OFDM time-domain waveforms are chosen such that mutual orthogonality is ensured in the frequency domain. Time dispersion is easily handled by such systems because the substreams are essentially free of intersymbol interference (ISI). The baseband equivalent transmitted signal in a generic N-channel multicarrier system is shown in Figure 4.3 and is expressed as

$$s(t) = \sum_{k=-\infty}^{+\infty} \sum_{l=0}^{N-1} a_{k,l}\phi_{k,l}(t) = \sum_{k=-\infty}^{+\infty} \sum_{l=0}^{N-1} a_{k,l}\phi_l(t - kT_s), \tag{4.7}$$

where T_s is the symbol period and $a_{k,l}$ is the information-bearing symbol, $\phi_{k,l}(t) = \phi_l(t - kT_s)$, $l = 0, 1, ..., N-1$ are the fundamental basis waveforms. The transmitted signal $s(t)$ is linearly distorted by the multipath fading channel operator $\mathbf{H}$ as in

$$y(t) = (\mathbf{H}s)(t) = \sum_{k=-\infty}^{+\infty} \sum_{l=0}^{N-1} a_{k,l} f_{k,l}(t)$$

$$= \sum_{k=-\infty}^{+\infty} \sum_{l=0}^{N-1} a_{k,l} f_l(t - kT_s) \tag{4.8}$$

where $f_l(t) = (\mathbf{H}\phi_l)(t)$. Observe that (4.7) is essentially mimicking (4.4). The fundamental problem is to select the transmission basis $\phi_{k,l}(t) = \phi_l(t - kT)$ in such a way that the projection of the signal onto the signal set $\phi_{k,l}(t)$ in absence of noise gives the transmitted symbols up to a complex gain β_l

$$z_{k,l} = \int_{-\infty}^{+\infty} y(t)\phi^*_{k,l}(t)dt = \int_{-\infty}^{+\infty} y(t)\phi^*_l(t - kT_s)dt = \beta_l a_{k,l}.$$

This condition implies not only relative simplicity of the receiver but also robustness to additive white Gaussian noise in the sense of a capacity-achieving design. The transmission basis employed in multicarrier systems is

$$\phi_{k,l}(t) \quad = \quad g(t - kT_s)e^{j2\pi lFt}, \tag{4.9}$$

where F is the carrier frequency spacing and $g(t)$ is a shaping window. The use of pulses as in (4.9) results in a rectangular *tiling* of the time-frequency plane. The product $T_sF \geq 1$ defines the time-frequency product of each independent function in the signal set. In the OFDM case the pulse $g(t)$ in (4.9) is a rectangular window of duration T_s and $F = 1/T_s$. A *coarsification* of the time-frequency grid is typically employed using a guard-time between temporal adjacent symbols for mitigation of the time-dispersive characteristic of a frequency selective channel. However, properly shaping the basic symbols in each subchannel by using a pulse different from the rectangular one mitigates frequency dispersion effects of the channel caused by Doppler spreads. If the channel is perfectly static with Fourier transform $H(f)$,

$$g(t) = \begin{cases} \frac{1}{\sqrt{T_s}} & 0 \leq t \leq T_s \\ 0 & \text{elsewhere} \end{cases},$$

and the guard-time is long enough to cover for the support of the channel, one obtains

$$\begin{aligned} z_{k,l} &= \int_{-\infty}^{+\infty} y(t)g(t - kT)e^{-j2\pi lFt}dt \\ &= H(lF)a_{k,l} + n_{k,l} \end{aligned} \tag{4.10}$$

for $l = 0, 2, ..., N-1$ and $k = -\infty, ..., 0, ... + \infty$. At any given k arranging N samples in N-vectors gives

$$\mathbf{z}(k) = \Lambda_H \mathbf{a}(k) + \mathbf{n}(k) \tag{4.11}$$

where Λ_H is a diagonal matrix with generic lth diagonal element $H(lF)$, and the organization of $z_{k,l}$, $a_{k,l}$, and $n_{k,l}$ in the vectors $\mathbf{z}(k)$, $\mathbf{a}(k)$, and $\mathbf{n}(k)$, respectively, should be clear from the context. Multicarrier transmission with N subcarriers is supposed to asymptotically approach C as subcarrier spacing $B/N = F$ decreases and N increases. Assuming that $P_s^{(o)}(f)$ and $\frac{|H(f)|^2}{N(f)}$ are flat within F, at each carrier f_i, the capacity of the generic ith subchannel is

$$C_i = F \log_2 \left[1 + P_s^{(o)}(f_i) \frac{|H(f_i)|^2}{N(f_i)} \right],$$

so that the aggregate rate is $\sum_{i=1}^{N} C_i$, and $\sum_i C_i \Rightarrow C$, as $N \to \infty$. The superscript $^{(o)}$ indicates that the power assigned to the particular subcarrier obeys the water-filling solution. In practice the projection operations are implemented by DFT-based transformations. This is exactly the point that makes OFDM an attractive practical technique. Assume that $T_s = RT_S$ for R positive integer and write (4.7) as

$$s(t) = \sum_{k=-\infty}^{+\infty} \sum_{l=0}^{N-1} a_{k,l} \phi_l(t - kRT_S).$$

Sampling at $T_e \leq T_S$, we obtain

$$s_k(n) = \sum_{m=-\infty}^{+\infty} \sum_{l=0}^{N-1} a_{m,l} \phi_l \left((nN - mR)T_S - kT_e \right)$$

with $s_k(n) = s\left((nR-k)T_e\right)$, $k = 0, 1, ..., N-1$ where R is such that $RT_e = NT_S$. The particular case $R = N$ and $T_e = T_S$, $\phi_l(t) = g(t)e^{2\pi j f_l t}$, $g(t) = \begin{cases} \frac{1}{\sqrt{NT_S}} & 0 \leq t \leq NT_S \\ 0 & \text{elsewhere} \end{cases}$ and $f_l = \frac{l}{NT_S}$ collapses to an FFT-based multicarrier system

$$\begin{aligned} s_k(n) &= \sum_{m=-\infty}^{+\infty} \sum_{l=0}^{N-1} a_{m,l} g\left((n-m)NT_S - kT_S\right) \exp\left\{\frac{j2\pi lk}{N}\right\} \\ &= \frac{1}{\sqrt{NT_S}} \sum_{l=0}^{N-1} a_{n-1,l} \exp\left\{\frac{j2\pi lk}{N}\right\}. \end{aligned} \tag{4.12}$$

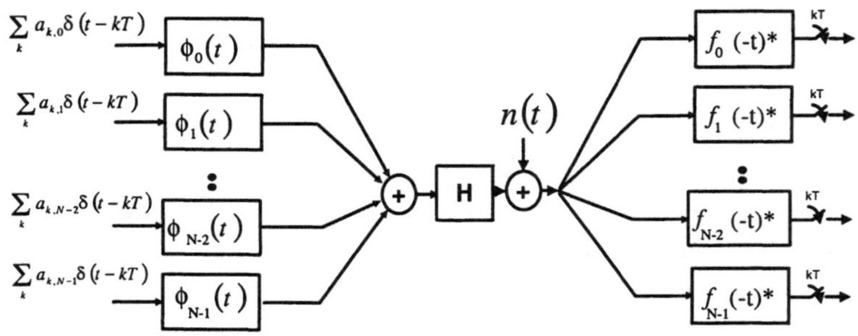

Figure 4.3 General transmitter/receiver realization in a multicarrier system.

If we define

$$\mathbf{a}(n) = [a_{n-1,0}, a_{n-1,1}, ..., a_{n-1,N-1}]^T$$

and

$$\mathbf{s}(n) = [s_0(n), s_1(n), ..., s_{N-1}(n)]^T,$$

(4.12) can be rewritten in vector form as

$$\mathbf{s}(n) = \mathbf{F}\mathbf{a}(n), \tag{4.13}$$

where $\mathbf{F}$ represents the (orthonormal) mapping[1] of the inverse Fourier transform and we have assumed without loss of generality a unitary sample period T_S. Similarly at the receiver we can collect the received baseband samples in a vector $\mathbf{r}(n)$ and apply the transformation $\mathbf{F}^H$. It is possible to show that if data are properly cyclically prefixed the channel convolution will appear as a cyclic convolution and diagonalization of the channel is achieved, just like the ideal multicarrier scheme. The modeling assumptions we have described can be summarized as in Figure 4.4.

4.1.3 Approaching Capacity with Single-Carrier Transmission

An alternative to multicarrier transmission is serial or single-carrier (SC) transmission. The channel $h(t)$ can be converted to

1. The k, l element of $\mathbf{F}$ is $\frac{1}{\sqrt{N}}\exp\left\{\frac{j2\pi lk}{N}\right\}$.

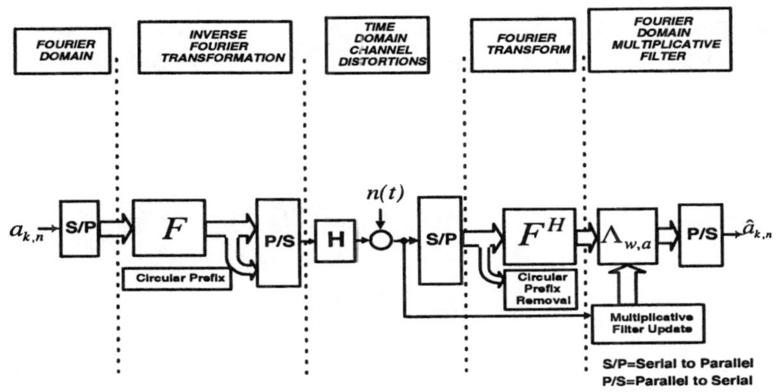

Figure 4.4 Functional block diagram of the baseband signal processing of an OFDM system that employs IFFT/FFT. The transformation $\mathbf{F}$ is the basic DFT matrix.

an equivalent discrete-time channel with ISI by using a transmit filter that shapes the transmit spectrum to the optimal water-filling form and a whitening matched filter (WMF) in the receiver. First, we assume that the capacity achieving band B is a single positive-frequency region of the spectrum with width W and focus on the minimum bandwidth passband transmission of complex symbols at rate $1/T_s = W$. The complex baseband transmitted signal is

$$s(t) = \sum_n a_n p_T(t - nT_s)$$

where a_n are complex symbols with variance σ_a^2, $p_T(t)$ with Fourier transform $P_T(f)$, the transmit-shaping filter selected to make the transmitted spectrum equal to the water-filling spectrum

$$\frac{\sigma_a^2}{T_s}|P_T(f)|^2 = P_s^o(f).$$

The receive filter is $P_R(f)$ is such that $|P_R(f)|^2 = 1/N(f)$ over B where $N(f)$ is the one-sided power spectral of the additive noise $n(t)$. The overall complex signal representation is

$$\tilde{r}(t) = \sum_n a_n v(t - nT_s) + w(t)$$

where $v(t) = p_T(t) * h(t) * p_R(t)$ and $w(t)$ is AWGN with normalized power spectral density. A filter matched to $v(t)$ [with response $v^*(t)$] is applied to $\tilde{r}(t)$ and is sampled at symbol rate to obtain

$$\tilde{z}_n = \int \tilde{r}(t)v^*(t - nT_s)dt.$$

The overall receive filter has response $P_R^{MF}(f) = \frac{P_T^*(f)H^*(f)}{N(f)}$

$$\tilde{z}_n = \sum_k q_k a_{n-k} + n_n,$$

where

$$q_k = q(kT_s) = \int df \frac{|P_T^*(f)H^*(f)|^2}{N(f)} e^{j2\pi ft} = \int df Q(f)e^{j2\pi ft}\bigg|_{t=kT_s}.$$

It is important to observe that the Fourier transform of the sequence q_k is the $1/T_s$-aliased spectrum of $Q(f)$,

$$\tilde{Q}(f) = \frac{1}{T_s}\sum_m Q(f + mT_s),$$

the so-called "folded spectrum." Focusing on the minimum-bandwidth (Nyquist) spectrum signaling implies that $\tilde{Q}(f) = \frac{1}{T_s}Q(f)$. Using the D-transform of sequences[2]

$$\tilde{z}(D) = q(D)a(D) + n(D),$$

and the spectral factorization $q(D) = h^*(D^{-1})A^2h(D)$ gives

$$1/A^2h^*(D^{-1})\tilde{z}(D) = z(D) = h(D)a(D) + n(D)$$

where $\log A^2 = T_s \int_W \log\tilde{Q}(f)df$. The sequence $z(D)$ is the T_s-sampled output of the WMF with response

$$P_R^{WMF}(f) = \frac{P_T^*(f)H^*(f)\tilde{H}(f)}{N(f)\tilde{Q}(f)},$$

and it can be optimally processed by a maximum likelihood sequence detector (MLSD). Such a canonical approach is capacity-achieving.

2. The D-transform of a sequence x_n is defined as $x(D) = \sum_n x_n D^n$, where D is the delay operator.

The complexity of the MLSD strategy may be prohibitive in some practical cases, so the designer has the option of using decision feedback equalization (DFE). Price [7] proved that at high SNR ISI compensated with DFE does not contribute signifcantly to a loss of capacity. The direct consequence of this important observation is that tail-canceling equalization and coding can approach capacity. The minimum mean squared error form of this result is obtained in [8].

4.2 The Multiantenna Channel

We have seen in Chapter 3 that linear space-time models can be expressed as

$$s(t,\vec{r}) = \int_T \int_\Gamma h(t,\tau,\vec{r},\vec{r}')x(\tau,\vec{r}')d\vec{r}'d\tau \tag{4.14}$$

where $h(t,\tau,\vec{r},\vec{r}')$ is the time-variant/space-variant wavefield response of the channel (assumed space-time and frequency selective), T is the time interval of interest, and Γ is the spatial volume of interest. The wavefield $x(t,\vec{r})$ is the wavefield generated by the transmitter. In the general case of N_{tx} transmit antennas, we can write

$$x(t,\vec{r}) = \int d\omega X(\omega,\vec{r},\vec{r}')e^{j\omega t},$$

where $X(\omega,\vec{r})$ is expressed as

$$X(\omega,\vec{r}) = \sum_{n=1}^{N_{\text{tx}}} q_n(\omega)\delta(\vec{r}-\vec{r}_n)$$

and N_{tx} is the number of elements at the transmit array, $q_n(\omega)$ are the N_{tx} information bearing signals transmitted at the N_{tx} elements of the transmit array, $\vec{r}_n$, $n = 1,2,...,N_{\text{tx}}$ are the locations of the elements in the transmit array [$\delta(\vec{r})$ is simply the spatial delta function]. Observe that $\vec{r}$ is a 3-D or 2-D parameterization of the spatial domain. Generally $h(t,\tau,\vec{r},\vec{r}')$ describes the space-time response of the system to an impulse generated at time t (as measured at delay τ) and spatial location $\vec{r}'$ (as measured at location $\vec{r}$).

4.3 The Spatio-Temporal Modes

If the channel is time-invariant $[h(t, \tau, \vec{r}, \vec{r}') = h(t - \tau, \vec{r}, \vec{r}')\ ]$, we obtain a simple model for the wavefield of interest

$$S(\omega, \vec{r}) = \int_{\Gamma} H(\omega, \vec{r}, \vec{r}')X(\omega, \vec{r}')d\vec{r}', \tag{4.15}$$

where $x(t, \vec{r}) = \int d\omega X(\omega, \vec{r}, \vec{r}')e^{j\omega t}$, $h(t, \vec{r}, \vec{r}') = \int d\omega H(\omega, \vec{r}, \vec{r}')e^{j\omega t}$. With noise we have

$$R(\omega, \vec{r}) = S(\omega, \vec{r}) + N(\omega, \vec{r}) \tag{4.16}$$

For simplicity of notation from now on we will drop the explicit dependence on ω. It is evident that the spatio-temporal model (4.16) can be manipulated similarly to the time-only case (4.1). Define

$$R_h(t, \tau, \vec{r}, \vec{r}') = \int_T \int_{\Gamma} h(t', t, \vec{r}'', \vec{r})h^*(t', \tau, \vec{r}'', \vec{r}')dt'd\vec{r}'' \tag{4.17}$$

and consider

$$\int_T \int_{\Gamma} R_h(t, u, \vec{r}, \vec{s})\phi(u, \vec{s})dud\vec{s} = \lambda\phi(t, \vec{r}) \quad t \in [0, T],\ \vec{r} \in \Gamma. \tag{4.18}$$

The $\mathcal{R}_{st}$ solutions to this integral eigenequation describe the $\mathcal{R}_{st}$ modes of the channel

$$\phi_1(t, \vec{r}),\ \phi_2(t, \vec{r}),\ \ldots, \phi_{\mathcal{R}_{st}}(t, \vec{r})$$

with the associated eigenvalues

$$\lambda_1,\ \lambda_2,\ \ldots, \lambda_{\mathcal{R}_{st}}.$$

If noise is temporally and spatially white

$$\sigma^2\delta(t - t')\delta(\vec{r} - \vec{r}') = E\left\{n(t, \vec{r})n^*(t', \vec{r}')\right\}$$

then signaling as $x(t, \vec{r}) = \sum_{n=1}^{\mathcal{R}_{st}} a_n\phi_n(t, \vec{r})$ processing as in Figure 4.5 has the potential of achieving capacity with Gaussian symbols a_n. In a typical wireless channel

$$\mathcal{R}_{st} >> \mathcal{R},$$

given the same bandwidth and power constraints. In other words, the spatio-temporal multiantenna channel has many more dimensions than the single-antenna channel. This was somehow already predicted in Chapter 3.

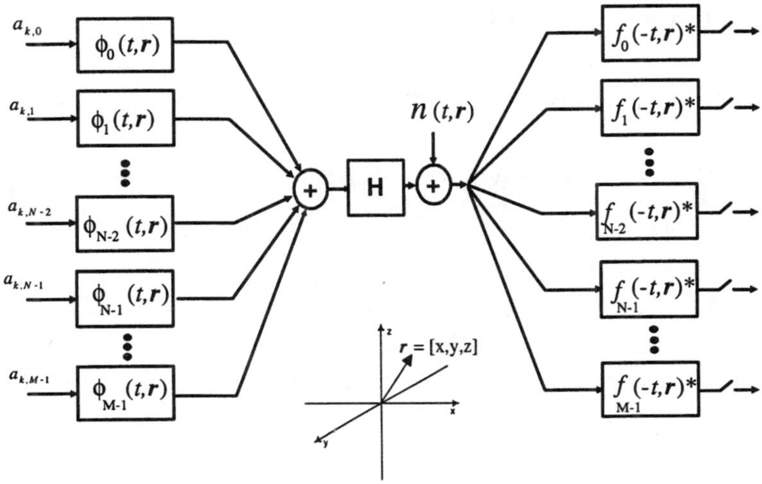

Figure 4.5 Capacity-achieving transmitter/receiver realization for a time-varying, space-varying channel. The waveforms $\phi_i(t)$ diagonalize the space-time correlation kernel for $h(t,\tau,\vec{r},\vec{s})$ and are solutions of an integral eigenequation.

4.3.1 Achievable Throughput of MIMO Systems

A common model for multiantenna systems is a multiple-input multiple-output (MIMO) system. This model was already expressed in the last example of Section 3.5.2 as (with noise)

$$\mathbf{r}(t) = \int_{-\infty}^{t} \mathbf{H}(\tau)\mathbf{s}(t-\tau) + \mathbf{n}(t),$$

or in the frequency domain

$$\mathbf{R}(\omega) = \mathbf{H}(\omega)\mathbf{S}(\omega) + \mathbf{N}(\omega),$$

where $\mathbf{N}(\omega)$ is the Fourier transform of the vector $[n_1(t), ..., n_{N_{rx}}(t)]^T$. Consider the transmitted signals at the N_{tx} antennas expressed in the time domain as

$$s_i(t) \quad = \quad \sum_l a_i(l)p_i(t-lT_s) \tag{4.19}$$

where $a_k(l)$ is a QAM symbol, T_s is the symbol period and $p_i(t)$ a pulse-shaping filter of the Nyquist type (refer to Chapter 2). The achievable rate, assuming that the signaling bandwidth is exactly limited to the Nyquist bandwidth, is obtained as in [9]

$$C = \frac{T_s}{2\pi} \int_{-\pi/T_s}^{\pi/T_s} \log det \left[\mathbf{I} + \mathbf{H}(\omega)\Sigma(\omega)\mathbf{H}(\omega)^H \mathbf{N}(\omega)^{-1} \right] d\omega \quad (4.20)$$

where $\Sigma(\omega) = E\left\{\mathbf{A}(\omega)\mathbf{A}(\omega)^H\right\}$ with $\mathbf{A}(\omega)$ as the Fourier transform of the vector $[a_1(n), ..., a_{N_{tx}}(n)]^T$ is the water-filling power spectral density with a power constraint $\frac{T_s}{2\pi}\int_{-\pi/T_s}^{\pi/T_s} \Sigma(\omega)d\omega \leq P$. If the transmitter has no knowledge of the channel, the symbols $a_l(n)$ are identically and independently distributed, the same unity power is transmitted at each transmit antenna we obtain

$$C = \frac{T_s}{2\pi} \int_{-\pi/T_s}^{\pi/T_s} \log det \left[\mathbf{I} + \mathbf{H}(\omega)\mathbf{H}(\omega)^H \mathbf{N}(\omega)^{-1} \right] d\omega.$$

With a finite excess bandwidth bounded by $[-J\pi/T_s, J\pi/T_s]$, with J as an integer, we have

$$\begin{aligned} C &= \frac{T_s}{2\pi} \int_{-\pi/T_s}^{\pi/T_s} d\omega \\ &\quad \log det \left[\left(\tilde{\mathbf{N}}(\omega) + \sum_{i=1}^{N_{\text{tx}}} \tilde{\mathbf{H}}_i(\omega)\tilde{\mathbf{H}}_i(\omega)^H \right) \tilde{\mathbf{N}}(\omega)^{-1} \right], \end{aligned} \quad (4.21)$$

where

$$\tilde{\mathbf{H}}_k(\omega)^T = \left[H_{k,1}(\omega - 2\pi J/T_s), ..., H_{k,N_{\text{rx}}}(\omega + 2\pi J/T_s) \right],$$

while the folded spectra of the noise processes at each receive antenna are organized in the matrix [10] $\tilde{\mathbf{N}}(\omega)$ as

$$\tilde{\mathbf{N}}(\omega) = \begin{bmatrix} \mathbf{N}(\omega - 2\pi J/T_s) & \mathbf{0} & \cdots \\ \mathbf{0} & \mathbf{N}(\omega - 2\pi(J+1)/T_s) & \cdots \\ \mathbf{0} & \cdots & \mathbf{0} \\ \cdots & \mathbf{0} & \mathbf{N}(\omega + 2\pi J/T_s) \end{bmatrix},$$

with

$$\mathbf{N}(\omega - 2\pi m/T_s) = \begin{bmatrix} N_1(\omega - 2\pi m/T_s) & 0 & \cdots \\ 0 & \cdots & \cdots \\ 0 & \cdots & 0 \\ \cdots & 0 & N_{N_{\text{rx}}}(\omega - 2\pi m/T_s) \end{bmatrix}.$$

In Appendix 4A we report more details regarding capacity computations.

4.4 The Flat Multiantenna Channel

The general treatment of previous sections is extremely relevant in practice but somehow obscure at a first glance. To simplify the concepts to their essential form we consider the *flat* channel, that is the nonfrequency selective channel. If the channel is not frequency-selective, $\mathbf{H}(\omega) \triangleq \mathbf{H}$ and

$$\mathbf{r}(t) = \mathbf{H}\mathbf{s}(t) + \mathbf{n}(t),$$

with $\mathbf{r}(t) = [r_1(t), ..., r_{N_{\text{rx}}}(t)]$, $\mathbf{s}(t) = \left[s_1(t), ..., s_{N_{\text{tx}}}(t)\right]$, and $\mathbf{n}(t) = [n_1(t), ..., n_{N_{\text{rx}}}(t)]$. We also assume that noise is spatially white, that is, that $E\left\{\mathbf{n}(t)\mathbf{n}(s)^H\right\} = \tilde{\sigma}_n^2\delta(t-s)\mathbf{I}$. If $s_i(t)$ is given in (4.19) and $p_i(t)$ is a Nyquist pulse with zero-excess bandwidth (a sinc pulse), the discrete-time symbol-spaced equivalent of the MIMO system is

$$y_k(n) \;=\; \tilde{r}_k(t)|_{t=nT} = \sum_{i=1}^{N_{\text{tx}}} h_{k,i}a_i(n) + \eta_k(n) \quad k = 1, 2, ..., N_{\text{rx}} \tag{4.22}$$

where $a_i(n)$ is the complex M-ary symbol transmitted at the ith transmit antenna, and $\eta_k(n)$ is white Gaussian noise with variance σ_n^2. We can compress (4.22) in vector notation as

$$\mathbf{y}(n) = \mathbf{H}\mathbf{a}(n) + \mathbf{n}(n) \tag{4.23}$$

where the kth element of the N_{rx}-vector $\mathbf{y}(n)$ is $y_k(n)$, the k, i element of the $N_{\text{rx}} \times N_{\text{tx}}$ matrix $\mathbf{H}$ is $h_{k,i}$, the ith element of the N_{tx}-vector $\mathbf{a}(n)$ is $a_i(n)$, and the kth element of the N_{rx}-vector $\mathbf{n}(n)$ is $\eta_k(n)$. We assume that $\sum_{i=1}^{N_{\text{tx}}} |h_{k,i}|^2 = N_{\text{tx}}$. The capacity (when the transmitter is not allowed to know the channel) in this case is found solving the eigenvalue problem

$$\mathbf{H}\mathbf{H}^H\mathbf{v} = \lambda\mathbf{v}$$

which gives

$$C = \log_2 det\left[\mathbf{I} + \frac{\text{SNR}}{N_{\text{tx}}}\mathbf{H}\mathbf{H}^H\right] = \sum_{i=1}^{N_{tx}} \log_2\left(1 + \frac{\text{SNR}}{N_{\text{tx}}}\lambda_i\right) \quad \text{in bps/Hz} \qquad (4.24)$$

where λ_i is the ith eigenvalue of $\mathbf{HH}^H$. Fundamentally (4.24) shows that we can regard a Gaussian vector N_{tx}-channel with signal-to-noise ratio $\frac{\text{SNR}\lambda_i}{N_{\text{tx}}}$. Channels with large λ_i will give a large contribution to the capacity.

Example

In this example we will find the maximum and minimum capacity of a multiantenna channel. For $N_{\text{tx}} = N_{\text{rx}}$, the highest capacity is obtained when $\lambda_i = N_{\text{tx}}$

$$C = N_{\text{tx}} \log_2 (1 + \text{SNR}) .$$

The channel is then $\mathbf{H} = N_{\text{tx}}\mathbf{I}$, with $\mathbf{I}$ the identity matrix. In this case the channel has N_{tx} dimensions (communication modes). At the other extreme when $\lambda_1 = N_{\text{tx}}^2$ and $\lambda_i = 0$, $i \neq 1$, the MIMO channel *degenerates* to

$$C = \log_2 (1 + N_{\text{tx}}\text{SNR}) ,$$

and in this case the channel has only one dimension (communication mode).

In the following two examples we normalize the propagation matrix $\mathbf{H}$ so that $\sum_{i=1}^{N_{\text{tx}}} |h_{k,i}|^2 = 1$ and that the degenerate MIMO channel is exactly equal to $\log_2(1 + \text{SNR})$, the capacity of a one-transmit, one-receive antenna system.

Example

Consider $N_{\text{tx}} = N_{\text{rx}} = 2$ antennas in free space and communication at a carrier with frequency 10 GHz arranged as in Figure 4.6. The antenna separation is d and the distance between transmitter

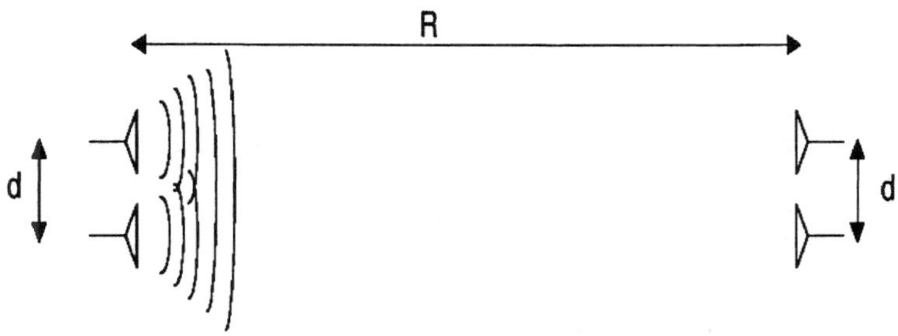

Figure 4.6 Two transmitters and two receivers in free space.

and receiver is d. We have explained in Chapter 3 that between two objects in free space (when the far-field assumption is valid), even though Maxwell's equations dictate a complicated propagation mechanism, we can always think of a single plane wave traveling in the direction of the line that connects the two objects. Moreover, we have explained that the propagation path of such a wave will be afflicted by an attenuation and a phase shift proportional to the distance between the two objects. We know from Chapter 3 that the propagation characteristics in this case results, for the model (4.23), in

$$\mathbf{H} = \frac{1}{\sqrt{2}} \begin{bmatrix} \Gamma_s(R)\exp\left\{-j\frac{2\pi}{\lambda}R\right\} & \Gamma_s(\tilde{R})\exp\left\{-j\frac{2\pi}{\lambda}\tilde{R}\right\} \\ \Gamma_s(\tilde{R})\exp\left\{-j\frac{2\pi}{\lambda}\tilde{R}\right\} & \Gamma_s(R)\exp\left\{-j\frac{2\pi}{\lambda}R\right\} \end{bmatrix} \qquad (4.25)$$

where $\lambda = c/f_c = 3e8/10e9$ is the wavelength and

$$\tilde{R} = \sqrt{d^2 + R^2}.$$

We can assume $\Gamma_s(\tilde{R}) \simeq \Gamma_s(R) \simeq 1$ without loss of generality. Consider $R = 1$ km and $d = 3$m. The eigenvalues of $\mathbf{HH}^H$ are $\lambda_1 = 1.58$ and $\lambda_2 = 0.41$, so the capacity formula gives us

$$\log_2\left(1 + \frac{\text{SNR}}{2}\lambda_1\right) + \log_2\left(1 + \frac{\text{SNR}}{2}\lambda_2\right) \quad \text{bps/Hz}.$$

Observe that a single transmitter and single receiver system would have the capacity

$$\log_2(1 + \text{SNR}) \quad \text{bps/Hz}.$$

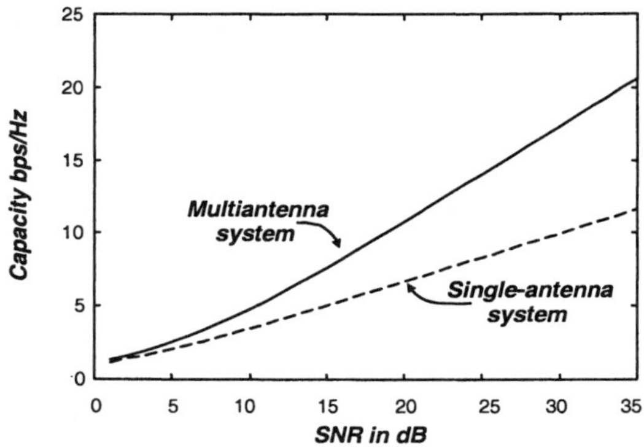

Figure 4.7 Capacities of the systems in the example when $R = 1$ km, $d = 15$m.

The capacities are plotted in Figure 4.7. It is instructive to see that if $d = 0.5$m, $\lambda_1 = 1.99$, and $\lambda_2 = 0.0003$, so that the contribution to capacity of the second term $\log_2\left(1 + \frac{\text{SNR}}{2}\lambda_2\right)$ vanishes and capacities of multiantenna and single-antenna systems become equivalent.

Example

The previous example showed that when the antennas are closely spaced, capacity collapses to single-antenna capacity. In this example we will attempt an explanation of why multipath reflections remove this restriction. Consider Figure 4.8. We refer again to Chapter 3 and consider that the reflector will create the presence of two plane waves traveling from the two transmitters to the reflector and two plane waves traveling from the reflector to the receivers. Assuming that the reflected waves have a reflection attenuation equal to 60% the attenuations of the nonreflected waves, we can characterize the propagation matrix **H** as

$$\begin{aligned}\mathbf{H}' &= \mathbf{H} \\ &+0.6\begin{bmatrix} \exp\left\{-j\frac{2\pi}{\lambda}(P1+P3)\right\} & \exp\left\{-j\frac{2\pi}{\lambda}(P1+P4)\right\} \\ \exp\left\{-j\frac{2\pi}{\lambda}(P2+P3)\right\} & \exp\left\{-j\frac{2\pi}{\lambda}(P2+P4)\right\} \end{bmatrix}\end{aligned} \tag{4.26}$$

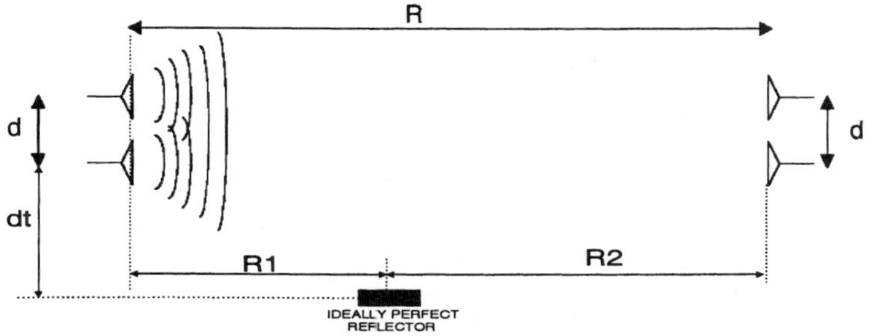

Figure 4.8 Two transmitters and two receivers in free space with one reflector.

where **H** is given in (4.25) and $P1 = \sqrt{R2^2 + dt^2}$, $P3 = \sqrt{R1^2 + dt^2}$, $P2 = \sqrt{R2^2 + (dt + d)^2}$, and $P4 = \sqrt{R1^2 + (dt + d)^2}$. Consider $R =$ 1 km and $d = 0.5$m, $R1 = R2 = R/2$, and $dt = 100$m. If we normalize $\mathbf{H}'$ so that $\sum_{i=1}^{2} |h'_{k,i}|^2 = 1$, we have $\lambda_1 = 1.6$ and $\lambda_2 = 0.39$. The capacity of the multiantenna system is again significantly higher (see Figure 4.9) than the single-antenna system even though the antennas are spaced only $d = 0.5$m. In its simplicity this example captures all of the important concepts related to multiantenna technology.

When the signal is reflected in an unpredictable way, the elements of the matrix **H** are modeled as Rayleigh distributed independent processes. Interestingly, it is shown in [11] that if the elements of **H** account for a Rayleigh propagation environment, a lower bound to the capacity is

$$C > \sum_{k=N_{\text{rx}}-N_{\text{tx}}+1}^{N_{\text{tx}}} \log_2 \left[1 + \frac{\text{SNR}}{N_{\text{tx}}} \chi_{2k}^2\right], \tag{4.27}$$

where χ_{2k}^2 is a chi-squared [12] random variable with $2k$ degrees of freedom. Moreover, for $N_{\text{rx}} = N_{\text{tx}}$

$$C > \sum_{k=1}^{N_{\text{tx}}} \log_2 \left[1 + \frac{\text{SNR}}{N_{\text{tx}}} \chi_{2k}^2\right]. \tag{4.28}$$

The achievable maximum data rate for $N_{\text{rx}} = N_{\text{tx}} = 1, 2, 3, 4$ is plotted in Figure 4.10. The data rate increase over traditional technology is remarkable.

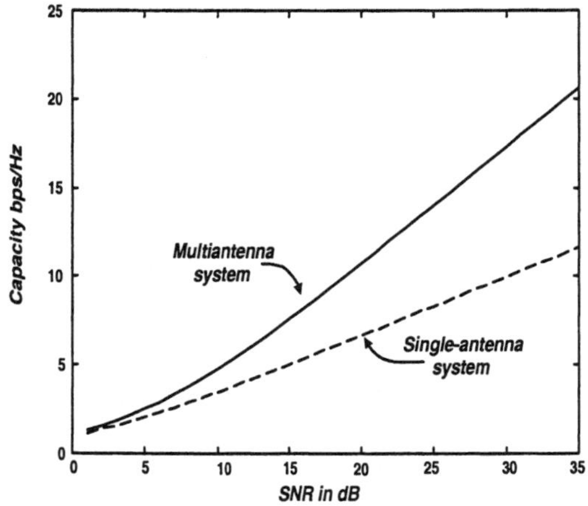

Figure 4.9 Capacities of the system in the example with one reflector and $R = 1$km and $d = 0.5$m, $R1 = R2 = R/2$, $dt = 100$m.

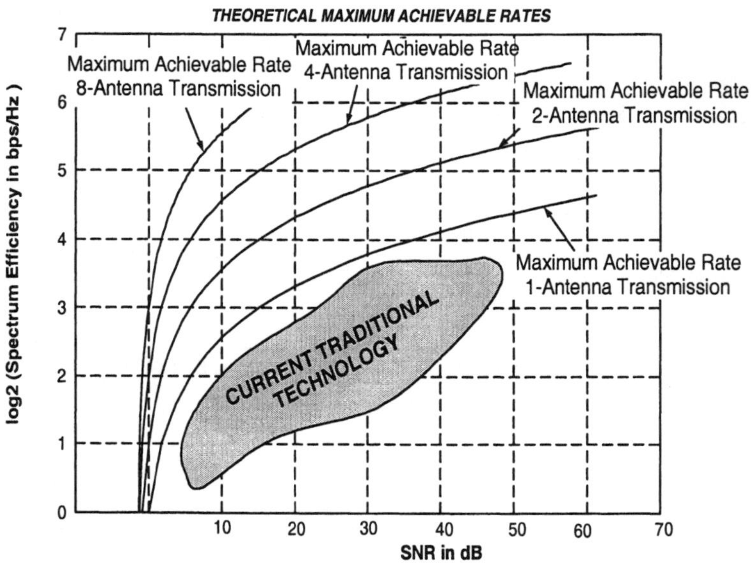

Figure 4.10 The capacity of a multiantenna system in nonfrequency selective Rayleigh fading environment with $N_{rx} = N_{tx}$.

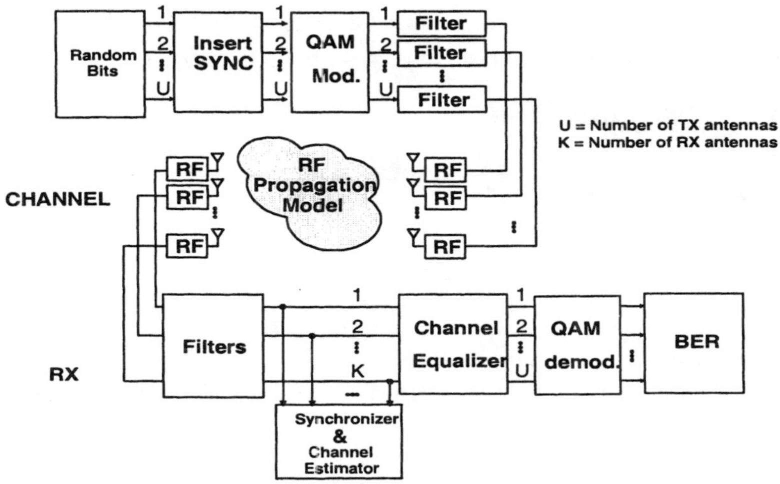

Figure 4.11 Block diagram of the system developed by WJ Communications Inc. in 1998 to validate the theory.

4.4.1 Practical Issues

The reader at this point might wonder if the impressive data rates promised by theory are in fact achievable by realistic modems. In the sections that follow we report the analytical and experimental results of a simple approach to the demodulation of multiantenna signals. For simplicity of notation let $K = N_{\text{rx}}$ and $U = N_{\text{tx}}$ and consider the system in Figure 4.11. The algorithm of the simplest modem is:

- Estimate the channel $\mathbf{H}$, using N training symbols;
- Multiply the received vector by the pseudo-inverse of $\mathbf{H}$.

Exploiting a sequence of NU known training symbols (N symbols per transmit antenna), the basic algorithm estimates the array transfer function $\mathbf{H}$, defined $\hat{\mathbf{H}}$ using the maximum likelihood principle. Since the noise is white Gaussian, the symbols are then demodulated by means of

$$\text{slice}\left\{\hat{\mathbf{a}}(m)\right\} \tag{4.29}$$

where

$$\hat{\mathbf{a}}(m) = \left[\hat{\mathbf{H}}^H\hat{\mathbf{H}}\right]^{-1}\hat{\mathbf{H}}^H\mathbf{y}(m). \tag{4.30}$$

Because of the noise, the imperfect estimate of the steering vector, and the residual interference, we have the event that $\hat{\mathbf{a}}(m) \neq \mathbf{a}(m)$ and the demodulator makes wrong symbol decisions.

4.4.2 The Output SNIR

Assume that the signal to be demodulated is $a_U(n)$ and partition $\mathbf{H}$ and $\hat{\mathbf{H}}$ as

$$\mathbf{H} = [\mathbf{D}, \mathbf{v}] \tag{4.31}$$

$$\hat{\mathbf{H}} = \left[\hat{\mathbf{D}}, \hat{\mathbf{v}}\right] \tag{4.32}$$

so that $\mathbf{v}$ and $\hat{\mathbf{v}}$ are the last columns of $\mathbf{H}$ and $\hat{\mathbf{H}}$, respectively. Define the matrices

$$\mathbf{P}_1 \triangleq \mathbf{D}\left[\mathbf{D}^H\mathbf{D}\right]^{-1}\mathbf{D}^H \tag{4.33}$$

$$\mathbf{P} \triangleq \mathbf{I} - \mathbf{P}_1 \tag{4.34}$$

$$\mathcal{D} \triangleq \left[\mathbf{D}^H\mathbf{D}\right]^{-1}\mathbf{D}^H \tag{4.35}$$

$$\hat{\mathbf{P}}_1 \triangleq \hat{\mathbf{D}}\left[\hat{\mathbf{D}}^H\hat{\mathbf{D}}\right]^{-1}\hat{\mathbf{D}}^H \tag{4.36}$$

$$\hat{\mathbf{P}} \triangleq \mathbf{I} - \hat{\mathbf{P}}_1 \tag{4.37}$$

$$\hat{\mathcal{D}} \triangleq \left[\hat{\mathbf{D}}^H\hat{\mathbf{D}}\right]^{-1}\hat{\mathbf{D}}^H. \tag{4.38}$$

Using these definitions, we have

$$\left[\mathbf{H}^H\mathbf{H}\right]^{-1}\mathbf{H}^H = \frac{1}{\mathbf{v}^H\mathbf{P}\mathbf{v}}\begin{bmatrix} \mathcal{D}\left(\mathbf{v}^H\mathbf{P}\mathbf{v}\mathbf{I} - \mathbf{v}\mathbf{v}^H\mathbf{P}\right) \\ \mathbf{v}^H\mathbf{P} \end{bmatrix} \tag{4.39}$$

and

$$\hat{a}_U(n) = \frac{\hat{\mathbf{v}}^H\hat{\mathbf{P}}\mathbf{y}(n)}{\hat{\mathbf{v}}^H\hat{\mathbf{P}}\hat{\mathbf{v}}} = \hat{\beta}_U a_U(n) + \sum_{k=1}^{U-1}\hat{\beta}_k a_k(n) + \hat{\mathbf{d}}^H\mathbf{n}(n), \tag{4.40}$$

where

$$\hat{\beta}_U \triangleq \frac{\hat{\mathbf{v}}^H\hat{\mathbf{P}}\mathbf{v}}{\hat{\mathbf{v}}^H\hat{\mathbf{P}}\hat{\mathbf{v}}}$$

$$\hat{\beta}_n \triangleq \frac{\hat{\mathbf{v}}^H\hat{\mathbf{P}}\mathbf{H}_n}{\hat{\mathbf{v}}^H\hat{\mathbf{P}}\hat{\mathbf{v}}}$$

$$\hat{\mathbf{d}}^H \triangleq \frac{\hat{\mathbf{v}}^H \hat{\mathbf{P}}}{\hat{\mathbf{v}}^H \hat{\mathbf{P}} \hat{\mathbf{v}}}$$

with $\mathbf{H}_n$ as the nth column of $\mathbf{H}$. The measure of signal quality we adopt is the signal-to-noise plus interference ratio

$$\frac{\text{Signal Power}}{\text{Noise Power} + \text{Interference Power}}$$

that can be expressed as

$$\text{SNIR} = [\text{SIR}^{-1} + \text{SNR}^{-1}]^{-1}$$

where

$$\text{SIR} = \frac{p_U E\left\{|\beta_U|^2\right\}}{\sum_{n=1}^{U-1} p_n E\left\{|\beta_n|^2\right\}}$$

is the signal-to-interference ratio and

$$\text{SNR} = \frac{p_U E\left\{|\beta_U|^2\right\}}{\sigma_n^2 E\left\{||\hat{\mathbf{d}}||^2\right\}}$$

is the signal-to-noise ratio (p_n as the power of the nth signal). As in [13], we consider that the errors in the estimate $\mathbf{H}$ are random and perform a sensitivity analysis of $\hat{\beta}_n$. A first-order Taylor expansion of $\hat{\beta}_n$ gives

$$\hat{\beta}_n \simeq \sum_{p=1}^{K} \sum_{q=1}^{U} \frac{\partial \overline{\mathbf{H}}}{\partial \overline{h}_{p,q}(0)} \delta \overline{h}_{p,q}(0) + \sum_{p=1}^{K} \sum_{q=1}^{U} \frac{\partial \tilde{\mathbf{H}}}{\partial \tilde{h}_{p,q}(0)} \delta \tilde{h}_{p,q}(0) \quad (4.41)$$

where $\overline{\mathbf{H}} = \text{Real}\{\mathbf{H}\}$, $\tilde{\mathbf{H}} = \text{Imag}\{\mathbf{H}\}$, $\overline{h}_{p,q}(0) = \text{Real}\{h_{p,q}(0)\}$, $\tilde{h}_{p,q}(0) = \text{Imag}\{h_{p,q}(0)\}$. It is possible to show that

$$\frac{\partial \beta_n}{\partial \overline{\mathbf{H}}_{p,q}} = -\delta_{q,n} \frac{\mathbf{v}^H \mathbf{P} \mathbf{e}_p}{\mathbf{v}^H \mathbf{P} \mathbf{v}}$$

$$\frac{\partial \beta_n}{\partial \tilde{\mathbf{H}}_{p,q}} = -j \delta_{q,n} \frac{\mathbf{v}^H \mathbf{P} \mathbf{e}_p}{\mathbf{v}^H \mathbf{P} \mathbf{v}},$$

where $\mathbf{e}_n$ is the nth column of $\mathbf{I}$, the identity matrix. Defining the vectors

$$\mathbf{c} \triangleq \begin{bmatrix} \text{vec}\{\overline{\mathbf{H}}^T\} \\ \text{vec}\{\tilde{\mathbf{H}}^T\} \end{bmatrix}$$

$$\mathbf{g}_n \triangleq \begin{bmatrix} \mathbf{g}_n^{(1)} \\ \mathbf{g}_n^{(2)} \end{bmatrix}$$

with

$$\mathbf{g}_n^{(1)} = \frac{\partial \beta_n}{\partial \text{vec}\{\overline{\mathbf{H}}^T\}} = -\frac{\mathbf{P}^* \mathbf{v}^* \otimes \mathbf{e}_n}{\mathbf{v}^H \mathbf{P} \mathbf{v}}$$

$$\mathbf{g}_n^{(2)} = \frac{\partial \beta_n}{\partial \text{vec}\{\tilde{\mathbf{H}}^T\}} = -j\frac{\mathbf{P}^* \mathbf{v}^* \otimes \mathbf{e}_n}{\mathbf{v}^H \mathbf{P} \mathbf{v}}$$

we can express (4.41) as

$$\hat{\beta}_n \simeq \mathbf{g}_n^T \delta \mathbf{c}$$

and

$$E\left\{|\hat{\beta}_n|^2\right\} = \mathbf{g}_n^T E\{\delta \mathbf{c} \delta \mathbf{c}^H\} \mathbf{g}_n^*.$$

An expression for $E\{\delta \mathbf{c} \delta \mathbf{c}^H\}$ can be found from the Cramer-Rao bound for the estimate of parameters of linear systems with known signals, which gives

$$E\{\delta \mathbf{c} \delta \mathbf{c}^H\} = \frac{\sigma_n^2}{2N} \mathbf{I}_K \otimes \left[\overline{\mathbf{R}}_s - \tilde{\mathbf{R}}_s \overline{\mathbf{R}}_s^{-1} \tilde{\mathbf{R}}_s\right]^{-1},$$

where $\mathbf{R}_s = \frac{1}{N} \sum_{m=1}^N \mathbf{a}(n) \mathbf{a}(n)^H$, $\overline{\mathbf{R}}_s = \text{Real}\{\mathbf{R}_s\}$ and $\tilde{\mathbf{R}}_s = \text{Imag}\{\mathbf{R}_s\}$. So ultimately we obtain

$$E\left\{|\hat{\beta}_n|^2\right\} \simeq \frac{\sigma_n^2}{N \mathbf{v}^H \mathbf{P} \mathbf{v}} \mathbf{e}_n^T \left[\overline{\mathbf{R}}_s - \tilde{\mathbf{R}}_s \overline{\mathbf{R}}_s^{-1} \tilde{\mathbf{R}}_s\right]^{-1} \mathbf{e}_n \qquad n \neq U.$$

Following a similar procedure one finds

$$E\left\{|\hat{\beta}_U|^2\right\} \simeq 1 + \frac{\mathbf{v}^H \mathbf{P} E\left\{\overline{\mathbf{v}}\overline{\mathbf{v}}^T + \tilde{\mathbf{v}}\tilde{\mathbf{v}}^T\right\} \mathbf{P} \mathbf{v}}{\left(\mathbf{v}^H \mathbf{P} \mathbf{v}\right)^2}$$

where again $\tilde{\mathbf{v}} = \text{Imag}\{\mathbf{v}\}$ and $\overline{\mathbf{v}} = \text{Real}\{\mathbf{v}\}$. If we assume uncorrelated training signals $\mathbf{R}_s = \text{diag}\{p_1, p_2, ..., p_N\}$, we have $E\left\{\overline{\mathbf{v}}\overline{\mathbf{v}}^T + \tilde{\mathbf{v}}\tilde{\mathbf{v}}^T\right\} \simeq \frac{\sigma_n^2}{p_U N} \mathbf{I}$. For reasonable N $(N > 1)$ we get

$$\text{SNRO} \simeq \frac{p_U}{\sigma_n^2} \mathbf{v}^H \mathbf{P} \mathbf{v}$$

and

$$\text{SIRO} \simeq N \frac{p_U}{\sigma_n^2(U-1)} \mathbf{v}^H \mathbf{P} \mathbf{v}.$$

We finally obtain

$$\text{SNIR} \simeq \frac{N p_U}{\sigma_n^2(N+U-1)} \mathbf{v}^H \mathbf{P} \mathbf{v} \triangleq \gamma.$$

4.4.3 The Distribution of γ

In obstructed environment the elements of $\mathbf{H}$ are independent Gaussian complex random variables with a variance σ_f^2. The pdf of γ is evidently found from the pdf of $\mathbf{v}^H\mathbf{P}\mathbf{v}$. Define the singular value decomposition of $\mathbf{P} = \mathbf{U}\Sigma\mathbf{U}^H$ where Σ has $K-U+1$ diagonal elements different from zero (equal exactly to one) and the rest of the elements equal to zero. We can write

$$\mathbf{v}^H\mathbf{P}\mathbf{v} = \mathbf{v}^H\mathbf{U}\Sigma\mathbf{U}^H\mathbf{v} = \mathbf{m}^H\Sigma\mathbf{m} = \sum_{n=1}^{K-U+1} |m_n|^2$$

where $\mathbf{m} \triangleq \mathbf{U}^H\mathbf{v}$ with the nth element equal to m_n. Since the elements of $\mathbf{v}$ are Gaussian, the elements of $\mathbf{m}$ are Gaussian and

$$E\{\mathbf{m}\mathbf{m}^H\} = E\{\mathbf{U}\mathbf{v}\mathbf{v}^H\mathbf{U}^H\} = \sigma_f^2\mathbf{I}.$$

This means that the elements of $\mathbf{m}$ are uncorrelated and therefore that $\mathbf{v}^H\mathbf{P}\mathbf{v}$ is chi-square distributed with $2(K-U+1)$ degrees of freedom.

4.4.4 Error Probabilities

Define

$$\rho_n = \frac{p_n}{\sigma_n^2}.$$

Since

$$\gamma = \rho_U \frac{N}{N+U-1} \sum_{n=1}^{K-U+1} |m_n|^2$$

is chi-square distributed with $2(K-U+1)$ degrees of freedom, the results from [12] provide the pdf of γ as

$$p_\Gamma(\gamma) = \frac{\gamma^{K-U}}{(K-U)!\,\overline{\gamma}^{K-U+1}} \exp\left\{-\frac{\gamma}{\overline{\gamma}}\right\} \tag{4.42}$$

where

$$\overline{\gamma} \triangleq \rho_U \left(\frac{N}{N+U-1} \right) \sigma_f^2.$$

Observe that

$$\rho_U = \frac{p_n}{\sigma_n^2} = \frac{E_s}{N_0} = \log_2(M)\frac{E_b}{N_0}.$$

where $\frac{E_b}{N_0}$ (and $\frac{E_s}{N_0}$) is the average SNR per bit (and per symbol) per antenna in the Nyquist band, while M is the number of points in the constellation of the digital modulation scheme employed. Assuming a Gaussian distribution for the interference terms affecting the decision on a generic symbol, the probability of error can be derived according to

$$P_b = \int_0^\infty P_b(\gamma) p_\Gamma(\gamma) d\gamma \tag{4.43}$$

where $P_b(\gamma)$ is the probability of error expression for any particular digital modulation scheme in AWGN given an SNR equal to γ.

Example

Consider that $\pi/4$ DQPSK [14] is a variation of 4-DPSK (or DQPSK) [12] used in the North-american standard for digital communication IS-136. In $\pi/4$ DQPSK the symbols

$$a_l(m) = \tilde{a}_m^{(l)} + j\tilde{b}_m^{(l)}$$

are obtained as

$$\tilde{a}_m^{(l)} = \tilde{a}_{m-1}^{(l)} \cos[\Delta\phi_m^{(l)}] - \tilde{b}_{m-1}^{(l)} \sin[\Delta\phi_m^{(l)}],$$

$$\tilde{b}_m^{(l)} = \tilde{a}_{m-1}^{(l)} \sin[\Delta\phi_m^{(l)}] + \tilde{b}_{m-1}^{(l)} \cos[\Delta\phi_m^{(l)}],$$

where $\Delta\phi_m^{(l)} = \pi/4$ if $\text{bit}_{1,m}^{(l)} = 0$ and $\text{bit}_{2,m}^{(l)} = 0$, $\Delta\phi_m^{(l)} = 3\pi/4$ if $\text{bit}_{1,m}^{(l)} = 1$ and $\text{bit}_{2,m}^{(l)} = 0$, $\Delta\phi_m = -3\pi/4$ if $\text{bit}_{1,m}^{(l)} = 1$ and $\text{bit}_{2,m}^{(l)} = 1$, $\Delta\phi_m = -\pi/4$ if $\text{bit}_{1,m}^{(l)} = 0$ and $\text{bit}_{2,m}^{(l)} = 1$. The $\pi/4$-DQPSK bit error rate for a nonfaded channel and using differential detection is [12]

$$P_b(\gamma) = Q(a_\gamma, b_\gamma) - 0.5 I_0(a_\gamma\, b_\gamma) \exp\left\{ -\frac{a_\gamma^2 + b_\gamma^2}{2} \right\}, \tag{4.44}$$

$$a_\gamma = \sqrt{0.5\gamma}\left(\sqrt{2+\sqrt{2}} - \sqrt{2-\sqrt{2}}\right), \tag{4.45}$$

$$b_\gamma = \sqrt{0.5\gamma}\left(\sqrt{2+\sqrt{2}} + \sqrt{2-\sqrt{2}}\right), \tag{4.46}$$

where $Q(a,b)$ is the Marcum function and $I_0(x)$ is the modified Bessel function of order zero. This expression is well approximated by

$$P_b(\gamma) \simeq Q\left(\sqrt{1.1716\frac{E_b}{N_0}}\right) = Q\left(\sqrt{\frac{1.1716}{2}\gamma}\right), \tag{4.47}$$

where $Q(x) = \frac{1}{2}erfc(x/\sqrt{2}) = \frac{1}{\sqrt{2\pi}}\int_0^\infty e^{-0.5\,t^2}dt$. Using this approximation, we can integrate and find a closed-form expression for the BER of $\pi/4$-DQPSK as in [12]

$$P_b \simeq \left[\frac{1}{2}(1-\mu)\right]^{K-U+1} \sum_{n=0}^{K-U+1} \binom{K-U+n}{n} \left[\frac{1}{2}(1+\mu)\right]^n \tag{4.48}$$

where

$$\mu = \sqrt{\frac{\tilde{\gamma}}{1+\tilde{\gamma}}}$$

and

$$\tilde{\gamma} = \frac{1.716}{4}\frac{E_b}{N_0}\left(\frac{N}{N+U-1}\right)\sigma_f^2 = \frac{1.716}{4}\frac{E_b}{N_0}\left(\frac{N}{N+U-1}\right)\sigma_f^2.$$

Example

Consider $M-QAM$ where $M = 2^Q$. It is easily seen that

$$P_b = \int_0^\infty P_{b,M-QAM}(\gamma)p_\Gamma(\gamma)d\gamma \tag{4.49}$$

where the well-known expression (see [12]) is

$$P_{b,M-QAM}(\gamma) = 1 - \left(1 - 2\left(1 - \frac{1}{\sqrt{M}}\right)Q\left[\sqrt{\frac{3\gamma}{(M-1)}}\right]\right)^2.$$

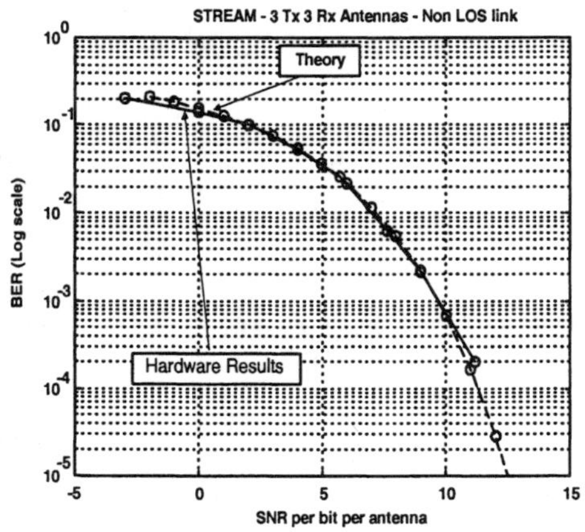

Figure 4.12 BER (analytical versus hardware results) of a multiantenna system in nonfrequency selective Rayleigh fading environment with three transmit antennas and three receive antennas.

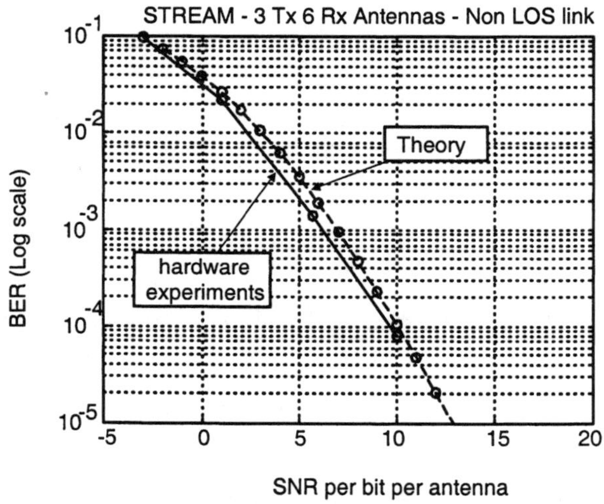

Figure 4.13 BER (analytical versus hardware results) of a multiantenna system in nonfrequency selective Rayleigh fading environment with three transmit antennas and six receive antennas.

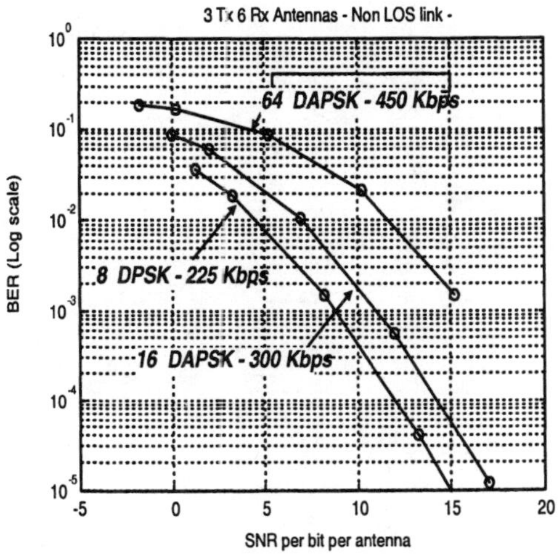

Figure 4.14 BER of a multiantenna system in nonfrequency selective Rayleigh fading environment with three transmit antennas and six receive antennas.

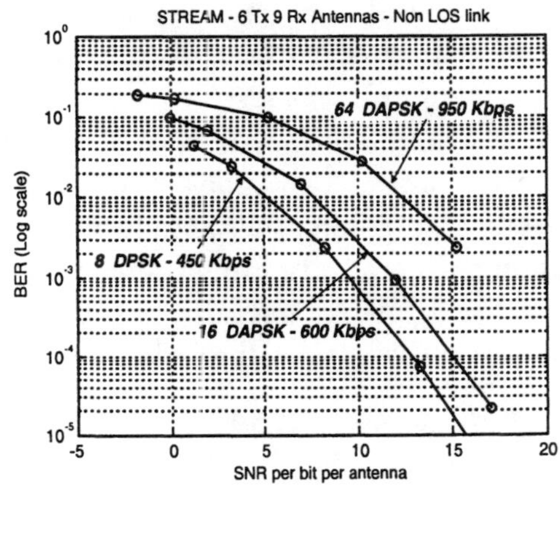

Figure 4.15 BER of a multiantenna system in nonfrequency selective Rayleigh fading environment with six transmit antennas and nine receive antennas.

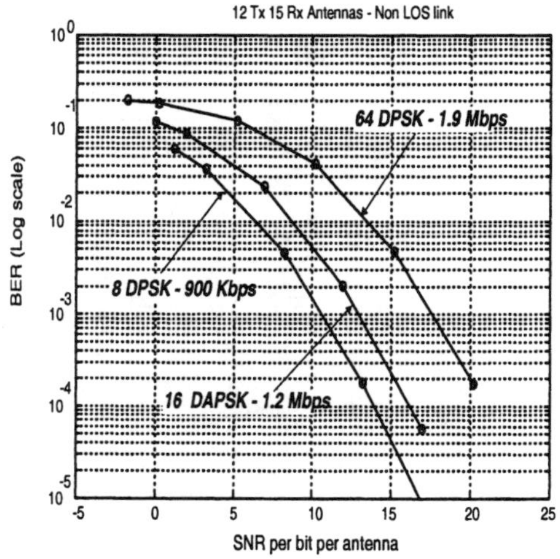

Figure 4.16 BER of a multiantenna system in nonfrequency selective Rayleigh fading environment with 12 transmit antennas and 15 receive antennas.

4.4.5 Numerical Results

All of the results are for an 835-MHz carrier with a bandwidth of about 30 kHz, $\alpha = 0.35$ and $\sigma_f^2 = 1$. We consider $N = 14$. Figures 4.12 and 4.13 show validation of the analytical approach by means of the measurements from hardware. Hardware experiments were performed at WJ Communications.

Figure 4.14 shows numerical results for higher-order modulation formats with three transmit antennas and six receive antennas. The modulation format differential amplitude phase shift keying (DAPSK) is a simple generalization of differential phase shift keying (DPSK) where not only phase but also amplitude is differentially encoded in two consecutive symbol periods.

- Using 8 DPSK, we have 225 Kbps.
- Using 16 DAPSK, we have 300 Kbps.
- Using 64 DAPSK, we have 450 Kbps.

Figure 4.15 shows numerical results for six transmit antennas and nine receive antennas:

- Using 8 DPSK, we have 450 Kbps.

- Using 16 DAPSK, we have 600 Kbps.
- Using 64 DAPSK, we have 950 Kbps.

Figure 4.16 shows numerical results for 12 transmit antennas and 15 receive antennas:

- Using 8 DPSK, we have 900 Kbps.
- Using 16 DAPSK, we have 1.2 Mbps.
- Using 64 DAPSK, we have 1.9 Mbps.

Observe that the diversity gain is essentially determined by the fixed parameter $K - U + 1$.

4.5 Key Points of This Chapter

In this chapter we have explored the theoretical limitations of the single-antenna wireless channel. We have derived capacity for a general (time-varying) frequency-selective channel and shown capacity-achieving schemes. A capacity-achieving transmission scheme is a scheme that coupled with *optimum coding* will get close to capacity. We have shown the fundamental theoretical equivalence of multicarrier and single-carrier modulations and addressed the spatio-temporal multiantenna channel. The impressive theoretical capacity achievable with a multiantenna system was derived and a few simple examples involving two transmit antennas and two receive antennas were provided. In the last part of the chapter we analyzed the performance of a straightforward approach to multiantenna signal demodulation. Some experimental results prove that most of the theoretical advantage of multiantenna technology can be obtained using simple approaches.

References

[1] T. Cover and J. Thomas, *Elements of Information Theory*, New York: Wiley 1991.

[2] H. J. Landau and H. O. Pollak, "Prolate Spheroidal Wave Functions, Fourier Analysis and Uncertainty-III: The Dimensions of the Space of Essentially Time and Band-Limited Signals," *Bell Systems Tech. J.*, Vol. 41, pp. 1295–1336, July 1962.

[3] R.W. Chang, "Orthogonal Frequency Division Multiplexing," U.S. Patent 3,488,445, filed 1966, issued Jan. 6, 1970.

[4] L. J. Cimini, "Analysis and Simulation of a Digital Mobile Channel Using Orthogonal Frequency Division Multiplexing," *IEEE Trans. Comm.*, Vol. 33, pp. 665–675, July 1985.

[5] B. Hirosaki, "An Orthogonally Multiplexed QAM System Using the Discrete Fourier Transform," *IEEE Trans. Comm.*, Vol. 29, pp. 982–989, July 1981.

[6] J. S. Chow, J. C. Tu, and J. M. Cioffi, "A Discrete Multitone Transceiver System for HDSL Application," *IEEE J. Sel. Areas Comm.*, Vol. 9, No. 6, pp. 895–908, Aug. 1991.

[7] R. Price, "Nonlinearly Feedback-Equalized PAM Versus Capacity for Noisy Filter Channel," *Proc. 1972 Internat. Conf. Comm.*, pp. 22.12–22.17, June 1972.

[8] J. M. Cioffi et al., "MMSE Decision Feedback Equalizer and Coding, Part I and II," *IEEE Trans. Commun.*, Vol. 43, pp. 2582–2604, Oct. 1995.

[9] R. S. Cheng and S. Verdu, "Gaussian Multiaccess Channels with ISI: Capacity Region and Multiuser Water-Filling," *IEEE Trans. on Information Theory*, Vol. 39, N. 3, pp. 773–785, May 1993.

[10] S. L. Ariyavisitakul, J. H. Winters, and I. Lee, "Optimum Space-Time Processors with Dispersive Interference: Unified Analysis and Required Filter Span," *IEEE Trans. Comm.*, Vol. 47, pp. 1073–1083, July 1999.

[11] G. J. Foschini, "Layered Space-Time Architecture for Wireless Communication in a Fading Environment When Using Multi-Element Antennas," *Bell Labs Tech. J.*, Vol. 1, No. 2, pp. 41–59, 1996.

[12] J. G. Proakis, *Digital Communications*, New York: McGraw-Hill, 1989.

[13] A. Swindlehurst and T. Kailath, "A Performance Analysis of Subspace-Based Methods in the Presence of Model Errors-Part I: The MUSIC Algorithm," *IEEE Trans. Signal Processing*, Vol. 40, pp. 1758–1774, July 1992.

[14] TIA/EIA/IS-136.1-A, "TDMA Cellular/PCS–Radio Interface–Mobile Station–Base Station Compatibility–Digital Control Channel," Oct. 1996.

[15] G. J. Foschini and M. J. Gans, "On Limits of Wireless Communication Systems in a Fading Environment When Using Multiple Antennas," *Wireless Pers. Comm.*, Vol. 6, No. 3, pp. 311–335, March 1998.

Appendix 4A: The Mechanics of MIMO Capacity Computations

4A.1 System Model

We have shown that an N_{rx}-receive-antenna receiver N_{tx}-transmit-antenna system ($N_{\text{tx}} \leq N_{\text{rx}}$) can be represented as a MIMO model

$$r_i(t) = \sum_{k=1}^{N_{\text{tx}}} \sum_{l=-\infty}^{+\infty} a_k(l) h_{k,i}(t - lT_s) + \tilde{n}_i(t), \quad i = 1, ..., N_{\text{rx}} \quad (4\text{A}.1)$$

where $a_l(n) = \tilde{a}_n^{(l)} + j\, \tilde{b}_n^{(l)}$ are the complex symbols defining the signal constellation, $h_{k,i}(t)$ is the channel response between kth transmit antenna and ith receive antenna. Sampling at R times the symbol rate $1/T_s$, the filtered, downconverted, and base-banded signal at the output of the kth antenna can be written as

$$y_k(n) = \tilde{r}_k(t)|_{t=\frac{nT_s}{R}} = \sum_{i=1}^{N_{\text{tx}}} \sum_{l} h_{k,i}(n - lR) a_i(l) + \eta_k(n)$$
$$k = 1, 2, ..., N_{\text{rx}} \quad (4\text{A}.2)$$

where $h_{k,i}(n) = \tilde{h}_{k,i}(t)\Big|_{t=\frac{nT_s}{R}}$, $\eta_k(n)$ is sampled Gaussian noise. The multichannel representation of fractionally sampled signals can be obtained defining

$$y_k^{(i)}(n) \triangleq y_k(nR + i)$$

$$h_{k,l}^{(i)}(n) \triangleq h_{k,l}(nR + i)$$

and

$$\eta_k^{(i)}(n) \triangleq \eta_k(nR + i).$$

We can then write (4A.2) as

$$y_k^{(i)}(n) = \sum_{m=1}^{N_{\text{tx}}} \sum_{l=0}^{L} g_{k,m}^{(i)}(l) a_m(n - l) + \eta_k^{(i)}(n)$$
$$k = 1, 2, ..., N_{\text{rx}}, \quad i = 1, 2, ..., R. \quad (4\text{A}.3)$$

Observe that we have considered finite support $L+1$ for $h_{k,m}^{(i)}(l)$ because we assume a truncation of the pulse-shaping filter and finite duration of the multipath fading channel (finite delay spread). Now (4A.3) can be expressed in vector form as

$$\tilde{\mathbf{y}}(n) = \mathbf{H}\mathbf{x}(n) + \tilde{\mathbf{z}}(n), \tag{4A.4}$$

where

$$\tilde{\mathbf{y}}(n) = \left[\mathbf{y}_1^{(1)}(n)^T, ..., \mathbf{y}_1^{(R)}(n)^T, ..., \mathbf{y}_{N_{\text{rx}}}^{(1)}(n)^T, ..., \mathbf{y}_{N_{\text{rx}}}^{(R)}(n)^T\right]^T,$$

$$\mathbf{y}_k^{(i)}(n) = \left[y_k^{(i)}(n-N+1), ..., y_k^{(i)}(n+L)\right]^T$$

$$\tilde{\mathbf{z}}(n) = \left[\mathbf{n}_1^{(1)}(n)^T, ..., \mathbf{n}_1^{(R)}(n)^T, ..., \mathbf{n}_{N_{\text{rx}}}^{(1)}(n)^T, ..., \mathbf{n}_{N_{\text{rx}}}^{(R)}(n)^T\right]^T,$$

$$\mathbf{n}_k^{(i)}(n) = \left[\eta_k^{(i)}(n-N+1), ..., \eta_k^{(i)}(n+L)\right]^T$$

$$\mathbf{x}(n) = \left[a_1(n-N+1), ..., a_1(n), ..., a_{N_{\text{tx}}}(n-N+1), ..., a_{N_{\text{tx}}}(n)\right]^T$$

and

$$\mathbf{H} = \begin{bmatrix} \mathbf{H}_{1,1} & \cdots & \mathbf{H}_{1,N_{\text{tx}}} \\ \cdots & \cdots & \cdots \\ \mathbf{H}_{N_{\text{rx}},1} & \cdots & \mathbf{H}_{N_{\text{rx}},N_{\text{tx}}} \end{bmatrix} \Rightarrow (N+L)N_{\text{rx}} \times NN_{\text{tx}}$$

with

$$\mathbf{H}_{k,l} = \begin{bmatrix} \mathbf{H}_{k,l}^{(1)} \\ \mathbf{H}_{k,l}^{(2)} \\ \vdots \\ \mathbf{H}_{k,l}^{(R)} \end{bmatrix}$$

and

$$\left[\mathbf{H}_{k,l}^{(i)}\right]_{n,m} = \begin{cases} h_{k,l}^{(i)}(m-n) & 0 \le m-n \le L \\ 0 & m-n > L \text{ or } m-n < 0 \end{cases}.$$

Observe that the model consists in a transmission of NN_{tx} scalar symbols each n. Also, since we assume that the samples of the Gaussian noise form a white process, we have

$$E\{\tilde{\mathbf{z}}(n)\tilde{\mathbf{z}}(n)^H\} = \sigma^2 \mathbf{I}_{NN_{\text{tx}}}$$

Assumption: The rank of $\mathbf{H}$ is $\mathcal{R} = NN_{\text{tx}}$, that is, we assume that N and N_{tx} are chosen such that $\mathbf{H}$ is a "tall" matrix, with full column rank.

4A.2 Water-Filling the Eigenvalues of the Channel

To apply the parallel Gaussian channels model, we perform the singular value decomposition (SVD) to $\mathbf{H} = \mathbf{U}\Sigma\mathbf{V}^H$ and multiply each side of (4A.4) by $\mathbf{H}^\dagger = \mathbf{V}\Sigma^{-1}\mathbf{U}^H$ to obtain

$$\mathbf{y}(n) = \mathbf{x}(n) + \mathbf{z}(n), \tag{4A.5}$$

where $\mathbf{y}(n) = \mathbf{H}^\dagger \tilde{\mathbf{y}}(n)$, $\mathbf{z}(n) = \mathbf{H}^\dagger \tilde{\mathbf{z}}(n)$. Now to simplify notation, rename the components of the vectors $\mathbf{x}(n) = \left[X_1, X_2, ..., X_{NN_{\text{tx}}}\right]^T$, $\mathbf{y}(n) = \left[Y_1, Y_2, ..., Y_{NN_{\text{tx}}}\right]^T$, and $\mathbf{z}(n) = \left[Z_1, Z_2, ..., Z_{NN_{\text{tx}}}\right]^T$. The capacity of this model [1] is given by

$$C = \max_{\sum_{i=1}^{NN_{\text{tx}}} E\{|X_i|^2\} \leq P} I\left(X_1, X_2, ..., X_{NN_{\text{tx}}}; Y_1, Y_2, ..., Y_{NN_{\text{tx}}}\right)$$

where $I(X;Y)$ is the mutual information between the random (vector) variables X and Y. We also have

$$I\left(X_1, ..., X_{NN_{\text{tx}}}; Y_1, ..., Y_{NN_{\text{tx}}}\right) = h\left(Y_1, ..., Y_{NN_{\text{tx}}}\right) - h\left(Z_1, ..., Z_{NN_{\text{tx}}}\right) \tag{4A.6}$$

where $h(X)$ is the differential entropy [1] for the random (complex vector) variable X. Define K_X, K_Y, K_Z as the covariance matrices of $\mathbf{x}(n) = \left[X_1, X_2, ..., X_{NN_{\text{tx}}}\right]^T$, $\mathbf{y}(n) = \left[Y_1, Y_2, ..., Y_{NN_{\text{tx}}}\right]^T$, and $\mathbf{z}(n) = \left[Z_1, Z_2, ..., Z_{NN_{\text{tx}}}\right]^T$, respectively. Since $\mathbf{x}(n)$ and $\mathbf{z}(n)$ are independent, we have $K_Y = K_X + K_Z$. Moreover the entropy $h\left(Y_1, Y_2, ..., Y_{NN_{\text{tx}}}\right)$ is maximized when $(Y_1, Y_2, ..., Y_{NN_{\text{tx}}})$ is circularly symmetric complex Gaussian, which means that $(X_1, X_2, ..., X_{NN_{\text{tx}}})$ has to be circularly symmetric complex Gaussian. In this case we obtain [1]

$$h\left(Y_1, Y_2, ..., Y_{NN_{\text{tx}}}\right) = \log\left[(2\pi e)^{NN_{\text{tx}}}\det(K_X + K_Z)\right],$$

$$h\left(Z_1, Z_2, ..., Z_{NN_{\text{tx}}}\right) = \log\left[(2\pi e)^{NN_{\text{tx}}}\det(K_X)\right].$$

Imposing a power constraint at the transmitter $\sum_{i=1}^{NN_{\text{tx}}} E\{|X_i|^2\} \leq P$, we then obtain that the maximum achievable rate is

$$C = \max_{\sum_{i=1}^{NN_{\text{tx}}} E\{|X_i|^2\} \leq P} \log \left[\frac{\det(K_X + K_Z)}{\det(K_Z)} \right].$$

This maximization problem can be solved as in [1] by water-filling the eigenvalues of K_Z, λ_i, $i = 1, 2, ..., NN_{\text{tx}}$. This gives

$$C = \sum_{i=1}^{NN_{\text{tx}}} \log \left[1 + \frac{(\xi + \lambda_i)^+}{\lambda_i} \right]$$

where $(x)^+ = \max(x, 0)$ and ξ is chosen so that

$$\sum_{i=1}^{NN_{\text{tx}}} (\xi - \lambda_i)^+ = P.$$

The eigenvalues λ_i are easily found from

$$\begin{aligned} E\{\mathbf{z}(n)\mathbf{z}(n)^H\} &= \mathbf{H}^{\dagger} E\{\mathbf{z}(n)\mathbf{z}(n)^H\} \left(\mathbf{H}^{\dagger}\right)^H \\ &= \mathbf{V}\Sigma^{-1}\mathbf{U}^H \sigma^2 \mathbf{I}_{NN_{\text{tx}}} \mathbf{U}\Sigma^{-1}\mathbf{V}^H = \sigma^2 \mathbf{V}\Sigma^{-2}\mathbf{V}^H \end{aligned}$$

which clearly shows that $\lambda_i = \frac{\sigma^2}{|\lambda_{h,i}|^2}$ where $\lambda_{h,i}$ is the ith element of the diagonal matrix Σ. We can then write

$$C = \sum_{i=1}^{NN_{\text{tx}}} \log_2 \left[1 + \frac{\left(\xi + \frac{\sigma^2}{|\lambda_{h,i}|^2}\right)^+}{\frac{\sigma^2}{|\lambda_{h,i}|^2}} \right] \quad \text{bits/transmission,} \qquad (4A.7)$$

where ξ is chosen so that

$$\sum_{i=1}^{NN_{\text{tx}}} \left(\xi - \frac{\sigma^2}{|\lambda_{h,i}|^2} \right)^+ = P.$$

Observe that the capacity has been derived according to the time-invariant and deterministic channel assumption. In reality the above expression should be averaged with respect to the random nature of the channel and its eigenvalues. More correctly one should compute

$$\begin{aligned} C &= \int_{\lambda_{h,1}=-\infty}^{+\infty} \int_{\lambda_{h,2}=-\infty}^{+\infty} \cdots \int_{\lambda_{h,NN_{\text{tx}}}=-\infty}^{+\infty} \\ & \sum_{i=1}^{NN_{\text{tx}}} \log_2 \left[1 + \frac{\left(\xi + \frac{\sigma^2}{|\lambda_{h,i}|^2}\right)^+}{\frac{\sigma^2}{|\lambda_{h,i}|^2}} \right] \\ & \cdot p_{\Lambda_{h,1},\ldots,\Lambda_{h,i}} \left(\lambda_{h,1}, \ldots, \lambda_{h,NN_{\text{tx}}}\right) d\lambda_{h,1}, \ldots, d\lambda_{h,NN_{\text{tx}}}, \end{aligned}$$

where $p_{\Lambda_{h,1},\ldots,\Lambda_{h,i}} \left(\lambda_{h,1}, \ldots, \lambda_{h,NN_{\text{tx}}}\right)$ is the joint distribution of the eigenvalues of the randomly varying channel $\mathbf{H}$ [15].

The expression (4A.7) can be equivalently written for the non-white noise case

$$E\{\tilde{\mathbf{z}}(n)\tilde{\mathbf{z}}(n)^H\} \neq \sigma^2 \mathbf{I}_{NN_{\text{tx}}}$$

as

$$C = \log\det\left[\mathbf{I} + \mathbf{H}\tilde{\mathbf{R}}_x \mathbf{H}\mathbf{R}_{\tilde{z}}^{-1}\right] \tag{4A.8}$$

where $\mathbf{R}_{\tilde{z}} = E\{\tilde{\mathbf{z}}(n)\tilde{\mathbf{z}}(n)^H\}$ and $\tilde{\mathbf{R}}_x$ is the autocovariance matrix of the transmission process which satisfies the water-filling principle. In the limit of an infinitely small sample period T_s/R for $R \to \infty$ it is possible to show that expression (4A.8) is equivalent to

$$C = \frac{T_s}{2\pi} \int_{-\pi/T_s}^{\pi/T_s} \log\det\left[\mathbf{I} + \mathbf{H}(\omega)\Sigma(\omega)\mathbf{H}(\omega)^H \mathbf{N}(\omega)^{-1}\right] d\omega \tag{4A.9}$$

in the zero-excess-bandwidth case and with the necessary assumptions to

$$\begin{aligned} C &= \frac{T_s}{2\pi} \\ & \cdot \int_{-\pi/T_s}^{\pi/T_s} \log\det\left[\left(\tilde{\mathbf{N}}(\omega) + \sum_{i=1}^{N_{\text{tx}}} \tilde{\mathbf{H}}_i(\omega)\tilde{\mathbf{H}}_i(\omega)^H\right) \tilde{\mathbf{N}}(\omega)^{-1}\right] d\omega. \end{aligned} \tag{4A.10}$$

4A.3 Practical Assumptions

First, we assume that transmission at each antenna is performed with constant power over the burst of N vector symbols and that each antenna transmits with the same power. Then we make the statistical assumptions that the symbols transmitted are identically and independently distributed (i.i.d.). We then have

$$\det(K_X) = \det(E\{\mathbf{x}(n)\mathbf{x}(n)^H\}) = \prod_{i=1}^{NN_{\text{tx}}} E\{|X_i|^2\} = P^{NN_{\text{tx}}},$$

and

$$\det(K_X + K_Z) = \prod_{i=1}^{NN_{\text{tx}}} \left(E\{|X_i|^2\} + \lambda_i\right) = \prod_{i=1}^{NN_{\text{tx}}} \left(P + \frac{\sigma^2}{|\lambda_{h,i}|^2}\right),$$

$$\det(K_Z) = \prod_{i=1}^{NN_{\text{tx}}} \lambda_i = \prod_{i=1}^{NN_{\text{tx}}} \frac{\sigma^2}{|\lambda_{h,i}|^2}.$$

We can find the maximum achievable rate under these assumptions as

$$\begin{aligned} \tilde{C} &= \log_2 \left[\frac{\prod_{i=1}^{NN_{\text{tx}}} \left(P + \frac{\sigma^2}{|\lambda_{h,i}|^2}\right)}{\prod_{i=1}^{NN_{\text{tx}}} \frac{\sigma^2}{|\lambda_{h,i}|^2}} \right] \\ &= \sum_{i=1}^{NN_{\text{tx}}} \log_2 \left[1 + \frac{P|\lambda_{h,i}|^2}{\sigma^2}\right] \end{aligned} \tag{4A.11}$$

Observe that $\tilde{C}$ is not the real capacity, but we can think of it as the practically achievable, without controlling the power at the transmitter antenna array. If we consider that W is the bandwidth used by the system and that $\sigma^2 = N_0/2$, we can write

$$\tilde{C}_W(P) = W \sum_{i=1}^{NN_{\text{tx}}} \log_2 \left[1 + \frac{P|\lambda_{h,i}|^2}{WN_0}\right] \tag{4A.12}$$

in bits per second per N symbols. This expression clearly reflects the effect of the multipath channel on the total achievable rate. For example, in the ideal case in which $|\lambda_{h,i}|^2 = 1$ for $i = 1, 2, ..., NN_{\text{tx}}$, (which means an ideal propagation environment) we can write

$$\widetilde{C}_W(P) = N_{\text{tx}} W \log_2 [1 + \text{SNR}] \quad \text{bps/Hz}, \tag{4A.13}$$

where SNR is the signal-to-noise ratio over the (Nyquist) bandwidth W.

5

Space-Time Modems Based on Traditional Principles

5.1 Introduction

In this chapter we will focus on effective, practical strategies for the detection of spatio-temporal digitally modulated signals when frequency-selective and possibly time-varying fading is severely affecting the bandwidth of interest. We give particular emphasis to MLSD (or estimators) because of their increased popularity in cellular communications. As an example of specific implementation, we detail two algorithms for fractionally spaced multiantenna sequence detection in cellular time division multiple access (TDMA) and report the results of a performance study involving hardware experiments. An approach based on decision feedback, minimum mean squared error (MMSE) processing is also described.

5.1.1 Traditional Space-Time Methods

If we exclude the multicarrier approach, which is treated in more detail in Chapter 6, there are today two major classes of detectors currently used in industry: maximum likelihood sequence estimators (MLSEs) and MMSE detectors (with and without feedback).

Assume linear modulation of N complex M-ary symbols at N_{tx} antennas and reception at N_{rx} antennas in a static environment. The model that we have already explained in the previous chapters is

$$r_k(t) = \sum_{i=1}^{N_{\mathrm{tx}}} \sum_{n=1}^{N} a_i(n) h_{k,i}(t - nT) + \eta_k(t) \qquad k = 1, 2, ..., N_{\mathrm{rx}}, \quad (5.1)$$

where $a_i(n)$ are complex symbols, $h_{k,i}(t)$ is the impulse response of the channel (eventually including transmit and receive filters), $\eta_k(t)$ is white Gaussian noise with two-sided spectral density equal to $N_0/2$. Detection theory gives the optimum approach to detection of the sequence of symbols $\mathbf{a} = \left[a_1(1), ..., a_{N_{\mathrm{tx}}}(N)\right]^T$ as

$$\hat{\mathbf{a}} = arg \min_{\mathbf{a} \in \mathcal{A}^{MN}} \sum_{k=1}^{N_{\mathrm{rx}}} \int_{t \in \mathcal{T}} |r_k(t) - \sum_{i=1}^{U} \sum_{n=1}^{N} a_i(n) h_{k,i}(t - nT)|^2 dt. \quad (5.2)$$

This detector for large temporal support of $h_{k,i}(t)$ (say approximately LT), for any practical N_{tx} and N_{rx}, is computationally prohibitive: The approximate complexity is $\mathcal{O}[M^{LN}\mathrm{tx}]$. Moreover the channel has to be known or estimated by means of training sequences. A different well-known and effective approach is the decision feedback (DF) (multichannel) filter, which is easily explained considering the frequency domain representation of (5.1)

$$R_k(f) = \sum_{i=1}^{N_{\mathrm{tx}}} A_i(f) H_{k,i}(f) + N_k(f), \quad (5.3)$$

where $R_k(f)$, $A_i(f)$, $H_{k,i}(f)$, $N_k(f)$ are the Fourier transforms of $r_k(t)$, $a_i(n)$, $h_{k,i}(t)$, $\eta_k(t)$, respectively. The DF processor output is expressed as

$$\begin{aligned} Z_i(f) &= \sum_{k=1}^{N_{\mathrm{rx}}} \sum_{m=-J}^{+J} R_k(f - m/T) C_{k,i}(f - m/T) \\ &\quad - \sum_{l=1}^{U} B_{l,i}(f) X_l(f), \end{aligned} \quad (5.4)$$

where $C_{k,i}(f)$ are the feed-forward filters and $B_{l,i}(f)$ are the feedback filters in the frequency domain. Further simplification $[B_{l,i}(f) = 0$

for any $l, i]$ of this detector gives the linear detector, which we will describe in detail. Observe that the summation with respect to m is due to spectrum folding and symbol rate sampling of the signal $\sum_{k=1}^{N_{\mathrm{rx}}} R_k(f)C_{k,i}(f)$. The factor J/T is large enough to cover for the excess bandwidth of $R_k(f)$. The filters are then designed to minimize the mean squared error

$$E\left\{|z_i(n) - a_i(n)|^2\right\} = T\int_{-1/2T}^{+1/2T} E\left\{|Z_i(f) - A_i(f)|^2\right\} df.$$

The MMSE solution for a linear multichannel filter is obtained as

$$\mathbf{W} = [\mathbf{R}_S + \mathbf{R}_I]^{-1}\,\mathbf{H}_i^*$$

where

$$\mathbf{W} = [W_1(f - J/T), ..., W_{N_{\mathrm{rx}}}(f - J/T), ..., W_{N_{\mathrm{rx}}}(f + J/T)]^T,$$

$$\mathbf{H}_i = [H_{1,i}(f - J/T), ..., H_{N_{\mathrm{rx}},i}(f - J/T), ..., H_{N_{\mathrm{rx}},i}(f + J/T)]^T,$$

$$\mathbf{R}_S = \mathbf{H}_i^*\mathbf{H}_i^T,$$

$$\mathbf{R}_I = \sum_l \mathbf{H}_l^*\mathbf{H}_l^T + N_0\mathbf{I}.$$

5.2 Optimum Sequence Detection

Consider a single transmit antenna system ($N_{\mathrm{tx}} = 1$). Sampling at R times the symbol rate we have the discrete-time system

$$y^{(k)}(n) = r_k(t)|_{t=\frac{nT}{R}} = \sum_l h^{(k)}(n - lR)a_l + \eta^{(k)}(n) \qquad (5.5)$$

where

$$\eta^{(k)}(n) = \eta_k(t)|_{t=\frac{nT}{R}}$$

is sampled Gaussian noise,

$$h^{(k)}(n) = h_{k,1}(t)|_{t=\frac{nT}{R}}$$

is the sampled impulse response of the channel between the transmit antenna and the kth receive antenna. Defining

$$y_i^{(k)}(n) = y^{(k)}(nR + i)$$

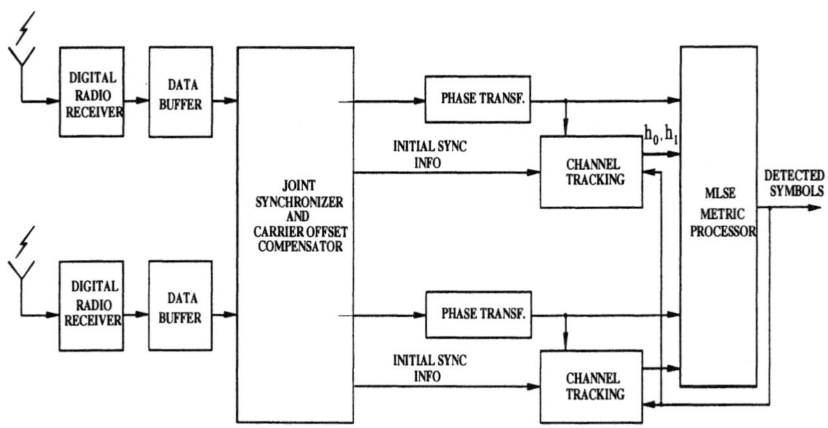

Figure 5.1 Block diagram of an equalizer strategy for two-antenna processing based on MLSE.

$$h_i^{(k)}(n) = h^{(k)}(nR + i)$$

and

$$\eta_i^{(k)}(n) = \eta^{(k)}(nR + i)$$

it is possible to see that (5.5) accepts an equivalent single input multiple output (SIMO) description given by

$$y_i^{(k)}(n) = \sum_l h_i^{(k)}(n - l)a_l + \eta_i^{(k)}(n) \tag{5.6}$$
$$i = 1, 2, ..., R, \quad k = 1, 2, ..., N_{\text{rx}}.$$

It is important to observe that while the random process $y^{(k)}(n)$ is wide-sense cyclostationary the subprocesses $y_i^{(k)}(n)$ are wide-sense stationary: In a sense we have $N_{\text{rx}}R$ virtual diversity channels. A block diagram of the receiver is depicted in Figure 5.1. Specializing our implementation to a case of high practical interest, we are interested in the case $N_{\text{rx}} = R = 2$ and to the case where the employed modulation is $\pi/4$-DQPSK (see [1]). We have already encountered $\pi/4$-DQPSK in Chapter 4, but we observe that the

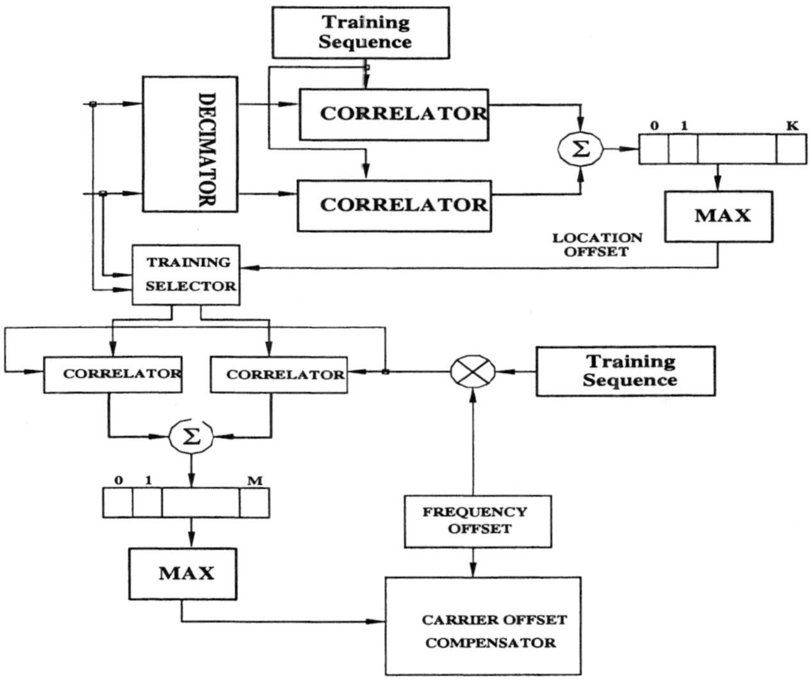

Figure 5.2 Block diagram of the front-end filtering and decimation stages in the IS-136 TDMA MLSE equalizer.

constellation of this modulation format is such that no phase transition through the origin of the complex plane is allowed. In Figure 5.2 we show front-end processing for the MLSE equalizer.

5.2.1 MLSE and the Viterbi Algorithm

Suppose we have the problem of detecting N symbols[1] a_n, $n = 1, 2, ..., N$. We assume that the length of the sequence N is large enough to disregard certain end effects. This concept will be made more precise when the need arises. An important assumption is that the noise samples $\eta^{(k)}(n)$ are spatially and temporally statistically independent. The other fundamental assumption is the truncation of

1. Note that there is a one-to-one mapping between each symbol in the a_n sequence and the actual phase shifts according to the $\pi/4$-DQPSK modulation.

the impulse responses $h_i^{(k)}(n)$, that is the assumption:

$$h_i^{(k)}(n) = 0 \quad \text{for } n < 0 \text{ and } n > D \quad i = 1,2 \text{ and } k = 1,2. \tag{5.7}$$

Arranging the channels in vector form

$$\mathbf{H}^T = \left[\mathbf{H}_1^{(1)^T}, \mathbf{H}_2^{(1)^T}, \mathbf{H}_1^{(2)^T}, \mathbf{H}_2^{(2)^T}\right]$$

where

$$\left[\mathbf{H}_i^{(k)}\right]_{n,m} = \begin{cases} h_i^{(k)}(m-n), & 0 \leq m-n \leq D \\ 0, & m-n > D \text{ or } m-n < 0 \end{cases}$$

and $\mathbf{h}_i^{(k)} = \left[h_i^{(k)}(0), h_i^{(k)}(1), ..., h_i^{(k)}(D)\right]^T$, the expression in (5.6) can be compactly expressed as

$$\mathbf{y} = \mathbf{Ha} + \mathbf{n} \tag{5.8}$$

where

$$\mathbf{y} = \left[\mathbf{y}_1^{(1)^T}, \mathbf{y}_2^{(1)^T} \mathbf{y}_1^{(2)^T}, \mathbf{y}_2^{(2)^T}\right]^T,$$

$$\mathbf{y}_i^{(k)} = \left[y_i^{(k)}(N), y_i^{(k)}(N-1), ..., y_i^{(k)}(1)\right]^T,$$

$$\mathbf{a} = \left[a_N, a_{N-1}, ..., a_1\right]^T,$$

$$\mathbf{n} = \left[\mathbf{n}_1^{(1)^T}, \mathbf{n}_2^{(1)^T} \mathbf{n}_1^{(2)^T}, \mathbf{n}_2^{(2)^T}\right]^T,$$

$$\mathbf{n}_i^{(k)} = \left[\eta_i^{(k)}(N), \eta_i^{(k)}(N-1), ..., \eta_i^{(k)}(1)\right]^T.$$

This allows the formulation of a hypothesis-testing problem

$$\begin{aligned} H_1 &: \mathbf{y} = \mathbf{Ha}^{(1)} + \mathbf{n} \\ H_2 &: \mathbf{y} = \mathbf{Ha}^{(2)} + \mathbf{n} \\ \cdots & \quad \cdots \\ H_{\tilde{M}} &: \mathbf{y} = \mathbf{Ha}^{(\tilde{M})} + \mathbf{n} \end{aligned}$$

where $\tilde{M} = 4^N$ and $\mathbf{a}^{(m)}$ is the $\pi/4$ DQPSK-modulated symbol sequence corresponding to the mth hypothesis H_m. The MAP criterion [2] states a decision in favor of the hypothesis that satisfies

$$\hat{\mathbf{a}} = arg \max_m \left\{p_{A|Y}\left(\mathbf{a}^{(m)}|\mathbf{y}\right)\right\}$$

where $p_{A|Y}\left(\mathbf{a}^{(m)}|\mathbf{y}\right)$ is the conditional probability that the data sequence $\mathbf{a}^{(m)}$ was transmitted given that $\mathbf{y}$ has been received. Using Bayes rule and assuming equiprobable sequences

$$\begin{aligned}\hat{\mathbf{a}} &= arg\max_{m}\left\{\frac{p_{Y|A}\left(\mathbf{y}|\mathbf{a}^{(m)}\right)p_A\left(\mathbf{a}^{(m)}\right)}{p_Y\left(\mathbf{y}\right)}\right\}\\ &= arg\max_{m}\left\{p_{Y|A}\left(\mathbf{y}|\mathbf{a}^{(m)}\right)\right\}\end{aligned}$$

we get the maximum likelihood rule (last equality) [2]. Since the vector noise process is assumed to have Gaussian zero-mean distribution with covariance

$$\mathbf{R}_n = E\left\{\mathbf{n}\mathbf{n}^H\right\} = \sigma_n^2\mathbf{I}, \tag{5.9}$$

we can write

$$p_N\left(\mathbf{n}\right) = \frac{1}{\left(\sqrt{2\pi}\right)^N\sqrt{\det\mathbf{R}_n}}e^{-\frac{1}{2}\mathbf{n}^H\mathbf{R}_n^{-1}\mathbf{n}}$$

and infer that also $\mathbf{y}$ conditioned to $\mathbf{a}^{(m)}$ is Gaussian distributed with mean

$$E_{Y/A}\left\{\mathbf{y}|\mathbf{a}^{(m)}\right\} = \mathbf{H}\mathbf{a}^{(m)}$$

and covariance

$$E_{Y/A}\left\{\left[\mathbf{y} - E_{Y/A}\left\{\mathbf{y}|\mathbf{a}^{(m)}\right\}\right]\left[\mathbf{y} - E_{Y/A}\left\{\mathbf{y}|\mathbf{a}^{(m)}\right\}\right]^H|\mathbf{a}^{(m)}\right\} = \mathbf{R}_n.$$

At this point we can easily write down the distribution $p_{Y|A}\left(\mathbf{y}|\mathbf{a}^{(m)}\right)$ as

$$p_{Y|A}\left(\mathbf{y}|\mathbf{a}^{(m)}\right) = \frac{1}{(\sqrt{2\pi})^N}\frac{1}{\sqrt{\det\mathbf{R}_n}}e^{-\frac{1}{2}\left[\mathbf{y}-\mathbf{H}\mathbf{a}^{(m)}\right]^H\mathbf{R}_n^{-1}\left[\mathbf{y}-\mathbf{H}\mathbf{a}^{(m)}\right]}.$$

The expression can be further simplified using the white noise assumption as

$$p_{Y|A}\left(\mathbf{y}|\mathbf{a}^{(m)}\right) = \text{constant}\times e^{-\frac{1}{2\sigma_n^2}\left\|\mathbf{y}-\mathbf{H}\mathbf{a}^{(m)}\right\|^2}. \tag{5.10}$$

Evidently the maximization of (5.10) with respect to m is equivalent to

$$\widehat{\mathbf{a}} = arg \min_{m} \left\{ \left\| \mathbf{y} - \mathbf{H}\mathbf{a}^{(m)} \right\|^2 \right\} \tag{5.11}$$

which can be also written as

$$\widehat{\mathbf{a}} = arg \min_{m} \left\{ \sum_{k=1}^{2} \sum_{i=1}^{2} \sum_{n=1}^{N} \left| y_i^{(k)}(n) - \sum_{l=0}^{D} h_i^{(k)}(l) a_{n-l}^{(m)} \right|^2 \right\} \tag{5.12}$$

where $a_n^{(m)}$ is the nth symbol of the mth hypothesized sequence $\mathbf{a}^{(m)}$ and we have assumed that $a_n^{(m)} = 0$ for $n > N$ and $n < 1$. The maximum likelihood estimate of the sequence would require the evaluation of

$$l_{\mathbf{a}} = \sum_{k=1}^{2} \sum_{i=1}^{2} \sum_{n=1}^{N} \left| y_i^{(k)}(n) - \sum_{l=0}^{D} h_i^{(k)}(l) a_{n-l} \right|^2$$

for every transmitted sequence $\mathbf{a}^{(m)}$ among the $\tilde{M}$ possible and decide for the one that gives the minimum value of $l_{\mathbf{a}}$. In other words the maximum likelihood rule requires $l_{\mathbf{a}}$ to be minimized over the whole set of possible sequences $\mathbf{a}$. There is a celebrated method known as the Viterbi algorithm (VA) [3] that performs this task efficiently. The significance of the VA is that the number of computations required for the detection of a sequence of length N grows *only linearly* with N. To apply the VA, we have to show that $l_{\mathbf{a}}$ can be reduced to a sum of terms where each term corresponds to the length of a branch in a suitable trellis diagram. It is convenient to define the variables

$$\mathcal{S}_n = [a_{n-1}, a_{n-2}, ..., a_{n-D}] \quad n = D, D+1, ..., N$$

and the quantities

$$U_{n+1}(\mathcal{S}_D, ..., \mathcal{S}_{n+1}) = \sum_{k=1}^{2} \sum_{i=1}^{2} \sum_{m=1}^{n} \left| y_i^{(k)}(m) - \sum_{l=0}^{D} h_i^{(k)}(l) a_{m-l} \right|^2$$

$$V_{n+1}(\mathcal{S}_n, \mathcal{S}_{n+1}) = \sum_{k=1}^{2} \sum_{i=1}^{2} \left| y_i^{(k)}(n) - \sum_{l=0}^{D} h_i^{(k)}(l) a_{n-l} \right|^2 .$$

It is easy to verify that

$$U_{n+1}(\mathcal{S}_D, ..., \mathcal{S}_{n+1}) = U_n(\mathcal{S}_D, ..., \mathcal{S}_n) + V_{n+1}(\mathcal{S}_n, \mathcal{S}_{n+1})$$
$$n = D, D+1, ..., N-1 \quad (5.13)$$

and that the recursive application of (5.13) yields

$$l_{\mathbf{a}} = U_D(\mathcal{S}_D) + V_{D+1}(\mathcal{S}_D, \mathcal{S}_{D+1}) + ...V_N(\mathcal{S}_{N-1}, \mathcal{S}_N)$$

which is the required decomposition.

The next step is to show a trellis such that we can associate to its branches the values taken by the functions $V_{n+1}(\mathcal{S}_n, \mathcal{S}_{n+1})$. Recall that we have assumed finite memory for the channel D. At any given time t the received signal $\tilde{r}^{(k)}(t)$ depends on a set of consecutive $D+1$ $\pi/4$ DQPSK-modulated symbols, say, $a_n, a_{n-1}, a_{n-2}, ..., a_{n-D}$, the last D of these has been defined as $\mathcal{S}_n$. This is the *state of the channel at time t*. We have a one-to-one correspondence between the sequence of transmitted symbols $a_1, a_2, ..., a_N$ and the sequence of states $\mathcal{S}_D, \mathcal{S}_{D+1}, ..., \mathcal{S}_N$. Therefore the problem of selecting the most likely symbols sequence is equivalent to that of selecting the most likely sequence of states.

For each value of the index n, for $n = D, D+1, ...N$ associate a set of 4^D nodes where each node corresponds to a value of $\mathcal{S}_n$. Each node has four branches stemming from it, one for each value taken by a_n. Also, the branches represent the transition from state $\mathcal{S}_n$ to the next state $\mathcal{S}_{n+1}$. The maximum likelihood detection problem has been reduced to the selection of a path through the trellis. The branches joining states $\mathcal{S}_n$ and $\mathcal{S}_{n+1}$ are assigned the values taken by the function $V_{n+1}(\mathcal{S}_n, \mathcal{S}_{n+1})$, usually referred to as the *metric*. The minimum-metric path corresponds to the most likely sequence of states and hence to the most likely sequence of symbols. The trellis diagram for a $\pi/4$ DQPSK system with two symbols memory is shown in Figure 5.3. The metric computation steps are reported in Figure 5.4:

- *Step 1:* Initialize. Let $n = D$, compute $U_D(\mathcal{S}_D)$ for every possible value of $\mathcal{S}_D$ and store them.
- *Step 2:* Extend survivor paths. Let $n = n+1$ and compute $V_{n+1}(\mathcal{S}_n, \mathcal{S}_{n+1})$ for each pair of states $\mathcal{S}_n, \mathcal{S}_{n+1}$ such that the transition from $\mathcal{S}_n$ to $\mathcal{S}_{n+1}$ is allowed by the trellis structure.

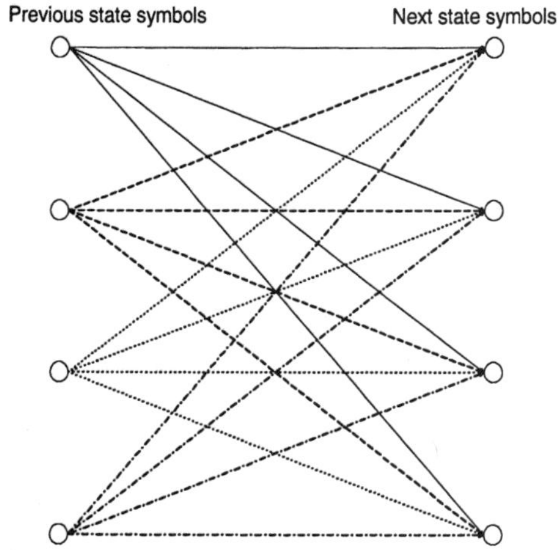

Figure 5.3 Trellis diagram for two-symbol memory detection of $\pi/4$-DQPSK detection using MLSE.

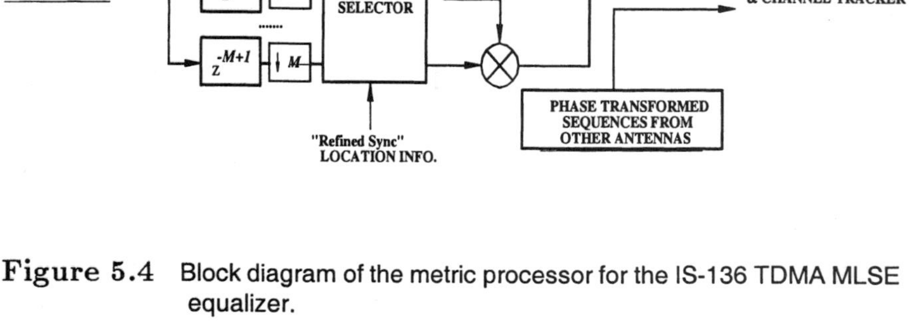

Figure 5.4 Block diagram of the metric processor for the IS-136 TDMA MLSE equalizer.

- *Step 3:* Drop paths. For each state $\mathcal{S}_{n+1}$ compute

$$u_{n+1}(\mathcal{S}_{n+1}) = \min_{\mathcal{S}_n} \{u_n(\mathcal{S}_n) + V_{n+1}(\mathcal{S}_n, \mathcal{S}_{n+1})\}$$

  where $u_D(\mathcal{S}_D) = U_D(\mathcal{S}_D)$ by definition and $u_{n+1}(\mathcal{S}_{n+1})$ is always recursively computed.
- *Step 4:* Trace-back procedure.
- *Step 5:* Iterate. Go to Step 2 if not end of slot.

The detected symbol is given at the output by means of the *trace-back* procedure, the procedure that outputs the estimated symbol as the one associated with the minimum euclidean distance path (going back in time Δ steps) among the retained paths (also defined *survivors*). Observe that Δ is the decision delay expressed in number of symbols.

5.2.2 Channel Estimation and Tracking

The described detection strategy requires a channel estimate of $h_i^{(k)}(n)$ $i = 1, 2$, $k = 1, 2$. Having available N_{sync} symbols, a_n coming from a known synchronization sequence is relatively easy to estimate the discrete-time channels $h_i^{(k)}(n)$. Defining

$$R_{y_i^{(k)}a}(m) = E\{y_i^{(k)}(n+m)a_n^*\},$$

$$R_{aa}(m) = E\{a_n^* a_{n+m}\},$$

we can write the well-known Wiener equation

$$R_{y_i^{(k)}a}(m) = \sum_{l=0}^{D} h_i^{(k)}(l) R_{aa}(m-l). \tag{5.14}$$

Organizing into vector notation

$$\mathbf{R}_{aa}\mathbf{h}_i^{(k)} = \mathbf{r}_{y_i^{(k)}a}$$

where $\mathbf{r}_{y_i^{(k)}a}$ is a $D+1 \times 1$ vector with

$$\left[\mathbf{r}_{y_i^{(k)}a}\right] = \left[R_{y_i^{(k)}a}(0), R_{y_i^{(k)}a}(1), \ldots, R_{y_i^{(k)}a}(D)\right]^T$$

and $\mathbf{R}_{aa}$ is a $D+1 \times D+1$ matrix with

$$\mathbf{R}_{aa} = \begin{bmatrix} R_{aa}(0) & R_{aa}(1)^* & \cdots & R_{aa}(q)^* \\ R_{aa}(1) & R_{aa}(0) & \cdots & R_{aa}(q-1)^* \\ \cdots & \cdots & \cdots & \cdots \\ R_{aa}(q) & R_{aa}(q-1) & \cdots & R_{aa}(0) \end{bmatrix}.$$

The unknown channel vector is

$$\mathbf{h}_i^{(k)} = \left[h_i^{(k)}(0), h_i^{(k)}(1), ..., h_i^{(k)}(D)\right]^T.$$

This system is easily solved in the least squares sense

$$\hat{\mathbf{h}}_i^{(k)} = \mathbf{R}_{aa}^{-1}\mathbf{r}_{y_i^{(k)}a} \tag{5.15}$$

and obviously $\hat{\mathbf{h}}_i^{(k)} = \left[\hat{h}_i^{(k)}(0), \hat{h}_i^{(k)}(1), ..., \hat{h}_i^{(k)}(D)\right]^T$. In the implementation one has to estimate statistical expectations by means of ensemble averages because obviously statistics of the processes are unknown. Then, (5.15) can be written as

$$\hat{\mathbf{h}}_i^{(k)} = \hat{\mathbf{R}}_{aa}^{-1}\hat{\mathbf{r}}_{y_i^{(k)}a}. \tag{5.16}$$

where

$$\left[\mathbf{r}_{y_i^{(k)}a}\right]_m = \sum_{n=D+1}^{N_{\text{sync}}} y_i^{(k)}(n)a_{n-m}^*, \qquad \text{for } 0 \le m \le D$$

$$\left[\hat{\mathbf{R}}_{aa}\right]_{l,m} = \sum_{n=D+1}^{N_{\text{sync}}} a_{n-m}^* a_{n-l}, \qquad \text{for } 0 \le l, m \le D.$$

In reality the impulse responses $h_i^{(k)}(l)$ will be time-varying so that it is necessary to track the variations of the channel.

The algorithm usually employed is a gradient approach [4]

$$\hat{\mathbf{h}}_i^{(k)}(n-\Delta) = \hat{\mathbf{h}}_i^{(k)}(n-\Delta-1) + \mu_1 \mathbf{a}(n-\Delta)^* \epsilon_i^{(k)}(n-\Delta)$$

where

$$\mathbf{a}(n) = [a_n, a_{n-1}, ..., a_{n-D}]^T,$$

$\hat{\mathbf{h}}_i^{(k)}(n) = \left[h_i^{(k)}(0,n), h_i^{(k)}(1,n), ..., h_i^{(k)}(D,n)\right]^T$, is the estimate of the channel at time step n,

$$\epsilon_i^{(k)}(n-\Delta) = y_i^{(k)}(n-\Delta) - \hat{\mathbf{h}}_i^{(k)}(n-\Delta-1)^T \mathbf{a}(n-\Delta),$$

μ_1 is the step size. It is useful to write the algorithm as

$$\hat{\mathbf{h}}_i^{(k)}(n-\Delta) = \hat{\mathbf{h}}_i^{(k)}(n-\Delta-1) - \mu_1 \widehat{\mathbf{g}}(n-\Delta) \qquad (5.17)$$

where $\widehat{\mathbf{g}}(n-\Delta) = -\mathbf{a}(n-\Delta)^* \widehat{\epsilon}_i^{(k)}(n-\Delta)$ is the estimate of the gradient vector at the nth iteration. Define

$$\mathbf{A}(n) = \begin{bmatrix} a_{n-\Delta} & a_{n-\Delta-1} & \cdots & a_{n-\Delta-D} \\ a_{n-\Delta-1} & a_{n-\Delta-2} & \cdots & \cdots \\ \vdots & \vdots & \vdots & \vdots \\ a_{n-\Delta-W+1} & a_{n-\Delta-W} & \cdots & a_{n-\Delta-W-D+1} \end{bmatrix}$$

and

$$\mathbf{y}_i^{(k)}(n-\Delta)$$
$$= \left[y_i^{(k)}(n-\Delta), y_i^{(k)}(n-\Delta-1), ..., y_i^{(k)}(n-\Delta-W+1)\right]^T.$$

5.2.3 Gradient-Based Channel Tracking

In principle one can interpret an adaptive filter as a feedback control system and use higher than first-order controllers. The idea is not exactly new; in fact it dates back to the 1970s. For example, the famous momentum LMS is one of these schemes. Here some specific rules to maintain stability are detailed together with analysis that supports the conclusion that this class of methods may not be appropriate when mean square error performance is of concern.

Traditional Gradient-Based Tracking Schemes

A well-known algorithm (see [4]) usually employed to track channel variations is a first-order gradient approach [4] (which is also the LMS algorithm)

$$\hat{\mathbf{h}}_i^{(k)}(n-\Delta) = \hat{\mathbf{h}}_i^{(k)}(n-\Delta-1) + \mu_1 \mathbf{a}(n-\Delta)^* \widehat{\epsilon}_i^{(k)}(n-\Delta)$$

where

$$\mathbf{a}(n) = [a_n, a_{n-1}, ..., a_{n-D}]^T$$

are the previously detected symbols in the absolute phase domain,

$$\hat{\mathbf{h}}_i^{(k)}(n) = \left[h_i^{(k)}(0,n), h_i^{(k)}(1,n), ..., h_i^{(k)}(D,n)\right]^T$$

is the estimate of the channel at time step n,

$$\widehat{\epsilon}_i^{(k)}(n-\Delta) = y_i^{(k)}(n-\Delta) - \hat{\mathbf{h}}_i^{(k)}(n-\Delta-1)^T \mathbf{a}(n-\Delta),$$

and μ_1 is the step size. Observe that we changed notation for $\epsilon_i^{(k)}(n-\Delta)$ with respect to previous sections as $\widehat{\epsilon}_i^{(k)}(n-\Delta)$ to emphasize the fact that only an *estimated* error is available.

It will also be useful for the following derivations to write the algorithm as

$$\hat{\mathbf{h}}_i^{(k)}(n-\Delta) = \hat{\mathbf{h}}_i^{(k)}(n-\Delta-1) - \mu_1 \widehat{\mathbf{g}}(n-\Delta) \tag{5.18}$$

where $\widehat{\mathbf{g}}(n-\Delta) = -\mathbf{a}(n-\Delta)^* \widehat{\epsilon}_i^{(k)}(n-\Delta)$ is the estimate of the gradient vector at the n-th iteration. The gradient estimate $\widehat{\mathbf{g}}(n-\Delta)$ may be noisy; that is why several variations of the LMS that use a filtered version of the gradient were proposed in the past.

Higher-Order Gradient Schemes

A second-order tracking scheme can be expressed as

- *Step 1:* $\Rightarrow$ $\widehat{\nabla}(n-\Delta) = \beta\widehat{\nabla}(n-\Delta-1) - \widehat{\mathbf{g}}(n-\Delta);$
- *Step 2:* $\Rightarrow$ $\hat{\mathbf{h}}_i^{(k)}(n+1-\Delta) = \hat{\mathbf{h}}_i^{(k)}(n-\Delta) + \mu\widehat{\nabla}(n-\Delta).$

It is evident the intention of the algorithm of filtering the gradient estimate along each component of $\hat{\mathbf{h}}_i^{(k)}(n-\Delta)$ with a one-pole filter $H(z)$ having a response

$$H(z) = \frac{1}{1-\beta z^{-1}}.$$

Note that $\beta = 0$ gives a first-order tracking loop identical to the traditional LMS.

Interpretation as a Feedback Control System and Stability Issues

The mean value of the recursive equations can be computed as

$$\begin{aligned} E\left\{\hat{\mathbf{h}}_i^{(k)}(n-\Delta+1)\right\} &= \mathbf{h}_i^{(k)}(n+1-\Delta) \\ &= \mathbf{h}_i^{(k)}(n-\Delta)+\mu\nabla(n-\Delta) \quad (5.19) \end{aligned}$$

and

$$\begin{aligned} E\left\{\hat{\nabla}(n-\Delta)\right\} &= \nabla(n-\Delta) \\ &= \beta\nabla(n-\Delta-1)+\mathbf{a}(n-\Delta)^*\epsilon_{i,\mathrm{true}}^{(k)}(n-\Delta), \end{aligned}$$

where $\mathbf{g}(n-\Delta) = -\epsilon_{i,\mathrm{true}}^{(k)}(n-\Delta)\mathbf{a}^*(n-\Delta)$ is the true error, not the estimated. Also observe that

$$\mathbf{a}(n-\Delta)^*\epsilon_{i,\mathrm{true}}^{(k)}(n-\Delta) = \mathbf{R}_{aa}\mathbf{h}_i^{(k)}(n-\Delta)-\mathbf{r}_{y_i^{(k)},a}. \quad (5.20)$$

These equations can be evidently interpreted as a feedback control system whose plant to be controlled is constituted by parallel discrete-time filters having response $G(z) = \frac{\mu}{z-1}$. The controller to be designed is $H(z)$.

The difference equation for the system is

$$\begin{aligned} \mathbf{h}_i^{(k)}(n+1-\Delta) &= \mathbf{h}_i^{(k)}(n-\Delta)(1+\beta) \\ &\quad -\beta\mathbf{h}_i^{(k)}(n-\Delta-1)-\mu\mathbf{g}(n-\Delta). \quad (5.21) \end{aligned}$$

Observe that we are interested in the conditions under which $\mathbf{g}(n-\Delta)$ converges to zero.

Diagonalize (that is, find the eigenvalue decomposition of) the Hessian[2] $\mathbf{R}_{aa}$ as $\mathbf{R}_{aa} = \mathbf{U}\mathbf{\Lambda}\mathbf{U}^H$ and define

$$\tilde{\mathbf{g}}(n-\Delta) = \mathbf{U}^H\mathbf{a}(n-\Delta)^*\epsilon_{i,\mathrm{true}}^{(k)}(n-\Delta),$$

$$\tilde{\mathbf{r}}_{y_i^{(k)},a} = \mathbf{U}^H\mathbf{r}_{y_i^{(k)},a}$$

$$\tilde{\mathbf{h}}_i^{(k)}(n-\Delta) = \mathbf{U}^H\mathbf{h}_i^{(k)}(n-\Delta).$$

2. In fact, the autocorrelation matrix is in this case also the Hessian.

Writing the recursion with the transformed parameters

$$\begin{aligned}\widetilde{\mathbf{h}}_i^{(k)}(n+1-\Delta) &= (1+\beta)\widetilde{\mathbf{h}}_i^{(k)}(n-\Delta) \\ &\quad -\mu\mathbf{\Lambda}\widetilde{\mathbf{h}}_i^{(k)}(n-\Delta) - \beta\widetilde{\mathbf{h}}_i^{(k)}(n-\Delta-1) \\ &\quad +\mu\widetilde{\mathbf{r}}_{y_i^{(k)},a} \end{aligned} \tag{5.22}$$

$$\begin{aligned}\widetilde{\mathbf{g}}(n-\Delta+1) - (1+\beta)\widetilde{\mathbf{g}}(n-\Delta) + \mu\mathbf{\Lambda}\widetilde{\mathbf{g}}(n-\Delta) \\ +\beta\widetilde{\mathbf{g}}(n-\Delta-1) = \mathbf{0}\end{aligned} \tag{5.23}$$

it is easy to see that as $n \to \infty$ we must have $\widetilde{\mathbf{g}}(n-\Delta) \to \mathbf{0}$ to have stability and this means that the characteristic polynomials in the z-domain in (5.22) and (5.23) must have roots for $\widetilde{\mathbf{g}}(n-\Delta)$ that lie inside the unit circle. In other words we must have

$$0 < \beta < 1 \tag{5.24}$$

$$0 < \mu < \frac{2(1+\beta)}{\lambda_{\max}} \tag{5.25}$$

where $\lambda_{\max}$ is the maximum eigenvalue of $\mathbf{R}_{aa}$. It is a well-known result [4] that to have stability in case of the traditional LMS we need

$$0 < \mu < \frac{2}{\lambda_{\max}}$$

in (5.18). By comparing (5.18)–(5.25), it is immediately evident that the second-order loop allows higher value of the step size μ, hence a faster convergence.

Variance of the Error

An important result for the first-order LMS is that the variance of the error with respect to the optimum Wiener solution is given by

$$\mathcal{E} = \mu\mathcal{E}_{\min}\sum_{l=1}^{D+1}\frac{\lambda_l}{2-\mu\lambda_l}, \tag{5.26}$$

where $\mathcal{E}_{\min}$ is the minimum error obtained as the Wiener equation is solved correctly at every time-step. Define $\mathbf{h}_i^{(k)}(n-\Delta)^{\text{Wiener}}$ as the optimum solution and observe that[3]

$$\mathcal{E} = E\left\{\left(\widehat{\mathbf{h}}_i^{(k)}(n-\Delta) - \mathbf{h}_i^{(k)}(n-\Delta)^{\text{Wiener}}\right)^H \cdot \mathbf{\Lambda}\left(\widehat{\mathbf{h}}_i^{(k)}(n-\Delta) - \mathbf{h}_i^{(k)}(n-\Delta)^{\text{Wiener}}\right)\right\}. \quad (5.27)$$

The result we get plugging in the difference equations obtained previously and computing the expectations is

$$\mathcal{E} = \frac{\mu \mathcal{E}_{\min}}{2(1-\beta)} \sum_{l=1}^{D+1} \frac{\lambda_l}{1 - \frac{\mu\lambda_l}{2(1+\beta)}}. \quad (5.28)$$

Equation (5.28) shows the main problem of this class of algorithms: *increased error variance.* So although higher-order tracking algorithms are very attractive, they must be used carefully because some applications may not tolerate a high error variance.

Extension to Third-Order Schemes

If we choose the controller as a two-pole filter

$$H(z) = \frac{1}{1 - \beta_1 z^{-1} - \beta_2 z^{-2}}, \quad (5.29)$$

we can follow a similar derivation and derive a third-order algorithm:

$$\widehat{\nabla}(n-\Delta) = \beta_1 \widehat{\nabla}(n-\Delta-1) + \beta_2 \widehat{\nabla}(n-\Delta-2) - \widehat{\mathbf{g}}(n-\Delta).$$

$$\hat{\mathbf{h}}_i^{(k)}(n+1-\Delta) = \hat{\mathbf{h}}_i^{(k)}(n-\Delta) + \mu\widehat{\nabla}(n-\Delta).$$

The region of stability is

$$0 < \mu < \frac{2(1+\beta_1+\beta_2)}{\lambda_{\max}} \quad (5.30)$$

$$-1 < \beta_2 < +1 \quad (5.31)$$

$$\beta_1 < \frac{1-\beta_2^2+\mu\beta_2\lambda_{\max}}{1+\beta_2}, \quad \beta_2 < 0 \quad (5.32)$$

$$\beta_1 < \frac{1-\beta_2^2+\mu\beta_2\lambda_{\min}}{1+\beta_2}, \quad \beta_2 > 0 \quad (5.33)$$

3. We transform all vectors by orthogonalization, that is, $\widetilde{\mathbf{v}} = \mathbf{U}^H\mathbf{v}$.

5.2.4 Least-Squares Methods for Channel Tracking

Consider the channel estimation method formulated at every step as in the following[4]

$$\min_{\hat{\mathbf{h}}_i^{(k)}(n-\Delta)} \left\| \mathbf{A}(n)\hat{\mathbf{h}}_i^{(k)}(n-\Delta) - \mathbf{y}_i^{(k)}(n-\Delta) \right\|^2 \tag{5.34}$$

where

$$\mathbf{A}(n) = \begin{bmatrix} a_{n-\Delta} & a_{n-\Delta-1} & \cdots & a_{n-\Delta-D} \\ a_{n-\Delta-1} & a_{n-\Delta-2} & \cdots & \cdots \\ \vdots & \vdots & \vdots & \vdots \\ a_{n-\Delta-W+1} & a_{n-\Delta-W} & \cdots & a_{n-\Delta-W-D+1} \end{bmatrix}$$

and

$$\mathbf{y}_i^{(k)}(n-\Delta) = \left[y_i^{(k)}(n-\Delta), y_i^{(k)}(n-\Delta-1), ..., y_i^{(k)}(n-\Delta-W+1) \right]^T$$

and the associated least-squares solution

$$\hat{\mathbf{h}}_i^{(k)}(n-\Delta) = \left[\mathbf{A}(n)^H \mathbf{A}(n) \right]^{-1} \mathbf{A}(n)^H \mathbf{y}_i^{(k)}(n-\Delta). \tag{5.35}$$

Consider now the problem at step $n+1$

$$\mathbf{A}(n+1)\hat{\mathbf{h}}_i^{(k)}(n+1-\Delta) = \mathbf{y}_i^{(k)}(n+1-\Delta)$$

and substitute $\hat{\mathbf{h}}_i^{(k)}(n+1-\Delta) = \partial\hat{\mathbf{h}}_i^{(k)}(n+1-\Delta) + \hat{\mathbf{h}}_i^{(k)}(n-\Delta)$ to obtain

$$\mathbf{A}(n+1)\partial\hat{\mathbf{h}}_i^{(k)}(n+1-\Delta) = \mathbf{y}_i^{(k)}(n+1-\Delta) - \mathbf{A}(n+1)\hat{\mathbf{h}}_i^{(k)}(n-\Delta).$$

A possible first-order solution to this least-squares problem would be

- *Step 1:* $\Rightarrow$

$$\partial\hat{\mathbf{h}}_i^{(k)}(n+1-\Delta) = \left[\mathbf{A}(n+1)^H \mathbf{A}(n+1) \right]^{-1} \mathbf{A}(n+1)^H \cdot \left[\mathbf{y}_i^{(k)}(n-\Delta) - \mathbf{A}(n+1)\hat{\mathbf{h}}_i^{(k)}(n-\Delta-1) \right];$$

4. Observe that the method is also the *sliding window* least-squares problem of [4] with window size W.

- *Step 2:* $\Rightarrow \hat{\mathbf{h}}_i^{(k)}(n+1-\Delta) = \mu\partial\hat{\mathbf{h}}_i^{(k)}(n+1-\Delta) + \hat{\mathbf{h}}_i^{(k)}(n-\Delta)$.

It should be evident that Step 2 can be interpreted as a gradient method and of course generalized to second-order schemes:

- *Step 1:* $\Rightarrow \partial\hat{\mathbf{h}}_i^{(k)}(n-\Delta)$
$= \left[\mathbf{A}(n)^H\mathbf{A}(n)\right]^{-1}\mathbf{A}(n)^H\left[\mathbf{y}_i^{(k)}(n-\Delta) - \mathbf{A}(n)\hat{\mathbf{h}}_i^{(k)}(n-1-\Delta)\right]$;

- *Step 2:* $\Rightarrow \widehat{\nabla}(n-\Delta) = \beta\widehat{\nabla}(n-\Delta-1) + \partial\hat{\mathbf{h}}_i^{(k)}(n-\Delta)$;

- *Step 3:* $\Rightarrow \hat{\mathbf{h}}_i^{(k)}(n-\Delta) = \hat{\mathbf{h}}_i^{(k)}(n-1-\Delta) + \mu\partial\widehat{\nabla}(n-\Delta)$.

and to third-order schemes:

- *Step 1:* $\Rightarrow \partial\hat{\mathbf{h}}_i^{(k)}(n-\Delta)$
$= \left[\mathbf{A}(n)^H\mathbf{A}(n)\right]^{-1}\mathbf{A}(n)^H\left[\mathbf{y}_i^{(k)}(n-\Delta) - \mathbf{A}(n)\hat{\mathbf{h}}_i^{(k)}(n-1-\Delta)\right]$;

- *Step 2:* $\Rightarrow \widehat{\nabla}(n-\Delta) = \beta_1\widehat{\nabla}(n-\Delta-1) + \beta_2\widehat{\nabla}(n-\Delta-2) + \partial\hat{\mathbf{h}}_i^{(k)}(n-\Delta)$;

- *Step 3:* $\Rightarrow \hat{\mathbf{h}}_i^{(k)}(n-\Delta) = \hat{\mathbf{h}}_i^{(k)}(n-1-\Delta) + \mu\partial\widehat{\nabla}(n-\Delta)$.

This last algorithm has proved to be remarkably robust in field-deployed base-station systems employing MLSE equalizers.

5.2.5 Simplifications for Metric Computations

We specialize to the case $D = 1$, which accounts in IS-136 for approximately 41.2 μsec of delay spread. The metric expression in this case is given by

$$V_{n+1}(\mathcal{S}_n, \mathcal{S}_{n+1}) = \sum_{i=1}^{2}\left|y_i^{(1)}(n) - \sum_{l=0}^{1} h_i^{(1)}(l)a_{n-l}\right|^2 + \sum_{i=1}^{2}\left|y_i^{(2)}(n) - \sum_{l=0}^{1} h_i^{(2)}(l)a_{n-l}\right|^2. \tag{5.36}$$

Now if we consider that the expression for

$$a_n = e^{j\frac{\pi}{4}\left(\sum_{m=1}^{n-1} x_m + x_n\right)}; \quad x_m \in \{+1, -1, +3, -3\} \tag{5.37}$$

and impose $x'_m = x_n - 1$ so that $x_m \in \{0, +2, -2, +4\}$, we have the transformation

$$a'_n = a_n\, e^{-jn\frac{\pi}{4}}; \quad a_n = a'_n\, e^{jn\frac{\pi}{4}} \tag{5.38}$$

that has the advantage $a'_n \in \{\pm 1, \pm j\}$. Now substituting the a_n expression into (5.36) and then multiplying by $|e^{-jn\frac{\pi}{4}}|^2 = 1$, we obtain

$$\begin{aligned} V_{n+1}\left(\mathcal{S}_n, \mathcal{S}_{n+1}\right) &= \sum_{i=1}^{2} \left| y_i^{(1)'}(n) - h_i^{(1)}(0)a'_n - h_i^{(1)'}(1)a'_{n-1} \right|^2 \\ &+ \sum_{i=1}^{2} \left| y_i^{(2)'}(n) - h_i^{(2)}(0)a_n - h_i^{(2)'}(1)a'_{n-1} \right|^2 \end{aligned} \tag{5.39}$$

where $h_i^{(k)'}(l) = h_i^{(k)}(l)\; e^{-j\frac{\pi}{4}}$ and $y_i^{(k)'}(n) = y_i^{(k)}(n)e^{-jn\frac{\pi}{4}}$. The metric as computed in (5.39) has the advantage of requiring only 16 multiplications. Now consider $W = 4$ so that

$$\mathbf{A}(n)' = \begin{bmatrix} a'_n & a'_{n-1} \\ a'_{n-1} & a'_{n-2} \\ a'_{n-2} & a'_{n-3} \\ a'_{n-3} & a'_{n-4} \end{bmatrix}$$

and $\mathbf{A}(n)^{H'}\mathbf{A}(n)' = \begin{bmatrix} 4 & \tilde{x}_n \\ \tilde{x}_n^* & 4 \end{bmatrix}$ where $\tilde{x}_n = \sum_{i=0}^{3} a'_{n-1-i}a^{*'}_{n-i}$, due to the fact that $a'_n \in \{\pm 1, \pm j\}$. We also have that

$$\left[\mathbf{A}(n)^{H'}\mathbf{A}(n)'\right]^{-1} = \begin{bmatrix} 4u_n & -\tilde{x}_n u_n \\ -\tilde{x}_n^* u_n & 4u_n \end{bmatrix}$$

with $u_n = \frac{1}{16-|\tilde{x}_n|^2}$. Define also

$$\begin{aligned} \epsilon_i^{(k)'}(n-\Delta) &= y_i^{(k)'}(n-\Delta) - \hat{h}_i^{(k)}(0, n-\Delta-1)a'_{n-\Delta} \\ &\quad -\hat{h}_i^{(k)'}(1, n-\Delta-1)a'_{n-\Delta-1}. \end{aligned} \tag{5.40}$$

The algorithm is summarized component-wise as in the following.

- *Step 1:*
  - $\tilde{x}_n = \sum_{i=0}^{3} a'_{n-1-i} a^{*'}_{n-i}$
  - $u_n = \frac{1}{16-|\tilde{x}_n|^2}$

- *Step 2:* $\Rightarrow$ <u>If $|\frac{\tilde{x}_n}{4}| = 1$ or $|\tilde{x}_n| = 4$ go to Step 3</u>
  - $\partial \hat{h}_i^{(k)}(0, n-\Delta) = 4u_n \sum_{i=0}^{3} \epsilon_i^{(k)'}(n-\Delta-i)a^{*'}_{n-i} - \tilde{x}_n u_n \sum_{i=0}^{3} \epsilon_i^{(k)'}(n-\Delta-i)a^{*'}_{n-i-1}$
  - $\partial \hat{h}_i^{(k)}(1, n-\Delta)' = -\tilde{x}_n u_n \sum_{i=0}^{3} \epsilon_i^{(k)'}(n-\Delta-i)a^{*'}_{n-i} + 4u_n \sum_{i=0}^{3} \epsilon_i^{(k)'}(n-\Delta-i)a^{*'}_{n-i-1}$
  - $\widehat{g}(0, n-\Delta) = -\partial \hat{h}_i^{(k)}(0, n-\Delta)$
  - $\widehat{g}(1, n-\Delta)' = -\partial \hat{h}_i^{(k)}(1, n-\Delta)'$.

- *Step 3:* $\Rightarrow$ <u>If $|\frac{\tilde{x}_n}{4}| \neq 1$ or $|\tilde{x}_n| \neq 4$ go to Step 4</u>
  - $\widehat{g}(0, n-\Delta)' = -\epsilon_i^{(k)'}(n-\Delta)a^{*'}_{n-\Delta}$.
  - $\widehat{g}(1, n-\Delta)' = -\epsilon_i^{(k)'}(n-\Delta)a^{*'}_{n-\Delta-1}$

- *Step 4:* $\Rightarrow$

$$\left[\widehat{\nabla}(n-\Delta)'\right]_1 = \beta_1 \left[\widehat{\nabla}(n-\Delta-1)'\right]_1 + \beta_2 \left[\widehat{\nabla}(n-\Delta-2)'\right]_1 - \widehat{g}(0, n-\Delta)'.$$

- *Step 5:* $\Rightarrow$

$$\left[\widehat{\nabla}(n-\Delta)'\right]_2 = \beta_1 \left[\widehat{\nabla}(n-\Delta-1)'\right]_2 + \beta_2 \left[\widehat{\nabla}(n-\Delta-2)'\right]_2 - \widehat{g}(1, n-\Delta)'.$$

- *Step 6:* $\Rightarrow$ $\hat{h}_i^{(k)}(0, n+1-\Delta) = \hat{h}_i^{(k)}(0, n-\Delta) + \mu \left[\widehat{\nabla}(n-\Delta)'\right]_1$.

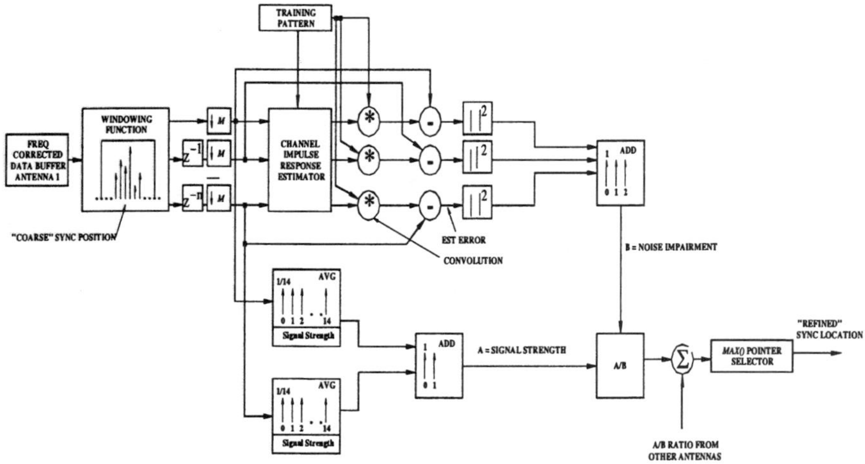

Figure 5.5 Block diagram of the synchronization stage and frequency offset compensation for the IS-136 TDMA MLSE equalizer.

- *Step 7:* $\Rightarrow$

$$\hat{h}_i^{(k)}(1, n+1-\Delta)' = \hat{h}_i^{(k)}(1, n-\Delta)' + \mu \left[\widehat{\nabla}(n-\Delta)'\right]_2 .$$

A block diagram of the channel tracking scheme stage of the detector we have described is depicted in Figure 5.5. A block diagram of the time-frequency synchronization stage of the detector we have described is depicted in Figure 5.6.

5.2.6 Performance Results

We report the results of some measurements performed on a base-station transceiver architecture developed by WJ Communications. The signaling period in IS-136 is 41.2 μsec and the worst case channel characteristics are described by a two-ray Rayleigh channel where the two rays have the same average energy. The variation of the channel can cause Doppler spreads up to 80 Hz, corresponding

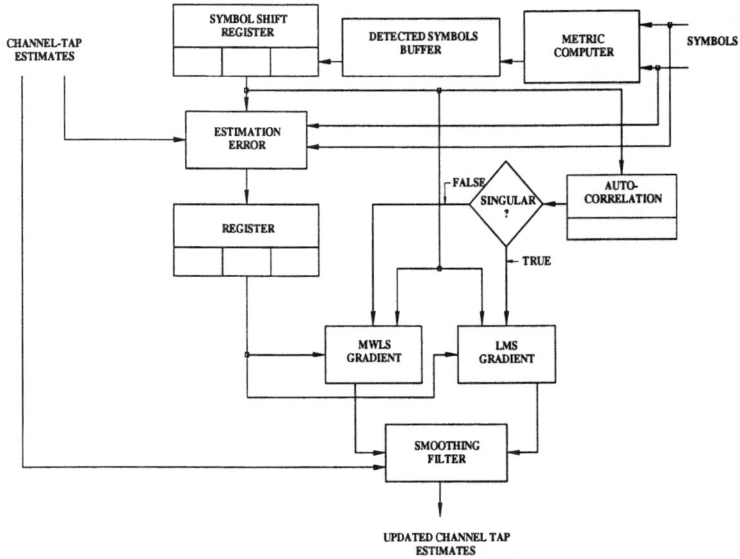

Figure 5.6 Block diagram of the channel tracking data flow for the IS-136 TDMA MLSE equalizer.

to a mobile speed of approximately 65 miles per hour. Figure 5.7 illustrates the remarkable performance of the space-time MLSE algorithm in a rapidly varying channel environment. In Figure 5.8 the space-time MLSE algorithm is analyzed with respect to the rate of change of the time-varying channel. In Figure 5.9 the space-time MLSE algorithm is analyzed with respect to the delay spread (or frequency-selectivity) of the time-varying channel. The robustness and simplicity of implementation of this scheme, associated with the fact that a complexity of 5–10 millions of instructions per second (MIPS) can be achieved, have all contributed to make the described algorithm the most popular approach to time-domain equalization in cellular communications.

5.3 Cross-Validated Signal Processing

We have seen that, since the channel is a priori unknown, a symbol decision-directed method must be used to perform channel estimation. However, this scheme implicitly assumes reliable symbol

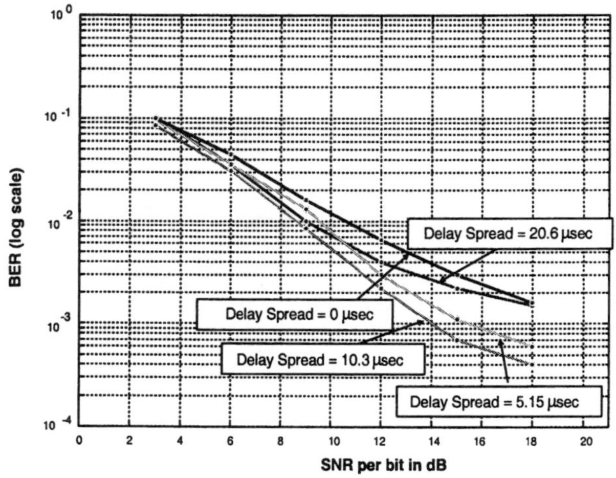

Figure 5.7 BER performance for the IS-136 TDMA MLSE equalizer when the channel experiences a rate of change consequent to the mobile traveling at a speed of 65 miles per hour.

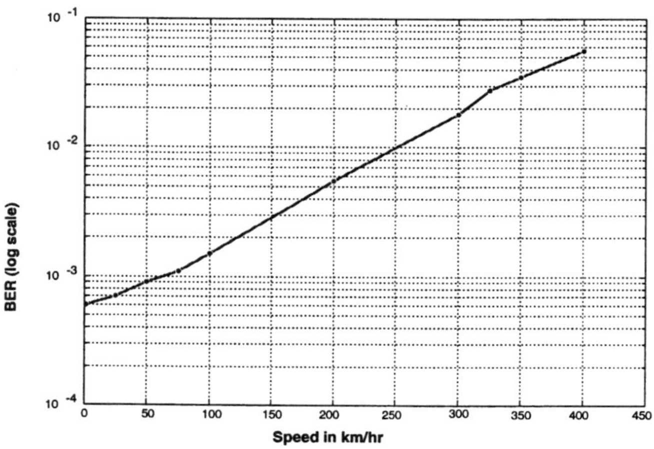

Figure 5.8 BER performance versus speed of the mobile with 20.6 microseconds spacing of the two Rayleigh fading taps.

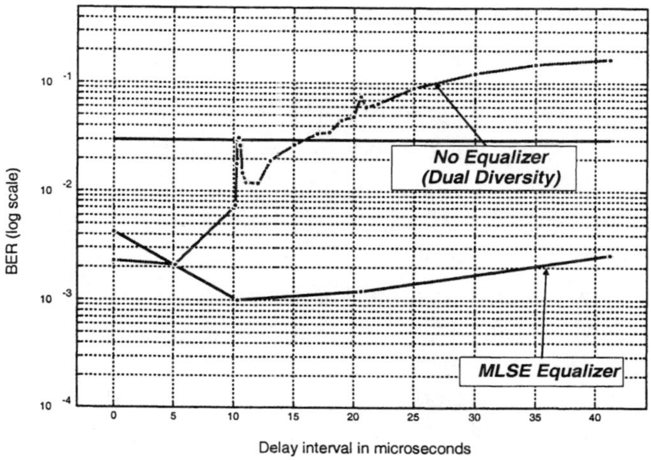

Figure 5.9 BER performance for the IS-136 TDMA MLSE versus delay spread as compared to a maximal ratio combining scheme with no equalizer.

decisions which may not be available at low SNR and requires that the estimate is performed with a certain delay (the decision delay) with respect to the current state. An alternative solution is to accomplish independent branch coefficients computation (with no delay) for each survivor path (per survivor processing, or PSP) [5] within the VA. However, in this case, the channel estimation algorithm loses optimality because there is a clear model mismatch (the assumed sequence for all but one survivor is not the transmitted one). In addition, the computational effort in the channel estimation of the PSP technique is practically unfeasible if a large support ISI channel is induced by significant delay spread of the frequency selective fading channel. The minimum survivor processing (MSP) technique of [6] uses at every step only one channel estimate, but this estimate is derived using the sequence relative to the survivor with the best metric value. However, the channel coefficients are affected by large variance because of the following facts:

- If the survivor with the best metric at step $n + 1$ is different from the survivor with the best metric at step n, the symbol sequence fed back to the channel estimator at step $n + 1$ may differ significantly from the symbol sequence used in the previous step. We define this event a "survivor swap."

- The survivor with the best metric value may contain a number of errors in the sequence even if a survivor swap did not occur.

In [6] it was suggested that the introduction of a small delay in the channel estimation procedure mitigates the two problems. The use of this delayed channel estimate in the metric is not technically correct and again significant performance degradation occurs. We describe in this section an MSP method that allows tracking of the discrete-time channel response with no delay with respect to the current state of the VA [7]. A regularization procedure forces the channel vector to lie in the appropriate a priori known subspace. The well-known variance-bias trade-off [8] is then automatically adjusted by means of a cross-validation "shrinkage" estimator which is at the same time optimal in a "small-sample" predictive sum of squares (PRESS) sense and asymptotically (model) mean squared error optimal. The method will be shown more effective than the PSP algorithm, the MSP approach and the traditional decision-directed approach.

5.3.1 A Different MLSE Model

In this section we will work out a slightly different baseband model for MLSE detection. This will give us more insight and suggest an improved strategy. We know that a complex baseband modulated signal can be represented as $\tilde{s}(t) = \sum_n a_n \, p_g(t - nT)$, where a_n are the complex symbols defining the signal constellation used for the particular digital modulation scheme. The filter $p_g(t)$ is a shaping filter (usually a square root raised cosine filter), and T is the signaling interval. The baseband continuous time representation of this signal at the receiver is

$$\tilde{r}(t) = \sum_n a_n \tilde{h}(t, t - nT) + \tilde{n}(t) \tag{5.41}$$

where $\tilde{h}(t, \tau)$ is the convolution of $\tilde{f}(t, \tau)$, the impulse response of a frequency selective fading multipath channel, and $\tilde{g}(\tau) = \int_{-\infty}^{+\infty} p_g(\beta) p_g(\tau - \beta) d\beta$, the overall shaping filter (in [1] it is a raised cosine function), and $\tilde{n}(t)$ is AWGN. The received signal (5.41) is the input of an ideal low-pass filter with bandwidth W Hz such that the bandwidth of $\tilde{r}(t)$ W_r satisfies $W > W_r$. Sampling the output of the ideal low-pass filter at $R/T = 2W$ rate (R is an integer) we have the discrete-time system whose output samples form a set of sufficient statistics for MLSE

$$y_n = \sum_{l=0}^{NR-1} h(n, n-l)\tilde{a}_l + \eta_n \tag{5.42}$$

where $\tilde{a}_n = \sum_l \delta(n - lR)a_l$ is the sequence of N symbols interleaved with $R-1$ zeroes and $\delta(n) = \begin{cases} 1 & n = 0 \\ 0 & n \neq 0 \end{cases}$, $h(n,k) = \tilde{h}(\frac{nT}{R}, \frac{kT}{R})$ and η_n are samples of a white Gaussian discrete-time process with $E\{\eta_n\eta_n^*\} = RN_0/T$ if N_0 is the two-sided power spectral density of the channel noise. Assuming that the effective support of $\tilde{g}(t)$ is $[-KT, KT]$ and that JT is the maximum duration of the fading channel delay profile, the effective time span of $\tilde{h}(t,\tau)$ is $(D+1)T/R = (J + 2K + 1)T$, which basically allows us to consider finite ISI over $D+1$ fractionally sampled samples

$$y_n = \mathbf{a}_n^T \mathbf{h}_n + \eta_n \quad n = 0, 1, ..., RN-1 \tag{5.43}$$

with $\mathbf{a}_n = [\tilde{a}_n, ...\tilde{a}_{n-D}]^T$ and $\mathbf{h}_n = [h(n,0), ..., h(n,D)]^T$.

Notations $[\mathbf{M}]_{l,m}$ and $[\mathbf{v}]_k$ stand for the l,m element of matrix $\mathbf{M}$ and the kth element of vector $\mathbf{v}$, respectively. We also use the notation $||\mathbf{v}|| = \sqrt{\sum_{i=1}^{M} |v_i|^2}$ for the 2-norm of the complex M-vector $\mathbf{v} = [v_1, ..., v_M]^T$, and $\mathbf{I}_M$ for the $M \times M$ identity matrix.

According to the maximum likelihood principle, the optimal metric corresponding to the hypothesis that the sequence $[\alpha_1, ..., \alpha_{RN-1}]^T$ was transmitted is

$$L_N = \sum_{n=0}^{NR-1} l(\boldsymbol{\alpha}_n, \mathbf{h}_n) = \sum_{n=0}^{NR-1} |y_n - \boldsymbol{\alpha}_n^T \mathbf{h}_n|^2, \tag{5.44}$$

where $\boldsymbol{\alpha}_n = [\alpha_n, ...\alpha_{n-D}]^T$, and $l(\boldsymbol{\alpha}_n, \mathbf{h}_n)$ is the branch metric. The VA is widely used to efficiently perform the exhaustive search in the trellis of the hypotheses. We have seen in the previous developments that in practice the time-varying coefficients $\mathbf{h}_n$ are unknown and must be identified by means of an input/output system identification procedure. If the true transmitted sequence could be provided at any step by the VA, channel identification could be accomplished by minimization of the MSE $E\{|\epsilon_n(\mathbf{a}_n, \hat{\mathbf{h}}_n)|^2\}$ with respect to $\hat{\mathbf{h}}_n$, where $\epsilon_n(\mathbf{a}_n, \hat{\mathbf{h}}_n) = y_n - \mathbf{a}_n^T \hat{\mathbf{h}}_n$. The typical approaches to achieve real-time channel estimation are gradient-based methods and recursive least squares (RLS)-based methods, described in Section 5.2. The typical recursion in both cases is

$$\hat{\mathbf{h}}_{n+1} = \hat{\mathbf{h}}_n + \mathbf{p}_n(\mathbf{a}_n, \hat{\mathbf{h}}_n) \tag{5.45}$$

where $\mathbf{p}_n(\mathbf{a}_n, \hat{\mathbf{h}}_n) = \mu\epsilon_n(\mathbf{a}_n, \hat{\mathbf{h}}_n)\mathbf{a}_n^*$ if the algorithm of choice is the LMS and μ is a step-size parameter that controls the rate of adjustment.[5] However, since the true transmitted sequence at $\mathbf{a}_n$ is not available in practice, one is forced to use sequences formed by past decisions on symbols at the output of the Viterbi processor: Estimation of the channel is delayed by Δ where Δ is the decision delay of the VA.

An alternative solution is to accomplish independent branch coefficients computation for each survivor path (PSP) [5] within the VA. Denoting $\boldsymbol{\alpha}_n$ as the sequence relative to a generic survivor path, one updates the survivor channel as

$$\hat{\mathbf{h}}_{n+1} = \hat{\mathbf{h}}_n + \mathbf{p}_n(\boldsymbol{\alpha}_n, \hat{\mathbf{h}}_n). \tag{5.46}$$

In essence, this approach is an extension of the VA, because it implements basically the generalized likelihood detector [2]. An important criticism to the method is the fact that the channel estimation algorithm (gradient-based or least squares) used in this particular context loses its optimality. It is in fact evident that the model assumed by the recursive identification algorithm for the generic survivor is

$$y_n = \boldsymbol{\alpha}_n^T \mathbf{h}_n + w_n \tag{5.47}$$

where the disturbance term

$$w_n = \eta_n + (\mathbf{a}_n - \boldsymbol{\alpha}_n)^T \mathbf{h}_n \tag{5.48}$$

5. Observe that in the case of an RLS approach $\mathbf{p}_n(\mathbf{a}_n, \hat{\mathbf{h}}_n) = \lambda\epsilon_n(\mathbf{a}_n, \hat{\mathbf{h}}_n)\mathbf{K}_n^*$ where $\mathbf{K}_n = \frac{\mathbf{P}_{n-1}\mathbf{a}_n}{\lambda + \mathbf{a}_n^H \mathbf{P}_{n-1}\mathbf{a}_n}$ and $\mathbf{P}_n = \lambda^{-1}(\mathbf{P}_{n-1} - \mathbf{a}_n^H \mathbf{P}_{n-1}\mathbf{K}_n)$ with λ as the forgetting factor.

is evidently non-Gaussian [6]. As a consequence the adjustment term $\mathbf{p}_n(\boldsymbol{\alpha}_n, \hat{\mathbf{h}}_n)$ cannot be determined correctly to minimize the mean squared error. This loss of optimality is crucial because in the known-channel VA the paths diverge in the steps immediately preceding step n and quickly converge going back to the past. Errors in $\boldsymbol{\alpha}_n$ cause the estimated coefficients to be significantly different from $\mathbf{h}_n$. This in turn increases the probability of having path metrics in the vicinity of step n that are not distinguishable from the correct and maximum likelihood one. Moreover the computational complexity in the channel estimation of the PSP technique is impractical if a long ISI channel is caused by large delay spread of the frequency selective fading channel. The MSP technique of [6] uses at every step only one channel estimate, but this estimate is derived using the sequence $\boldsymbol{\alpha}_n$ relative to the survivor with the best metric value. However, $\hat{\mathbf{h}}_n$ is a large variance estimate because of the following facts:

- If the survivor with the best metric at step $n+1$ is different from the survivor with the best metric at step n, the sequence $\boldsymbol{\alpha}_{n+1}$ fed back to the channel estimator at step $n+1$ may differ significantly from $\boldsymbol{\alpha}_n$, the symbol sequence used in the previous step; we define this event as a "survivor swap."
- The survivor with the best metric value may contain a number of errors in the sequence $\boldsymbol{\alpha}_n$, even if a survivor swap did not occur.

In [6] it was suggested that the introduction of a delay $0 \leq d \leq \Delta$ in the channel estimation procedure mitigates the two problems. The use of $\hat{\mathbf{h}}_{n-d}$ in the metric function $l(\mathbf{a}_n, \hat{\mathbf{h}}_{n-d})$ is, however, not technically correct and again significant performance degradation occurs.

Prompted by these observations we look for a low-variance channel estimate by doing the following:

1. Forcing the channel estimates to lie in a known subspace: It is well-known that by doing so we trade variance for bias.
2. Using cross-validation as a regularization procedure to automatically (and optimally) adjust the trade-off between variance and bias in the channel estimate.

5.3.2 Optimal Channel Estimation Based on Cross-Validation

The recursive channel estimation can be seen as

$$\hat{\mathbf{h}}_{n+1} = \sum_{i=1}^{n} \mathbf{p}_i(\boldsymbol{\alpha}_i, \hat{\mathbf{h}}_i). \tag{5.49}$$

Now observe that since in practice $\hat{\mathbf{h}}_1$ is estimated using a known[6] preamble, each term in the summation (5.49) depends on the initial condition $\hat{\mathbf{h}}_1$. We can rewrite (5.49) as

$$\hat{\mathbf{h}}_{n+1} = \frac{1}{n} \sum_{i=1}^{n} \mathbf{h}_n(i) \tag{5.50}$$

with $\mathbf{p}_i(\boldsymbol{\alpha}_i, \hat{\mathbf{h}}_i) = \frac{1}{n}\mathbf{h}_n(i)$ and model the channel bias in each term as

$$\mathbf{h}_n(i) = \mathbf{h}_{n+1} + \boldsymbol{\xi}_n(i) \tag{5.51}$$

where $\boldsymbol{\xi}_n(i)$ is a random disturbance with *unknown* distribution. We will show that it is possible to design an estimator that is optimal for any "small" n in a predictive sum of squares sense, while being asymptotically optimal in the (model) mean squared sense.

5.3.3 An Orthonormal Basis for $\mathbf{h}_n$

The first step is to design an orthonormal basis for $\mathbf{h}_n$ exploiting the "structure" of the channel impulse response $\tilde{h}(t, \tau)$. Observe that

$$\tilde{h}(t, \tau) = \int_0^{NT} \tilde{f}(t, \beta)\tilde{g}(\tau - \beta)d\beta, \tag{5.52}$$

can be discretized as shown in [9] by

$$\tilde{h}(t, \tau) \simeq T_I \sum_{l=0}^{M} \tilde{f}(t, lT_I)\tilde{g}(\tau - lT_I). \tag{5.53}$$

The accuracy of the approximation is arbitrary[7] (see [9]). The (fractionally) sampled channel then becomes

$$h(n, k) \simeq T_I \sum_{l=0}^{M} f_l(n)g_l(k) \tag{5.54}$$

6. In the absence of a preamble, using a blind method, the initialization of $\hat{\mathbf{h}}_1$ is relatively arbitrary.
7. It must hold that $(M + 1)T_I = NT$. Accuracy is achieved by choosing a discretization interval T_I that is sufficiently small.

where $g_l(k) = \tilde{g}(k\frac{T}{R} - lT_I)$ and $f_l(n) = \tilde{f}(n\frac{T}{R}, lT_I)$. Neglecting the approximation error (which can be made arbitrarily small), it is possible to write (5.54) in vector form as

$$\mathbf{h}_n = \mathbf{G}\mathbf{f}_n \tag{5.55}$$

where $[\mathbf{G}]_{k+1,i+1} = T_I g_i(k)$ and $[\mathbf{f}_n]_{i+1} = f_i(n)$. Basically (5.55) states that $\mathbf{h}_n$ is in the subspace spanned by the columns of $\mathbf{G}$, which, being known, allows the selection of an appropriate orthonormal basis for the channel vector. Selection of a basis is one way of incorporating prior signal information into any estimation procedure. The value of estimating the signal as represented with respect to the basis is that the estimator is insensitive to disturbance components outside the space spanned by the basis, that is the known subspace.

Consider the SVD [4] of $\mathbf{G} = \overline{\mathbf{U}}\boldsymbol{\Sigma}\mathbf{V}^H$ where $\overline{\mathbf{U}}$ is $D+1 \times D+1$, $\boldsymbol{\Sigma}$ is $D+1 \times M+1$ and $\mathbf{V}$ is $M+1 \times M+1$, and write

$$\mathbf{h}_n = \overline{\mathbf{U}}\ \overline{\boldsymbol{\theta}}_n \tag{5.56}$$

where $\overline{\boldsymbol{\theta}}_n = \boldsymbol{\Sigma}\mathbf{V}^H\mathbf{f}_n$. Consider $\mathbf{U} = [\mathbf{u}_1, ..., \mathbf{u}_Q]^T$, the $D+1 \times Q$ matrix formed by the first Q columns (singular vectors) of $\overline{\mathbf{U}}$ corresponding to the significant singular values in $\boldsymbol{\Sigma}$. Applying the transformation

$$\boldsymbol{\theta}_n = \mathbf{U}^H\mathbf{h}_n \tag{5.57}$$

we define a "subspace parameter" $\boldsymbol{\theta}_n$. The Q orthonormal columns of $\mathbf{U}$ span a Q-dimensional subspace defined $\mathcal{U}_Q$ such that $\mathcal{U}_Q \subseteq \mathcal{C}^{D+1}$. Observe that if the transformation $\mathbf{U}$ is selected with $Q < D+1$, the following facts are true:

- The computational complexity is reduced because less parameters have to be estimated in the subspace parameter.
- More importantly, the estimate of $\boldsymbol{\theta}_n$ is insensitive to the disturbance components of $[\boldsymbol{\xi}_n(i)]_m$ for $m = Q+1, Q+2, ..., D+1$ and the resulting estimator has a smaller variance than the full space estimator of $\mathbf{h}_n$.

However, these two advantages are obtained at the expenses of estimation bias because in practice

$$\mathbf{h}_n = \mathbf{U}\boldsymbol{\theta}_n = \mathbf{P}_U\mathbf{h}_n \tag{5.58}$$

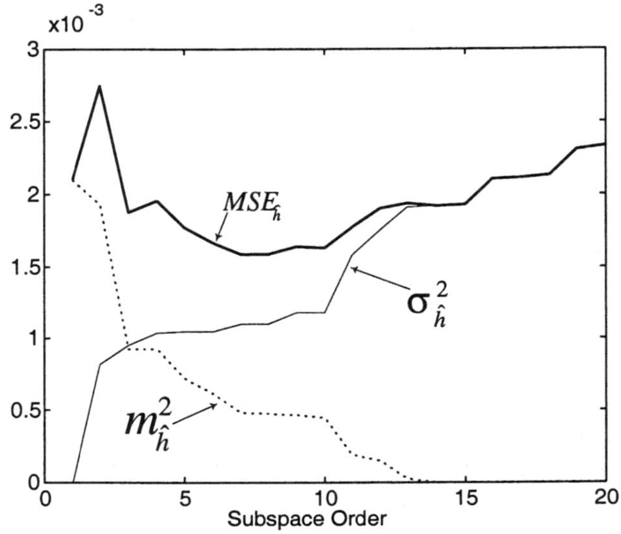

Figure 5.10 The evident trade-off bewteeen variance and bias in channel estimation. A 20-tap Rayleigh fading channel is estimated using recursive least squares in reduced order subspaces.

with $\mathbf{P}_U = \mathbf{U}\mathbf{U}^H$ holds only approximately true and while the variance increases with dimensions of the subspace the bias decreases [8]. Evidently the goal of selecting a transformation $\mathbf{U}$ is to optimize the trade-off between bias, variance, and computational load. The PRESS method can be seen as a method to automatically adjust the amount of bias and variance introduced by a given subspace estimation procedure [10]. In Figure 5.10 we show the evident variance-bias paradigm for a 20 taps channel fading at 8 km/hr estimated in a perfect training situation (that is, with completely known data), using recursive least squares, but using transformations projecting the parameters in subspaces with reduced dimensions.

The figure shows the variance and the squared bias of the estimated $\widehat{\mathbf{h}}_{n+1}$ using different orders of the subspace. Observe that the variance increases with the subspace order while the bias exhibits the opposite behavior. In this particular case the mean squared error of the coefficients exhibits a minimum, which allows advantages with respect to the full-rank estimator. In some other situations one may desire less variance, while a slight increase of MSE is still tolerable.

Remark: Observe that if the discretization interval T_I is selected such that $M+1 < D+1$ then the subspace $\mathcal{U}_Q$ with $Q = M+1$ gives a close to zero bias. However, this subspace selection is not always possible in general.

5.3.4 The PRESS Method

Since we have selected a proper basis for $\hat{\mathbf{h}}_n$, it is immediate to see that

$$\hat{\mathbf{h}}_{n+1} = \mathbf{U}\widehat{\boldsymbol{\theta}}_{n+1} \Rightarrow \widehat{\boldsymbol{\theta}}_{n+1} = \mathbf{U}^H\hat{\mathbf{h}}_{n+1} = \frac{1}{n}\sum_{i=1}^{n}\mathbf{U}^H\hat{\mathbf{h}}_n(i). \quad (5.59)$$

The estimator $\widehat{\boldsymbol{\theta}}_{n+1}$ can also be expressed as the solution of a least squares problem

$$\widehat{\boldsymbol{\theta}}_{n+1} = arg \min_{\boldsymbol{\theta}_{n+1}} \left|\left|\hat{\mathbf{h}}_{n+1} - \mathbf{U}\boldsymbol{\theta}_{n+1}\right|\right|^2. \quad (5.60)$$

In an attempt to regularize the least squares solution we formulate a "penalized" least squares (PLS) problem

$$\widetilde{\boldsymbol{\theta}}_{n+1} = arg \min_{\boldsymbol{\theta}_{n+1}} \left\{\left|\left|\hat{\mathbf{h}}_{n+1} - \mathbf{U}\boldsymbol{\theta}_{n+1}\right|\right|^2 + \boldsymbol{\theta}_{n+1}^H\boldsymbol{\Lambda}_n\boldsymbol{\theta}_{n+1}\right\} \quad (5.61)$$

where $\boldsymbol{\Lambda}_n$ is a diagonal matrix and $[\boldsymbol{\Lambda}_n]_{i,i} = \lambda_{n,i} \geq 0$ are the regularization parameters that control the amount of penalization in each component of the least squares estimate. The term $\boldsymbol{\theta}_{n+1}^H\boldsymbol{\Lambda}_n\boldsymbol{\theta}_{n+1}$ is the "penalizing functional." Increasing $\lambda_{n,i}$ increases the cost associated with channel estimates having a large component in the corresponding subspace of the transform space spanned by the columns of $\mathbf{U}$. The unique solution[8] to (5.61) is well-known [10]

$$\widetilde{\boldsymbol{\theta}}_{n+1}(\boldsymbol{\Gamma}_n) = (\mathbf{I} + \boldsymbol{\Lambda})^{-1}\widehat{\boldsymbol{\theta}}_{n+1} = \widehat{\boldsymbol{\theta}}_{n+1} - \boldsymbol{\Gamma}_n\widehat{\boldsymbol{\theta}}_{n+1} \quad (5.62)$$

where $\boldsymbol{\Gamma}_n$ is a diagonal matrix with $[\boldsymbol{\Gamma}_n]_{i,i} = \frac{\lambda_{n,i}}{1+\lambda_{n,i}} = \gamma_{n,i}$ and since $\lambda_{n,i}$ is a real non-negative number, we have $\gamma_{n,i} \in [0,1]$.

Remark: Since the channel estimates are already forced to lie in the subspace spanned by $\mathbf{U}$, it would appear that there is no

8. Observe that (5.62) is a "shrinkage" estimator because it shrinks the estimate close to the origin.

need for penalization. However, $\mathbf{h}_n = \mathbf{U}\boldsymbol{\theta}_n$ in our approach is only an efficient orthonormal representation for the parameter, not *the main mechanism to reduce variance.* Ideally we would like further reduction of variance by eliminating more columns from $\mathbf{U}$ and using constrained solutions of the form

$$\widehat{\boldsymbol{\theta}}'_{n+1} = arg \min_{\boldsymbol{\theta}_{n+1} \in \Theta_{\tilde{Q}}} \left\| \hat{\mathbf{h}}_{n+1} - \mathbf{U}\boldsymbol{\theta}_{n+1} \right\|^2 \tag{5.63}$$

with $\Theta_{\tilde{Q}} = \left\{ \boldsymbol{\theta}_{n+1} : \theta_{n+1,i} = 0, \quad \tilde{Q} \leq i \leq Q \right\}$, for some $\tilde{Q} < Q$. However, further reduction of the subspace size increases bias. The penalized least squares solution makes an optimal trade-off between these constrained estimators and the full subspace solution (5.60), in fact, we would obtain the estimator (5.63) from (5.62) if we selected $\gamma_{n,i} = 0$ for $i = 0, 1, .., \tilde{Q} - 1$ and $\gamma_{n,i} = 1$ for $i = \tilde{Q}, ..., Q$.

If we define

$$\widetilde{\boldsymbol{\theta}}_{n+1}(\boldsymbol{\Gamma}_n)_l = (\mathbf{I} - \boldsymbol{\Gamma}_n) \sum_{k=1, k \neq l}^{n} \mathbf{U}^H \mathbf{h}_n(k), \tag{5.64}$$

we can design a validation function to be minimized with respect to $0 \leq \gamma_{n,i} \leq 1$ as

$$\mathcal{V}(\boldsymbol{\Gamma}_n) = \sum_{l=1}^{n} \left\| \widetilde{\boldsymbol{\theta}}_{n+1}(\boldsymbol{\Gamma}_n)_l - \mathbf{U}^H \mathbf{h}_n(l) \right\|^2, \tag{5.65}$$

which is a PRESS. Basically $\widetilde{\boldsymbol{\theta}}_{n+1}(\boldsymbol{\Gamma}_n)_l$ is a PLS estimate of $\boldsymbol{\theta}_{n+1}$ with the lth vector omitted. Then this estimate is used to predict $\mathbf{U}^H \mathbf{h}_n(l)$. Summing the squares of the errors for any l one obtains the PRESS function $\mathcal{V}(\boldsymbol{\Gamma}_n)$, which is a small-sample optimality criterion. The objective is to find

$$\boldsymbol{\Gamma}_n^{(\text{PRESS})} = arg \min_{\boldsymbol{\Gamma}_n} \mathcal{V}(\boldsymbol{\Gamma}_n). \tag{5.66}$$

A closed form solution to the minimization problem can be found as in [10]. Define

$$\widehat{\boldsymbol{\theta}}_{n+1} = \left[\hat{\theta}_{n+1,1} \; \hat{\theta}_{n+1,2} \hat{\theta}_{n+1,Q} \right]^T,$$

$$\tilde{\boldsymbol{\theta}}_{n+1} = \left[\tilde{\theta}_{n+1,1}\ \tilde{\theta}_{n+1,2}\ \ldots\ \tilde{\theta}_{n+1,Q}\right]^T,$$

$$\hat{\theta}_{n+1,i} = \frac{1}{n}\sum_{l=1}^{n} \mathbf{u}_i^H \mathbf{h}_n(l),$$

and

$$\sigma_{n,i}^2 = \frac{1}{n-1}\sum_{l=1}^{n} |\mathbf{u}_i^H \mathbf{h}_n(l) - \hat{\theta}_{n+1,i}|^2.$$

Setting the derivative of $\mathcal{V}(\boldsymbol{\Gamma}_n)$ with respect to $\gamma_{n,i}$ equal to zero it is easy to obtain that

$$\gamma_{n,i}^{(\min)} = \frac{\widehat{\sigma}_{n,i}^2}{\frac{\widehat{\sigma}_{n,i}^2}{n} + (n-1)|\hat{\theta}_{n+1,i}|^2}.$$

It is evident that $\gamma_{n,i}^{(\min)} \geq 0$. If $0 \leq \gamma_{n,i}^{(\min)} \leq 1$, $\gamma_{n,i}^{(\min)}$ is the required minimizer of $\mathcal{V}(\boldsymbol{\Gamma}_n)$. If instead $\gamma_{n,i}^{(\min)} > 1$, we have that the required minimizer is equal to 1 because $\mathcal{V}(\boldsymbol{\Gamma}_n)$ is strictly decreasing in $0 \leq \gamma_{n,i} \leq 1$ whenever $\widehat{\sigma}_{n,i}^2 > \frac{\widehat{\sigma}_{n,i}^2}{n} + (n-1)|\hat{\theta}_{n+1,i}|^2$. The optimal choice for $\gamma_{n,i}$ is then

$$\gamma_{n,i}^{(\mathrm{PRESS})} = \mathcal{T}\left(\frac{\widehat{\sigma}_{n,i}^2}{\frac{\widehat{\sigma}_{n,i}^2}{n} + (n-1)|\hat{\theta}_{n+1,i}|^2}\right) \tag{5.67}$$

where

$$\mathcal{T}(x) = \begin{cases} x & x < 1 \\ 1 & x \geq 1 \end{cases}.$$

Observe that the PRESS estimator has an interesting statistical interpretation: It is asymptotically equivalent to the estimator that minimizes $E\left\{|\theta_{n+1,i} - \tilde{\theta}_{n+1,i}|^2\right\}$, the model mean squared error. In fact, modeling $\mathbf{h}_{n+1}$ as deterministic and $\boldsymbol{\xi}_n(i)$ as a zero-mean random disturbance with unknown distribution and $E\{\boldsymbol{\xi}_n(i)^H \boldsymbol{\xi}_n(i)\} < \infty$, one obtains that the value of $\gamma_{n,i}$ minimizing $E\left\{|\theta_{n+1,i} - \tilde{\theta}_{n+1,i}|^2\right\}$ is

$$\gamma_{n,i}^{(\mathrm{MSE})} = \frac{E\left\{|\hat{\theta}_{n+1,i}|^2\right\} - E\left\{\theta_{n+1,i}^* \hat{\theta}_{n+1,i}\right\}}{E\left\{|\hat{\theta}_{n+1,i}|^2\right\}}$$

$$= \frac{\sigma_{n,i}^2}{\sigma_{n,i}^2 + n|\theta_{n+1,i}|^2} \tag{5.68}$$

where $\sigma_{n,i}^2 = E\left\{\left|\mathbf{u}_i^H(\mathbf{h}_n(i) - \mathbf{h}_{n+1})\right|^2\right\}$. It is proved in [11] that asymptotically $\gamma_{n,i}^{(\text{PRESS})} \Rightarrow \gamma_{n,i}^{(\text{MSE})}$.

5.3.5 Adaptive Implementation

The transformation $\mathbf{U}$ allows the modification of the original data model into

$$y_n = \mathbf{z}(\mathbf{a}_n)^T \boldsymbol{\theta}_n + \eta_n \tag{5.69}$$

where $\mathbf{z}(\mathbf{a}_n) = \mathbf{U}^T\mathbf{a}_n$ is the transformed sequence. At every step the recursive procedure does the following:

- Obtains the estimate $\widehat{\boldsymbol{\theta}}_{n+1}$, adjusting the regularized estimate of the previous step;
- Computes the regularization parameters;
- Applies the regularization to $\widehat{\boldsymbol{\theta}}_{n+1}$.

The adaptation then becomes

$$\begin{aligned} \widehat{\boldsymbol{\theta}}_{n+1} &= \widetilde{\boldsymbol{\theta}}_n + \mathbf{U}^H \mathbf{p}_n(\boldsymbol{\alpha}_n, \widetilde{\boldsymbol{\theta}}_n) \\ &= \widetilde{\boldsymbol{\theta}}_n + \mu\epsilon(\boldsymbol{\alpha}_n, \widetilde{\boldsymbol{\theta}}_n)\mathbf{z}^*(\boldsymbol{\alpha}_n) \end{aligned} \tag{5.70}$$

where $\epsilon(\boldsymbol{\alpha}_n, \widetilde{\boldsymbol{\theta}}_n) = y_n - \mathbf{z}(\boldsymbol{\alpha}_n)^T\widetilde{\boldsymbol{\theta}}_n$. To obtain $\boldsymbol{\Gamma}_n$, we use a windowed estimate of $\widehat{\sigma}_{n,i}^2$

$$\widehat{\sigma}_{n,i}^2 = \frac{1}{q-1}\sum_{l=1}^{q}\left|\mu\mathbf{z}^*(\boldsymbol{\alpha}_{n-l+1})\epsilon(\boldsymbol{\alpha}_{n-l+1}, \widetilde{\boldsymbol{\theta}}_{n-l+1}) - \widehat{\boldsymbol{\theta}}_{n+1}\right|^2$$

for $i = 1, 2, ..., Q$, where q is the size of the sliding window. The algorithm is summarized in Table 5.1.

5.3.6 QR-Based Least Squares PRESS Tracking

RLS methods can be used to improve tracking performance. At each $n+1$ step the problem of estimating the channel can be expressed as

$$\min_{\boldsymbol{\theta}_{n+1}} \left\|\left[\begin{array}{c} \lambda\tilde{\mathbf{Z}}^{(n)} \\ \mathbf{z}(\boldsymbol{\alpha}_{n+1})^T \end{array}\right] \boldsymbol{\theta}_{n+1} - \left[\begin{array}{c} \lambda\mathbf{Y}^{(n)} \\ y_{n+1} \end{array}\right]\right\|^2 \tag{5.71}$$

Table 5.1
The Gradient PRESS-Regularized Algorithm

Input
$y_n,\ \boldsymbol{\alpha}_n,\ \widetilde{\boldsymbol{\theta}}_n,\ \mathbf{U},\ \mu,\ \beta,\ q,\ Q,$
Subspace projecton
$\mathbf{z}(\boldsymbol{\alpha}_n) = \mathbf{U}^T\boldsymbol{\alpha}_n$ $\epsilon(\boldsymbol{\alpha}_n, \widetilde{\boldsymbol{\theta}}_n) = y_n - \mathbf{z}(\boldsymbol{\alpha}_n)^T\widetilde{\boldsymbol{\theta}}_n$
Adaptive filtering
$\widehat{\boldsymbol{\theta}}_{n+1} = \widetilde{\boldsymbol{\theta}}_n + \mu\epsilon(\boldsymbol{\alpha}_n, \widetilde{\boldsymbol{\theta}}_n)\mathbf{z}^*(\boldsymbol{\alpha}_n)\ i = 1, 2, ..., Q$
Regularization
$\widehat{\sigma}^2_{n,i} = \frac{1}{q-1}\sum_{l=1}^{q}\left\lvert\mu\mathbf{z}^*(\boldsymbol{\alpha}_{n-l+1})\epsilon(\boldsymbol{\alpha}_{n-l+1}, \widetilde{\boldsymbol{\theta}}_{n-l+1}) - \widehat{\boldsymbol{\theta}}_{n+1}\right\rvert^2,\ i = 1, ..., Q$ $[\boldsymbol{\Gamma}_n]_{i,i} = \beta\,[\boldsymbol{\Gamma}_{n-1}]_{i,i} + \mathcal{T}\left(\frac{\widehat{\sigma}^2_{n,i}}{\widehat{\sigma}^2_{n,i}q^{-1} - (q-1)\lvert\hat{\theta}_{n+1,i}\rvert^2}\right)$ $\widetilde{\boldsymbol{\theta}}_{n+1} = (\mathbf{I} - \boldsymbol{\Gamma}_n)\widehat{\boldsymbol{\theta}}_{n+1}.$

where λ is the *forgetting factor* [4], $\tilde{\mathbf{Z}}^{(n)^T} = [\mathbf{z}(\boldsymbol{\alpha}_1), ..., \mathbf{z}(\boldsymbol{\alpha}_n)]\,\boldsymbol{\Lambda}(n)$ is the data matrix and $\mathbf{Y}^{(n)^T} = [y_1, y_2, ..., y_n]\,\boldsymbol{\Lambda}(n)$ with $\boldsymbol{\Lambda}(n) = diag[\lambda^{n-1}, \lambda^{n-2}, ..., 1]$, is equivalent to solving for $\partial\boldsymbol{\theta}_n = \boldsymbol{\theta}_{n+1} - \boldsymbol{\theta}_n$ the problem

$$\min_{\partial\boldsymbol{\theta}_n}\left\lVert\begin{bmatrix}\mathbf{V}^{(n+1)}\\ \mathbf{0}^T\end{bmatrix}\partial\boldsymbol{\theta}_n - \mathbf{Q}(n)\begin{bmatrix}\mathbf{0}\\ \epsilon_{n+1}\end{bmatrix}\right\rVert^2. \tag{5.72}$$

The orthogonal matrix $\mathbf{Q}(n)$ is such that $\mathbf{Q}(n)\tilde{\mathbf{Z}}^{(n+1)} = \begin{bmatrix}\mathbf{V}^{(n+1)}\\ \mathbf{0}^T\end{bmatrix}$ where $\mathbf{V}^{(n+1)}$ is upper triangular. As a consequence one can append a new data row $\mathbf{z}(\boldsymbol{\alpha}_{n+1})^T$ to $\mathbf{V}^{(n)}$, sweep the last row of the augmented matrix by means of an orthogonal transformation $\tilde{\mathbf{Q}}(n)$, apply the same transformation to the vector $\left[\mathbf{0}^T, \epsilon_{n+1}\right]^T$, and solve a $Q \times Q$ triangular system by back-substitution (for details, see [12]). $\tilde{\mathbf{Q}}(n)$ can be found as a product of Q Givens rotation matrices [4]. Further, one can avoid the back-substitution (implicitly a serial process) to enhance parallelism. The matrix factorization lemma [4] states that for any two matrices $\mathbf{A}_1$ and $\mathbf{A}_2$ related to $\mathbf{B}_1$ and $\mathbf{B}_2$ by orthogonal transformations, that is,

$$\tilde{\mathbf{Q}}\begin{bmatrix}\mathbf{A}_1 & \mathbf{A}_2\end{bmatrix} = \begin{bmatrix}\mathbf{B}_1 & \mathbf{B}_2\end{bmatrix} \tag{5.73}$$

where $\tilde{\mathbf{Q}}^H\tilde{\mathbf{Q}} = \mathbf{I}$, it holds the following fundamental relation

$$\mathbf{B}_1^H\mathbf{B}_2 = \mathbf{A}_1^H\mathbf{A}_2. \tag{5.74}$$

Fact 1: The inverse hermitian of $\mathbf{V}^{(n)}$ (indicated as $\mathbf{V}^{(n)^{-H}}$) can be recursively computed using

$$\tilde{\mathbf{Q}}(n)\begin{bmatrix} \mathbf{V}^{(n)^{-H}}\lambda^{-1} \\ \mathbf{0}^T \end{bmatrix} = \begin{bmatrix} \mathbf{V}^{(n+1)^{-H}} \\ \tilde{\mathbf{g}}(n+1)^T \end{bmatrix}. \tag{5.75}$$

Proof: Consider the factorization lemma with

$$\mathbf{A}_1 = \begin{bmatrix} \mathbf{V}^{(n)^{-H}}\lambda^{-1} \\ \mathbf{0}^T \end{bmatrix}$$

and

$$\mathbf{A}_2 = \begin{bmatrix} \lambda\mathbf{V}^{(n)} \\ \mathbf{z}(\boldsymbol{\alpha}_{n+1})^T \end{bmatrix}$$

and apply the orthogonal transformation to $\begin{bmatrix} \mathbf{A}_1 & \mathbf{A}_2 \end{bmatrix}$. Then it is possible to write

$$\tilde{\mathbf{Q}}(n)\begin{bmatrix} \mathbf{V}^{(n)^{-H}}\lambda^{-1} & \lambda\mathbf{V}^{(n)} \\ \mathbf{0}^T & \mathbf{z}(\boldsymbol{\alpha}_{n+1})^T \end{bmatrix} = \begin{bmatrix} \mathbf{V}^{(n+1)^{-H}} & \mathbf{V}^{(n+1)} \\ \tilde{\mathbf{g}}(n+1)^T & \mathbf{0}^T \end{bmatrix}.$$

It is then evident that writing down (5.74) for $\mathbf{B}_1 = \begin{bmatrix} \mathbf{V}^{(n+1)^{-H}} \\ \tilde{\mathbf{g}}(n+1)^T \end{bmatrix}$ and $\mathbf{B}_2 = \begin{bmatrix} \mathbf{V}^{(n+1)} \\ \mathbf{0}^T \end{bmatrix}$ we get

$$\begin{bmatrix} \mathbf{V}^{(n)^{-1}}\lambda^{-1} & \mathbf{0}^T \end{bmatrix}\begin{bmatrix} \lambda\mathbf{V}^{(n)} \\ \mathbf{z}(\boldsymbol{\alpha}_{n+1})^T \end{bmatrix}$$
$$= \begin{bmatrix} \mathbf{V}^{(n+1)^{-1}} & \tilde{\mathbf{g}}(n+1)^* \end{bmatrix}\begin{bmatrix} \mathbf{V}^{(n+1)} & \mathbf{0}^T \end{bmatrix}, \tag{5.76}$$

that is $\mathbf{V}^{(n)^{-1}}\mathbf{V}^{(n)} = \mathbf{V}^{(n+1)^{-1}}\mathbf{V}^{(n+1)} = \mathbf{I}_Q$.

Fact 2: $\partial\boldsymbol{\theta}_n$ is obtained as:

$$\partial\boldsymbol{\theta}_n = -\tilde{\mathbf{g}}(n+1)^* r_{n+1}, \tag{5.77}$$

where $\tilde{\mathbf{g}}(n+1)$ is obtained from (5.75) and r_{n+1} is obtained as shown in Table 5.2.

Proof: We use again the matrix factorization lemma with $\mathbf{A}_1$ and $\mathbf{B}_1$ defined as in the proof of fact 1 while we use now $\mathbf{A}_2 = \begin{bmatrix} \mathbf{0} \\ \epsilon_{n+1} \end{bmatrix}$ and $\mathbf{B}_2 = \begin{bmatrix} \overline{\mathbf{Y}}^{(n+1)} \\ r_{n+1} \end{bmatrix}$. Using again (5.74) (because obviously (5.73) is verified) we can write

$$\begin{bmatrix} \mathbf{V}^{(n)^{-1}}\lambda^{-1} & \mathbf{0} \end{bmatrix} \begin{bmatrix} 0 \\ \epsilon_{n+1} \end{bmatrix}$$
$$= \begin{bmatrix} \mathbf{V}^{(n+1)^{-1}} & \tilde{\mathbf{g}}(n+1)^* \end{bmatrix} \begin{bmatrix} \overline{\mathbf{Y}}^{(n+1)} \\ r_{n+1} \end{bmatrix}, \qquad (5.78)$$

which gives

$$\overbrace{\mathbf{V}^{(n+1)^{-1}}\overline{\mathbf{Y}}^{(n+1)}}^{\partial\boldsymbol{\theta}_n} + \tilde{\mathbf{g}}(n+1)^* r_{n+1}$$
$$= \lambda^{-1}\mathbf{V}^{(n)^{-1}}\mathbf{0} + \epsilon_{n+1}\mathbf{0} = \mathbf{0}. \qquad (5.79)$$

This last expression proves (5.77).

The algorithmic steps are summarized in Table 5.2. The regularization is obtained by replacing (5.70) with

$$\widehat{\boldsymbol{\theta}}_{n+1} = \widetilde{\boldsymbol{\theta}}_n - \tilde{\mathbf{g}}(n+1)^* r_{n+1}. \qquad (5.80)$$

To obtain $\boldsymbol{\Gamma}_n$ we use the estimate of $\widehat{\sigma}^2_{n,i}$

$$\widehat{\sigma}^2_{n,i} = \frac{1}{q-1}\sum_{l=1}^{q}\left|-\tilde{\mathbf{g}}(n-l+2)^* r_{n-l+2} - \widehat{\boldsymbol{\theta}}_{n+1}\right|^2$$

for $i = 1, 2, ..., Q$. In both methods the size q of the sliding window has to be designed in such a way to accommodate the time-varying characteristics of the channel; moreover, the regularization parameters are computed as

$$\gamma_{n,i}^{(\mathrm{PRESS})} = \mathcal{T}\left(\frac{\widehat{\sigma}^2_{n,i}}{\frac{\widehat{\sigma}^2_{n,i}}{q} - (q-1)|\hat{\theta}_{n+1,i}|^2}\right). \qquad (5.81)$$

Table 5.2
The QR-RLS PRESS-Regularized Algorithm

Input	
--	$y_n,\ \boldsymbol{\alpha}_n,\ \widetilde{\boldsymbol{\theta}}_n,\ \mathbf{U},\ \mathbf{V}^{(n)},\ \lambda,\ \beta,\ q,\ Q,$
Step 1 Step 2	$\mathbf{z}(\boldsymbol{\alpha}_n) = \mathbf{U}^T \boldsymbol{\alpha}_n$ $\epsilon_n = y_n - \mathbf{z}(\boldsymbol{\alpha}_n)^T \widetilde{\boldsymbol{\theta}}_n$
Step 1 Step 2 Step 3 for $l = 1, \ldots, Q$ do end Step 4: Step 5	$\tilde{\mathbf{V}}^{(n)} = \begin{bmatrix} \lambda \mathbf{V}^{(n)} & 0 & \mathbf{V}^{(n)^{-H}} \lambda^{-1} \\ \mathbf{z}(\boldsymbol{\alpha}_n)^T & \epsilon_n & \mathbf{0}^T \end{bmatrix}$ $\tilde{\mathbf{V}}_1^{(n)} = \tilde{\mathbf{V}}^{(n)}$ $a = \left[\tilde{\mathbf{V}}_l^{(n)}\right]_{l,l}, \quad b = \left[\tilde{\mathbf{V}}_l^{(n)}\right]_{Q+1,l}$ $c_l = \frac{a}{\sqrt{\|a\|^2+\|b\|^2}}, \quad s_l = \frac{b^*}{\sqrt{\|a\|^2+\|b\|^2}}$ $\tilde{\mathbf{Q}}^{(l)} = \mathbf{I}_{Q+1}$ $\left[\tilde{\mathbf{Q}}^{(l)}\right]_{Q+1,Q+1} = c_l, \quad \left[\tilde{\mathbf{Q}}^{(l)}\right]_{l,Q+1} = s_l,$ $\left[\tilde{\mathbf{Q}}^{(l)}\right]_{Q+1,l} = -s_l^*, \quad \left[\tilde{\mathbf{Q}}^{(l)}\right]_{l,l} = c_l,$ $\tilde{\mathbf{V}}_{l+1}^{(n)} = \tilde{\mathbf{Q}}^{(l)} \tilde{\mathbf{V}}_l^{(n)}$ $\tilde{\mathbf{V}}_{Q+1}^{(n)} \Rightarrow \begin{bmatrix} \mathbf{V}^{(n+1)} & \overline{\mathbf{Y}}^{(n+1)} & \mathbf{V}^{(n+1)^{-H}} \\ \mathbf{0}^T & r_{n+1} & \tilde{\mathbf{g}}(n+1)^T \end{bmatrix}.$ $\widehat{\boldsymbol{\theta}}_{n+1} = \widetilde{\boldsymbol{\theta}}_n - \tilde{\mathbf{g}}(n+1)^* r_{n+1}$
Step 1 $i = 1, \ldots, Q$ Step 2 $i = 1, \ldots, Q$ Step 3	$\widehat{\sigma}_{n,i}^2 = \frac{1}{q-1} \sum_{l=1}^{q} \left\| -\tilde{\mathbf{g}}(n-l+2)^* r_{n-l+2} - \widehat{\boldsymbol{\theta}}_{n+1} \right\|^2,$ $[\boldsymbol{\Gamma}_n]_{i,i} = \beta \, [\boldsymbol{\Gamma}_{n-1}]_{i,i} + \mathcal{T}\left(\frac{\widehat{\sigma}_{n,i}^2}{\widehat{\sigma}_{n,i}^2 q^{-1} - (q-1)\|\hat{\theta}_{n+1,i}\|^2} \right)$ $\widetilde{\boldsymbol{\theta}}_{n+1} = (\mathbf{I} - \boldsymbol{\Gamma}_n) \widehat{\boldsymbol{\theta}}_{n+1}.$

In rapidly fading environments the size of the window q has to be only a few samples, which causes the estimates of $\gamma_{n,i}^{(\mathrm{PRESS})}$ to be extremely noisy. A practical approach is to filter the estimated regularization parameters with a one pole filter having response $H(z) = \frac{1}{1-\beta z^{-1}}$. This gives the following update

$$\gamma_{n,i}^{(\mathrm{PRESS})} = \beta\gamma_{n-1,i}^{(\mathrm{PRESS})} + \mathcal{T}\left(\frac{\widehat{\sigma}_{n,i}^2}{\frac{\widehat{\sigma}_{n,i}^2}{q} - (q-1)|\hat{\theta}_{n+1,i}|^2}\right). \quad (5.82)$$

Observe that the "transformed" branch metric is

$$l(\mathbf{U}^T\boldsymbol{\alpha}_n, \widetilde{\boldsymbol{\theta}}_n) = |y_n - \mathbf{z}(\boldsymbol{\alpha}_n)^T\widetilde{\boldsymbol{\theta}}_n|^2, \quad (5.83)$$

which can be used directly in the VA.

5.3.7 Performance Analysis Results

We compare the following approaches:

- `CLMS`: Channel estimation with a single LMS channel estimator that is delayed $d = \Delta = 6$ symbols, which is the typical approach of [13];
- `CQRLS`: Channel estimation with a single QR-RLS channel estimator that is delayed $d = \Delta = 6$ symbols;
- `PSPLMS`: Channel estimation with one LMS channel estimator per survivor using the PSP technique with $d = 0$ [5];
- `PSPQRLS`: Channel estimation with one QR-RLS channel estimator per survivor using the PSP technique with $d = 0$;
- `MSPLMS`: Channel estimation with a single LMS channel estimator using the MSP technique with $d = 2$, which is the approach of [6];
- `MSPQRLS`: Channel estimation with a single QR-RLS channel estimator using the MSP technique with $d = 2$;
- `CVMSPLMS(Q)`: Channel estimation with a single regularized LMS channel estimator using the MSP technique with $d = 0$ and order of the subspace equal to Q;
- `CVMSPQRLS(Q)`: Channel estimation with a single regularized QR-RLS channel estimator using the MSP technique with $d = 0$ and an order of the subspace equal to Q.

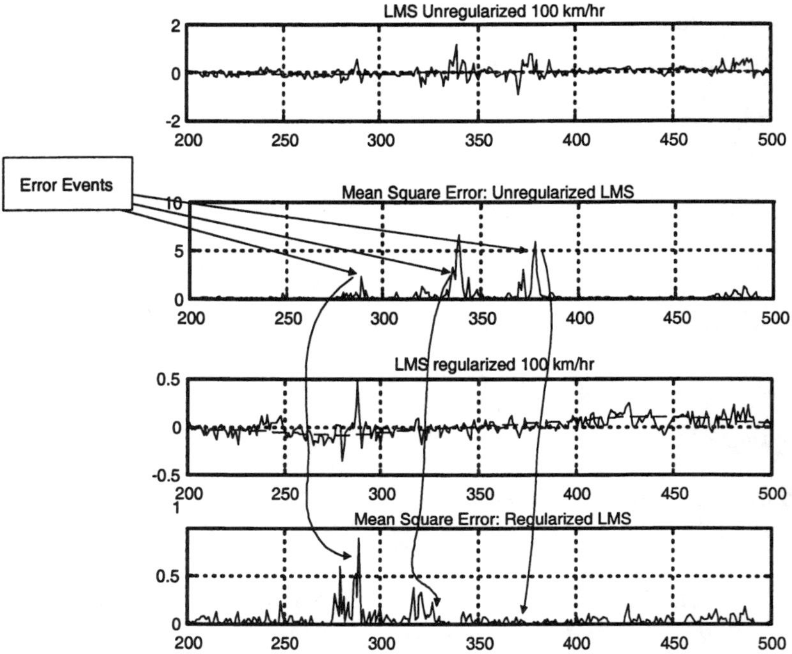

Figure 5.11 Tracking a two-ray Rayleigh faded channel at 100 km/hr with no delay between the VA and the channel estimator using the MSP technique.

The value of Q order of the "representation subspace" for the channel can be typically selected as the number of singular values in the SVD of $\mathbf{G}$ with magnitude larger than a small number ranging between 0.01 and 0.001. We report the performance of the method with different values of Q. The over-sampling factor is $R = 4$. Figure 5.11 shows experiments where we try to investigate the tracking capability of the method in a computer simulation experiment. The plots show the trajectory of the real part of the maximum energy fading coefficient compared to the estimated one. The speed of the mobile is 100 km/hr and the delay interval is 20.6 μsec. We compare `CVMSPLMS(4)` with `MSPLMS`. The SNR per bit is equal to 20 dB. The size of the sliding window for computation of the regularization parameters is $q = 5$. Observe the relative resistance to error events exhibited by the cross-validated

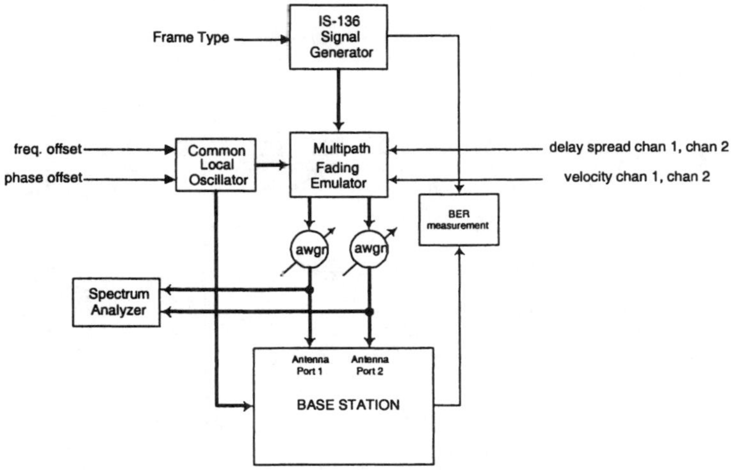

Figure 5.12 Hardware test setup for lab experiments.

subspace approach. The following figures describe only hardware implementation results. The hardware test setup is depicted in Figure 5.12. The IS-136 signal generator simulates transmission of DTC frames coming from three different mobiles. Additive Gaussian noise is injected on both diversity channels. Observe that the DSP modem receives a sampling rate of 80 kHz after RF conversion and digital quadrature downconversion. The word length used is 16 and the algorithm has been implemented using Analog Devices fixed-point DSP processors. A polyphase raised-cosine filter (roll-off factor equal to 0.35) concatenated with an adjacent interference rejection filter transforms the rate to $4/T = 97.2$ kHz. Automatic gain control is operated on a slot-by-slot basis. An open-loop synchronizer chooses the optimum positioning of the metric computer with a resolution of $4/T$ rate. In addition, as required by the specification [1], there is carrier frequency offset between the local oscillator and the transmitted carrier frequency of about 213 Hz. Frequency offset is estimated and removed using a second-order phase locked loop (PLL). The trellis has four states because the span of the ISI channel is two symbol periods. This means that it is assumed to be $D + 1 = 8$. The complexity of each algorithm is reported in Table 5.3.7 in millions of instructions per second (MIPS) for a two-antenna

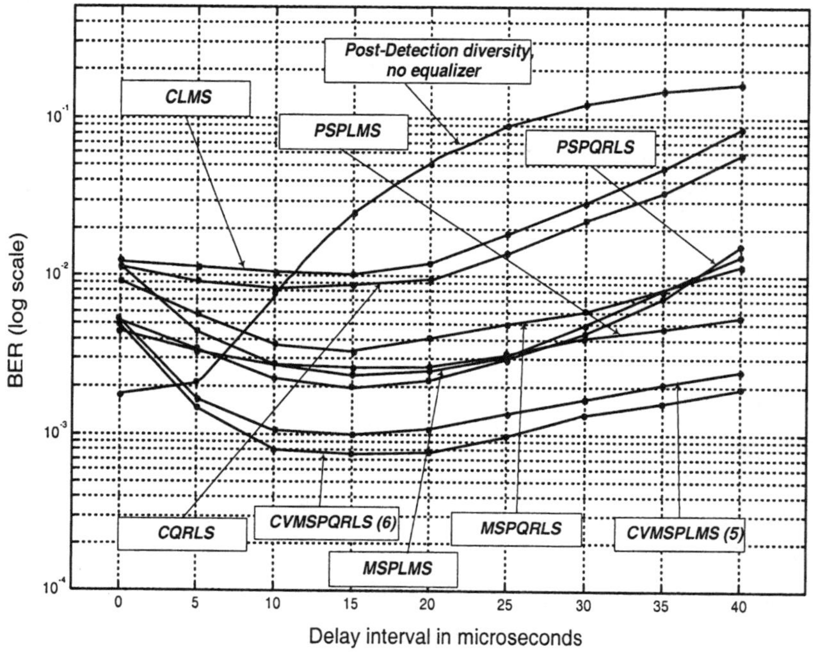

Figure 5.13 BER at SNR per bit equal to 25 dB and 100 km/hr, versus delay interval.

receiver. The complexity is relevant to the processing of one single reverse digital traffic channel (RDTC). The above algorithms have been implemented in Analog Devices ADSP-2181 processors (40-MHz clock rate, 16 bits fixed point arithmetic) with the exclusion of the PSP algorithms that have been implemented in simulated fixed-point *C*-language and executed on baseband samples collected from the same hardware platform. Figure 5.13 shows results at 100 km/hr versus delay interval, while the SNR per bit is 25 dB. Figure 5.14 is for delay interval equal to 20.6 μsec, varying the mobile speed and SNR per bit equal to 22 dB. It is evident that there is a significant advantage to using the cross-validated method. The complexity of the various approaches is reported in Table 5.3.

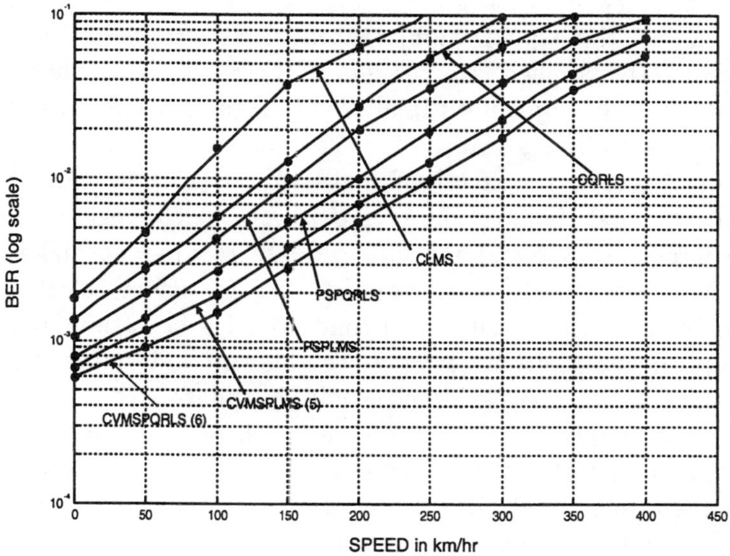

Figure 5.14 BER at SNR per bit equal to 22 dB and 20.6 μsec delay interval and versus mobile speed.

Table 5.3
Computational Complexity with Two Antennas, Four States VA Oversampling Factor Equal to Four, ISI Span of Two Symbols

Algorithm	MIPS
CLMS	20
CQRLS	34
PSPLMS	37
PSPQRLS	43
MSPLMS	26
MSPQRLS	32
CVMSPLMS(6)	27
CVMSPQRLS(6)	33

5.4 Space-Time QR-Based MMSE Equalization

Another very popular scheme for the detection of multiantenna systems is the equalization concept first proposed in [14] and subsequently applied to wireless TDMA in [15] where it was conjectured the implicit optimality of the scheme. The use of different feed-forward filters at the antennas and one single feedback filter was demonstrated to be effective because it was able to simultaneously combat signal fading, intersymbol, and cochannel interference. The implementation of the filter update requires special attention because low signal-to-noise ratio and fast frequency selective fading channels result generally in ill-conditioned adaptation. RLS based on QR decomposition [4] is a well-known and numerically well-behaved method to perform the filters update. However, the high computational complexity of the method has always been considered a remarkable problem. In this section we describe an algorithm based on an approximated QR factorization which improves in terms of computational efficiency over existing schemes mainly because square root operations are avoided. The approach uses a generalization of the scaled tangent rotation of [16] to update the Cholesky factor of the information matrix without needing to form it.

Consider a two-antenna receiver. The optimum equalization scheme has two feed-forward filtering sections (one per antenna) and one feedback filtering section (see Figure 5.15). Since the algorithm jointly optimizes the taps of these filters as to minimize the MSE, using samples of the received signal taken at different points in time and in space, this architecture is called space-time minimum mean square error (MMSE) equalizer. We can express the output of the space-time MMSE equalizer in vector notation as

$$z_n = \mathbf{y}(n)^T \mathbf{c}(n)$$

where[9]

$$\mathbf{y}(n) = \left[\mathbf{y}_{f_1}(n)^T, \mathbf{y}_{f_2}(n)^T, \mathbf{y}_b(n)^T\right]^T$$
$$\mathbf{c}(n) = \left[\mathbf{c}_{f_1}(n)^T, \mathbf{c}_{f_2}(n)^T, \mathbf{c}_b(n)^T\right]^T$$

9. Note that $\mathbf{c}_l^{(k)f}(n)$, $k = 1, 2$ are fractionally spaced taps, while $\mathbf{c}_l^b(n)$ are symbol spaced taps. Moreover, $\mathbf{c}(n)$ is updated once per symbol.

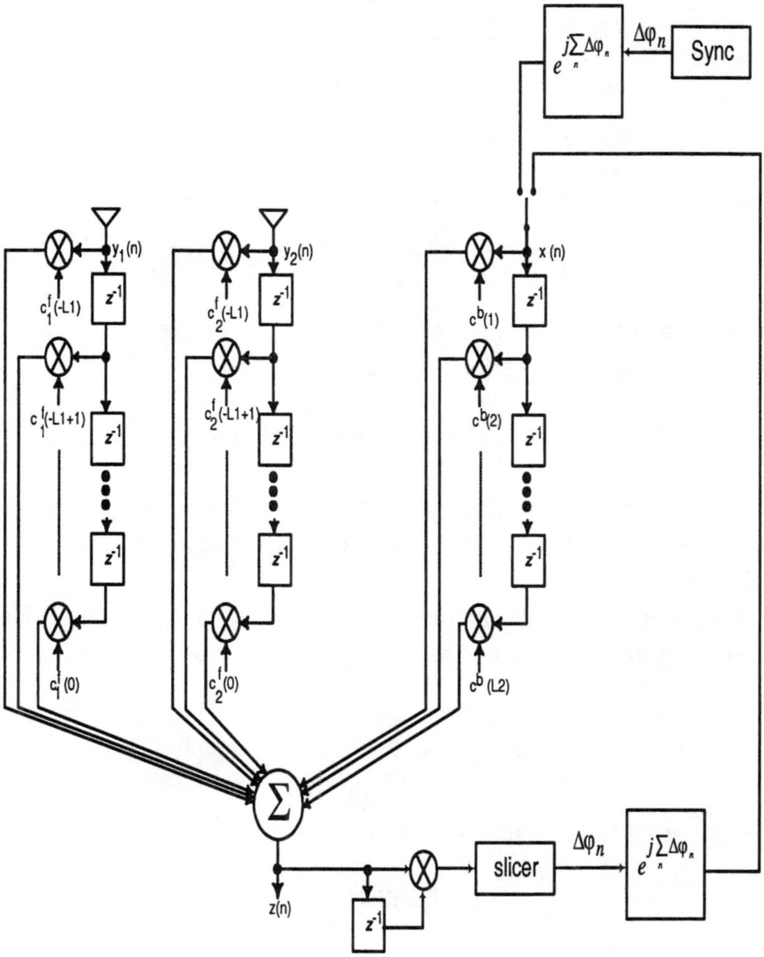

Figure 5.15 Block diagram of a space-time MMSE processor for differential modulations.

$$\mathbf{c}_{f_1}(n) = \left[c_{-L_1}^{(1)f}(n), c_{-L_1+1}^{(1)f}(n), ..., c_0^{(1)f}(n)\right]^T$$
$$\mathbf{c}_{f_2}(n) = \left[c_{-L_1}^{(2)f}(n), c_{-L_1+1}^{(2)f}(n), ..., c_0^{(2)f}(n)\right]^T$$
$$\mathbf{y}_{f_1}(n) = \left[y_{n+L_1}^{(1)}, y_{n+L_1-1}^{(1)}, ..., y_n^{(1)}\right]^T$$
$$\mathbf{y}_{f_2}(n) = \left[y_{n+L_1}^{(2)}, y_{n+L_1-1}^{(2)}, ..., y_n^{(2)}\right]^T$$
$$\mathbf{y}_b(n) = \left[x_{n-1}, x_{n-2}, ..., x_{n-L_2}\right]^T$$
$$\mathbf{c}_b(n) = \left[c_1^b(n), c_2^b(n), ..., c_{L_2}^b(n)\right]^T.$$

The adaptive algorithm minimizes the MSE defined as

$$\text{MSE} = E\left\{|\epsilon_n|^2\right\} = E\left\{|x_n - z_n|^2\right\},$$

and the goal of the adaptation process is to adjust $\mathbf{c}(n)$ [size $L = L_2 + 2(L_1 + 1)$] to converge toward the solution

$$\mathbf{c}_{\text{opt}} = E\left\{\mathbf{y}(n)\mathbf{y}(n)^H\right\}^{-1} E\left\{\mathbf{y}(n)^* x_n\right\}.$$

The sequence x_n is generated using the known training symbols (during training) and using past decisions (during data demodulation, in decision directed mode).

The QR approach is described as follows. The equalization problem can be reconducted to solving at each $n+1$ step the problem

$$\min_{\mathbf{c}} \left\| \begin{bmatrix} \lambda \tilde{\mathbf{Y}}^{(n)} \\ \mathbf{y}(n+1)^T \end{bmatrix} \mathbf{c} - \begin{bmatrix} \lambda \mathbf{X}^{(n)} \\ x_{n+1} \end{bmatrix} \right\|^2 \tag{5.84}$$

where λ is the *forgetting factor* [4],

$$\tilde{\mathbf{Y}}^{(n)^T} = [\mathbf{y}(1), \mathbf{y}(2), ..., \mathbf{y}(n)]\, \mathbf{\Lambda}(n)$$

is the data matrix and

$$\mathbf{X}^{(n)^T} = [x_1, x_2, ..., x_n]\, \mathbf{\Lambda}(n)$$

with $\mathbf{\Lambda}(n) = diag[\lambda^{n-1}, \lambda^{n-2}, ..., 1]$. The *normal equations* define the desired minimizer as

$$\tilde{\mathbf{Y}}^{(n+1)^H}\, \tilde{\mathbf{Y}}^{(n+1)}\, \mathbf{c} = \tilde{\mathbf{Y}}^{(n+1)^H}\, \mathbf{X}^{(n+1)}.$$

The use of orthogonal transformation to solve least squares problems is well-established, as is the inadvisability of using the normal equations [4]. Suppose that a matrix $\mathbf{V}^{(n)}$ is known from the previous step such that

$$\mathbf{Q}\,\tilde{\mathbf{Y}}^{(n)} = \begin{bmatrix} \mathbf{V}^{(n)} \\ \mathbf{0} \end{bmatrix},$$

with $\mathbf{Q}$ orthogonal and $\mathbf{V}^{(n)}$ upper triangular matrix, then the problem stated in (5.71) is equivalent to

$$\min_{\mathbf{c}(n+1)} \left\| \begin{bmatrix} \lambda \mathbf{V}^{(n)} \\ \mathbf{y}(n+1)^T \end{bmatrix} \mathbf{c}(n+1) - \begin{bmatrix} \lambda \hat{\mathbf{X}}^{(n)} \\ x_{n+1} \end{bmatrix} \right\|^2 \tag{5.85}$$

where

$$\mathbf{Q}\,\mathbf{X}^{(n)} = \begin{bmatrix} \hat{\mathbf{X}}^{(n)} \\ r \end{bmatrix},$$

because euclidean distance is preserved by orthogonal transformations. The traditional QR-RLS approach [4] obtains $\mathbf{c}(n+1)$ from the solution of the triangular linear system

$$\mathbf{V}^{(n+1)}\mathbf{c}(n+1) = \hat{\mathbf{X}}^{(n+1)}$$

where $\mathbf{V}^{(n+1)}$ is obtained by sweeping the row vector $\mathbf{y}(n+1)^T$ in (5.85) through orthogonal transformations represented by $\tilde{\mathbf{Q}}(n)$ and $\hat{\mathbf{X}}^{(n+1)}$ is obtained applying the same orthogonal transformations to

$$\left[\lambda \hat{\mathbf{X}}^{(n)^T}, x_{n+1}\right]^T.$$

5.4.1 CSTAR Transformations

The novelty of the method we present is described in the following two points:

1. The algorithm tracks the variation in $\mathbf{c}(n)$ from step n to step $n+1$ rather than $\mathbf{c}(n)$ itself. This saves multiplications.
2. The orthogonal transformations are approximated by scaled tangent rotations rather than the traditional Givens rotations. This avoids the need for square root operations.

Define $\partial\mathbf{c}(n) = \mathbf{c}(n+1) - \mathbf{c}(n)$. We know that $\tilde{\mathbf{Q}}(n)$ is such that

$$\tilde{\mathbf{Q}}(n)\begin{bmatrix} \lambda\mathbf{V}^{(n)} \\ \mathbf{y}(n+1)^T \end{bmatrix} = \begin{bmatrix} \mathbf{V}^{(n+1)} \\ \mathbf{0}^T \end{bmatrix}$$

with $\mathbf{V}^{(n+1)}$ upper triangular. Since

$$\tilde{\mathbf{Q}}(n)\left\{\begin{bmatrix} \lambda\hat{\mathbf{X}}^{(n)} \\ x_{n+1} \end{bmatrix} - \begin{bmatrix} \lambda\mathbf{V}^{(n)} \\ \mathbf{y}(n+1)^T \end{bmatrix}\mathbf{c}(n)\right\}$$
$$= \tilde{\mathbf{Q}}(n)\begin{bmatrix} \mathbf{0} \\ u_{n+1} \end{bmatrix}, \tag{5.86}$$

$\partial\mathbf{c}(n)$ is the solution of[10]

$$\min_{\partial\mathbf{c}(n)} \left\|\begin{bmatrix} \mathbf{V}^{(n+1)} \\ \mathbf{0}^T \end{bmatrix}\partial\mathbf{c}(n) - \tilde{\mathbf{Q}}(n)\begin{bmatrix} \mathbf{0} \\ u_{n+1} \end{bmatrix}\right\|^2 \tag{5.87}$$

where $u_{n+1} = x_{n+1} - \mathbf{y}(n+1)^T\mathbf{c}(n)$. Hence, $\partial\mathbf{c}(n)$ can be found by solving the triangular system

$$\mathbf{V}^{(n+1)}\,\partial\mathbf{c}(n) = \overline{\mathbf{X}}^{(n+1)} \tag{5.88}$$

where $\overline{\mathbf{X}}^{(n+1)}$ satisfies

$$\begin{bmatrix} \overline{\mathbf{X}}^{(n+1)} \\ r_{n+1} \end{bmatrix} = \tilde{\mathbf{Q}}(n)\begin{bmatrix} \mathbf{0} \\ u_{n+1} \end{bmatrix}.$$

It is then evident that all we need to solve the system (5.88) can be obtained by forming

$$\tilde{\mathbf{V}}^{(n)} = \begin{bmatrix} \lambda\mathbf{V}^{(n)} & \mathbf{0} \\ \mathbf{y}(n+1)^T & u_{n+1} \end{bmatrix}$$

and sweeping the bottom part of this matrix using plane rotations.[11] The orthogonal matrix $\tilde{\mathbf{Q}}(n)$ can be found as a product of $L = L_2 + 2(L_1+1)$ Givens rotation matrices [4]:

$$\tilde{\mathbf{Q}}(n) = \tilde{\mathbf{Q}}(n)^{(1)}\tilde{\mathbf{Q}}(n)^{(2)}...\tilde{\mathbf{Q}}(n)^{(L)}.$$

10. Just substitute $\mathbf{c}(n+1) = \mathbf{c}(n) + \partial\mathbf{c}(n)$ in (5.85) and use (5.86).
11. It should be clear now that by tracking $\partial\mathbf{c}(n)$ we have simplified the rotation step in $\tilde{\mathbf{V}}^{(n)}$ because the first L elements of the last column of $\tilde{\mathbf{V}}^{(n)}$ are equal to zero. We save $2L$ complex multiplications with respect to the QR-RLS of [4] and, more importantly, the dynamic range required to represent the elements of the last column of the augmented matrix $\tilde{\mathbf{V}}^{(n)}$ is reduced.

A single Givens rotation $\tilde{\mathbf{Q}}(n)^{(l)}$ annihilates the $L+1, l$-element $\tilde{v}^{(n)}_{L+1,l}$ $(l = 1, 2, ..., L)$ of $\tilde{\mathbf{V}}^{(n)}$ using $\tilde{v}^{(n)}_{l,l}$:

$$\tilde{\mathbf{Q}}(n)^{(l)}_{k,k} = 1 \quad k \neq l, L+1,$$

$$\tilde{\mathbf{Q}}(n)^{(l)}_{k,m} = 0,$$

$$\tilde{\mathbf{Q}}(n)^{(l)}_{L+1,L+1} = c_l,$$

$$\tilde{\mathbf{Q}}(n)^{(l)}_{l,L+1} = s_l,$$

$$\tilde{\mathbf{Q}}(n)^{(l)}_{L+1,l} = -s^*_l,$$

$$\tilde{\mathbf{Q}}(n)^{(l)}_{l,l} = c_l,$$

where

$$c_l = \frac{|\tilde{v}^{(n)}_{l,l}|}{\sqrt{|\tilde{v}^{(n)}_{l,l}|^2 + |\tilde{v}^{(n)}_{L+1,l}|^2}}$$

and

$$s_l = \frac{\tilde{v}^{(n)*}_{L+1,l}}{\sqrt{|\tilde{v}^{(n)}_{l,l}|^2 + |\tilde{v}^{(n)}_{L+1,l}|^2}} \frac{\tilde{v}^{(n)}_{l,l}}{|\tilde{v}^{(n)}_{l,l}|}.$$

The considerable drawback of the method is in the computation of the angles for the Givens rotations. Square-root computations are not easily implemented in DSP processors and are even more problematic in VLSI circuits. Usually they involve iterative procedures whose convergence is not always guaranteed. Scaled tangent rotations (STAR) were proposed in the context of RLS adaptive filtering for real time-series in [16]. We generalize the rotation to the complex domain but there is a difference that is important to note. In STAR [16] scaling is necessary to prevent instability caused by the fact that the tangent function may become infinite. Our modification for complex signals still contains a scaling operation but the factors *are not normalized to unity* as in [16] because this would involve again a square-root operation. The complex scaled tangent rotation (CSTAR) transformation $\mathbf{T}(n) = \mathbf{T}(n)^{(1)}\mathbf{T}(n)^{(2)}...\mathbf{T}(n)^{(L)}$

Table 5.4
The CSTAR Transformation

$\mathbf{T}(n)^{(l)}_{k,m} = 0 \; k, m = 1, ..., L+1$	
if $\left\| \frac{\tilde{v}^{(n)}_{L+1,l}\tilde{v}^{(n)*}_{l,l}}{\|\tilde{v}^{(n)}_{l,l}\|^2} \right\| \leq 1$	(Scaling is not required) $t = \frac{\tilde{v}^{(n)}_{L+1,l}\tilde{v}^{(n)*}_{l,l}}{\|\tilde{v}^{(n)}_{l,l}\|^2}$ $\mathbf{T}(n)^{(l)}_{L+1,L+1} = 1$ $\mathbf{T}(n)^{(l)}_{l,l} = 1$ $\mathbf{T}(n)^{(l)}_{L+1,l} = -t$ $\mathbf{T}(n)^{(l)}_{l,L+1} = t^*$
if $\left\| \frac{\tilde{v}^{(n)}_{L+1,l}\tilde{v}^{(n)*}_{l,l}}{\|\tilde{v}^{(n)}_{l,l}\|^2} \right\| > 1$	(Scaling is required) $t = \text{csign}\{\tilde{v}^{(n)}_{L+1,l}\tilde{v}^{(n)*}_{l,l}\}$, $z = \frac{\tilde{v}^{(n)}_{L+1,l}\tilde{v}^{(n)*}_{l,l}}{\|\tilde{v}^{(n)}_{l,l}\|^2} t^*$, $t_z = \frac{tz^*}{\|z\|^2}$, $\overline{z} = \frac{z^*}{\|z\|^2}$ $\mathbf{T}(n)^{(l)}_{L+1,L+1} = \overline{z}$ $\mathbf{T}(n)^{(l)}_{l,l} = 1$ $\mathbf{T}(n)^{(l)}_{L+1,l} = -t$ $\mathbf{T}(n)^{(l)}_{l,L+1} = t^*_z$
$\mathbf{T}(n)^{(l)}_{k,k} = 1 \;\; k \neq l, L+1$	

in analogy to the Givens rotation is defined in terms of each $\mathbf{T}(n)^{(l)}$ as in Table 5.4, where the complex sign function is

$$\text{csign}\{x\} = \frac{1}{\sqrt{2}}\text{sign}\{Real[x]\} + j\frac{1}{\sqrt{2}}\text{sign}\{Imag[x]\},$$

and the sweep is applied to the augmented matrix $\tilde{\mathbf{V}}^{(n)}$ (see Table 5.5). Observe that the Givens elementary complex rotation has two remarkable properties. First, it zeroes selectively one predetermined element of any complex matrix, which is indeed needed to triangularize the data matrix. Second, it is an orthogonal transformation; that is, $\tilde{\mathbf{Q}}(n)^{(l)}$ is an orthogonal matrix that

Table 5.5
The CSTAR Algorithm

Initialization	$\mathbf{c}(n) = \mathbf{0},\ \mathbf{V}^{(n)} = \gamma\mathbf{I}$
Input	$\mathbf{y}(n+1),\ \mathbf{c}(n),\ \lambda,\ x_n,\ \mathbf{V}^{(n)}$
Compute Prediction Error	$u_{n+1} = x_{n+1} - \mathbf{y}(n+1)^T\,\mathbf{c}(n)$
Form the matrix $\tilde{\mathbf{V}}^{(n)}$	$\tilde{\mathbf{V}}^{(n)} = \begin{bmatrix} \lambda\mathbf{V}^{(n)} & \mathbf{0} \\ \mathbf{y}(n+1)^T & u_{n+1} \end{bmatrix}$
Sweep $\mathbf{y}(n+1)^T$	$\mathbf{T}(n)\ \tilde{\mathbf{V}}^{(n)} \Rightarrow \begin{bmatrix} \mathbf{V}^{(n+1)} & \overline{\mathbf{X}}^{(n+1)} \\ \mathbf{0}^T & r_{n+1} \end{bmatrix}$
Solve by back-substitution	$\partial\mathbf{c}(n) = \left[\mathbf{V}^{(n+1)}\right]^{-1}\overline{\mathbf{X}}^{(n+1)}$
Obtain new taps	$\mathbf{c}(n+1) = \mathbf{c}(n) + \partial\mathbf{c}(n)$

preserves the original least squares problem. The CSTAR elementary transformation $\mathbf{T}(n)^{(l)}$ maintains the first property but it is not an orthogonal transformation. Note in the flow diagram of the transformation (Table 5.4) that whenever scaling is required, we obtain a nonorthogonal $\mathbf{T}(n)^{(l)}$ elementary matrix [and of course a nonorthogonal $\mathbf{T}(n)$]. So the CSTAR solution deviates from the optimum least squares solution.[12] However, it is possible to show along the same guidelines of [16] that the deviation from orthogonality is limited to the first few adaption steps because as new samples are processed, the scaling operation becomes more and

12. In fact at any given step

$$arg \min_{\partial\mathbf{c}(n)} \left\| \begin{bmatrix} \mathbf{V}_{GR}^{(n+1)} \\ \mathbf{0}^T \end{bmatrix} \partial\mathbf{c}(n) - \tilde{\mathbf{Q}}(n) \begin{bmatrix} \mathbf{0} \\ u_{n+1} \end{bmatrix} \right\|^2$$

$$\neq arg \min_{\partial\mathbf{c}(n)} \left\| \begin{bmatrix} \mathbf{V}_{CSTAR}^{(n+1)} \\ \mathbf{0}^T \end{bmatrix} \partial\mathbf{c}(n) - \mathbf{T}(n) \begin{bmatrix} \mathbf{0} \\ u_{n+1} \end{bmatrix} \right\|^2$$

where $\mathbf{V}_{GR}^{(n+1)}$ and $\mathbf{V}_{CSTAR}^{(n+1)}$ are the triangular matrices obtained from the sweep performed applying Givens rotations and CSTAR rotations, respectively.

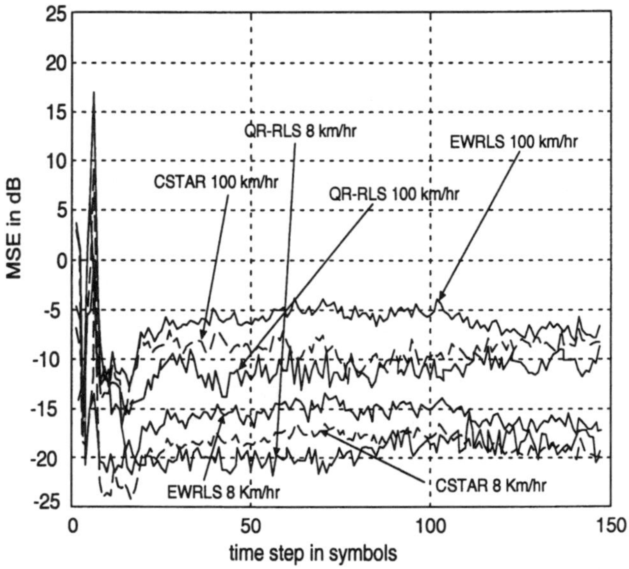

Figure 5.16 Performance of the adaptive schemes in terms of MSE.

more unnecessary. In other words, the algorithm has the property that

$$\lim_{n\to\infty} \left|\left|\mathbf{T}(n)^{(l)^H}\mathbf{T}(n)^{(l)} - \mathbf{I}\right|\right|_F = \mathbf{0}, \qquad l = 1, 2, ..., L \qquad (5.89)$$

where $\mathbf{I}$ is the identity matrix and we have used the notation $||\mathbf{M}||_F = \sqrt{\sum_{i=1}^{M}\sum_{j=1}^{M}|m_{i,j}|^2}$ for the Frobenius norm of a $M \times M$ complex matrix $\mathbf{M}$ whose generic i, j element is $m_{i,j}$. Experimental results show that the effect of this initial bias is negligible. The algorithm can be summarized as in Table 5.5. The described algorithm, the EWRLS algorithm (traditional RLS [17]), and the QR-RLS algorithm (traditional QR-based RLS [4]) have been implemented using 24 bits of resolution in the fixed-point arithmetic representation. Figure 5.16 shows the performance of the CSTAR algorithm compared to EWRLS and to QR-RLS in terms of MSE estimated and averaged over 100 runs. The delay interval for both diversity channels is equal to 41.2 μsec (one symbol period). The length of the feed-forward sections is 3; the length of the feedback section is 2. Sampling rate is $2/T$ ($R = 2$). The value of λ is 0.855

for 100 km/hr and 0.98 for 8 km/hr.[13]

5.5 Key Points of This Chapter

We have described several practical solutions to adaptive space-time reception in frequency selective multipath fading channels and analyzed their performance. Some traditional schemes (like the MLSE) have been enhanced by novel tracking schemes. Specializing the study to the current U.S. standard for digital TDMA cellular communications [1] and to specific hardware implementations, we presented the results of an experimental investigation. Moreover we have described a new method to update the digital filters of a MMSE space-time decision-feedback receiver. In some sense all of the techniques described in this chapter may be considered traditional because they simply generalize scalar approaches typically employed in single-antenna systems. Nevertheless, because of their significant robustness and simplicity, all of the algorithms are particularly suited for fixed-point arithmetic implementations.

References

[1] TIA/EIA/IS-136.1-A, "TDMA Cellular/PCS–Radio Interface–Mobile Station–Base Station Compatibility–Digital Control Channel," Oct. 1996.

[2] L. Van Trees, *Detection, Estimation, and Modulation Theory,* Part I, New York: Wiley, 1968.

[3] G. D. Forney, "Maximum Likelihood Sequence Estimation of Digital Sequences in the Presence of Intersymbol Interference," *IEEE Trans. Inform. Theory,* Vol. IT-18, pp. 363–378, May 1972.

[4] S. Haykin, *Adaptive Filter Theory,* Englewood Cliffs, NJ: Prentice-Hall, 1986.

[5] R. Raheli, A. Polydoros, and C.-K. Tzou, "Per-Survivor Processing: A General Approach to MLSE in Uncertain Environments," *IEEE Trans. Comm.,* Vol. 43, pp. 354–364, Feb./March/April 1995.

[6] G. Castellini et al., "A Continuously Adaptive MLSE Receiver for Mobile Communications: Algorithm and Performance," *IEEE Trans. Comm.,* Vol. 45, pp. 80–89, Jan. 1997.

[7] M. Martone, "Optimally Regularized Channel Tracking Techniques for Sequence Estimation Based on Cross-Validated Subspace Signal Processing," *IEEE Trans. Commun.,* Vol. 48, pp. 95–105, Jan. 2000.

---

13. In general, low speeds require larger values of λ to get the best performance out of the tracking scheme. However, there is marginal BER degradation if λ is kept fixed to 0.855.

[8] L. L. Scharf, *Statistical Signal Processing*, Reading, MA: Addison-Wesley, 1991.

[9] B. C. Ng, M. Cedervall, and A. Paulraj, "A Structured Channel Estimator for Maximum-Likelihood Sequence Detection," *IEEE Comm. Letters*, Vol. 1, pp. 52–55, March 1997.

[10] R. D. Nowak, "Penalized Least Squares Estimation of Volterra Filters and Higher-Order Statistics," *IEEE Trans. Sig. Proc.*, Vol.46, pp. 419–428, Feb. 1998.

[11] R.D. Nowak, *Optimal Signal Estimation Using Cross-Validation,* Rice University Tech. Rep., Rice University, Houston, TX, ECE-TR-9601, Jan. 1998.

[12] M. Martone, "Complex Scaled Tangent Rotations (CSTAR) for Fast Space-Time Adaptive Equalization of Wireless TDMA," *IEEE Trans. Commun.*, Vol. 46, pp. 1587–1591, Dec. 1998.

[13] M.-C. Chiu and C.-C. Chao, "Analysis of LMS-Adaptive MLSE Equalization on Multipath Fading Channels," *IEEE Trans. Comm.*, Vol. 44, pp. 1684–1692, Dec. 1996.

[14] P. Monsen, "MMSE Equalization of Interference on Fading Diversity Channels," *IEEE Trans. Comm.*, Vol. 34, pp. 5–12, Jan. 1984.

[15] C. Despins, D. Falconer, and S. Mahmoud, "Compound Strategies of Coding, Equalization, and Space Diversity for Wide-Band TDMA Indoor Wireless Channels," *IEEE Trans. Vehicular Tech.*, Vol. 41, pp. 369–379, Nov. 1992.

[16] K. J. Raghunath and K. K. Parhi, "Pipelined RLS Adaptive Filtering Using Scaled Tangent Rotations," *IEEE Trans. Signal Proc.*, Vol. 44, No. 10, pp. 2591–2604, Oct. 1996.

[17] J. G. Proakis, *Digital Communications*, New York: McGraw-Hill, 1989.

6

Spatio-Temporal Processing Using Higher-Order Statistics

6.1 Introduction

The space-time filtering approach in a multiantenna system results in a discrete-time MIMO model that must be deconvolved [1, 2]. Standard techniques for equalization described in the previous chapters are based on an equivalent minimum phase system modeling approach because they exploit only the second order statistics (SOS) of the signal (the MMSE criterion, for example). However, most of all real-world channels do not present the minimum phase (MP) condition. Motivated by these observations, the equalization of non-MP channels using higher-than-second-order-statistics (HOS) [3] has stimulated incredible interest during the last 10 years [4]. A blind algorithm well studied in its practical implementation [5] is the constant modulus algorithm (CMA) whose slow convergence under particular situations constitutes the most important objection. The methods proposed in this chapter are in part based on the same idea introduced in [6], where the super exponential algorithm of [7] was generalized to the multichannel case using third-order cumulants. Here we describe the application of the method to the reception of spatio-temporal signals using fourth-order cumulants and a new adaptive implementation based on a low-rank processing concept.

Low-rank/subspace processing is an extremely important branch of signal processing (see [8] for applications and algorithms). The space/time autocorrelation matrix of signals received at different elements of an array can be in practice very ill-conditioned. Any adaptive algorithm based on full-rank processing (for example, traditional RLS [9, 10]) is severely affected by the rank degeneracy problem. We describe an adaptive algorithm based on a low-rank approximation of the covariance matrix that is also modified for a fast acquisition by means of training sequences. In the second part of the chapter we show how the MMSE principle can be formulated in a MIMO context using fourth-order cumulants. Most of the works describing HOS-based algorithms for identification/deconvolution are blind and try not to exploit any input-signal knowledge. In practical systems the bit stream is typically organized into frames. Some known sequences are periodically transmitted in every slot, allowing the receiver to perform frame synchronization, symbol timing recovery, and equalizer training. The use of the known sequence is indispensable to fast start-up equalization. Another method proposed in this chapter is based on the same idea introduced in [11, 12] where high-order cumulants were used to derive MMSE methods for system identification. Here we modify the basic intuition so that it can be applied to the deconvolution of complex vector signals. Moreover, it is shown that it is indeed possible to derive a fully adaptive implementation based on a recursive Jacobi-type algorithm. Most of the theoretical developments in this chapter are validated by hardware experimentation. Refer to Appendix 6A for some introductory material on HOS.

6.1.1 Discrete-Time Model

Assuming that the transmission bandwidth is strictly band-limited to the Nyquist bandwidth and sampling at symbol rate T, we can compact the effect of the RF propagation channels in a U-transmit-antenna, K-receive-antenna as

$$y_k(n) = \sum_{l=1}^{U} \sum_{m} h_{k,l}(m) x_l(n-m) + \eta_k(n) \qquad k = 1, 2, ..., K \qquad (6.1)$$

where $\eta_k(n)$ is Gaussian white noise and $h_{k,l}(m)$ is the symbol-sampled[1] impulse responses $h_{k,l}(m)$. In the z-domain the transfer function (6.1) can be expressed as

$$\tilde{\mathcal{H}}(z) = \sum_{m=-\infty}^{\infty} \mathbf{H}(m) z^{-m} \tag{6.2}$$

where the organization of the $\mathcal{H}_{i,l}(z)$ polynomials (z-transforms of $h_{i,l}(m)$) in $\tilde{\mathcal{H}}(z)$ is given by

$$\tilde{\mathcal{H}}(z) = \begin{bmatrix} \mathcal{H}_{1,1}(z) & \cdots & \mathcal{H}_{1,U}(z) \\ \cdots & \cdots & \cdots \\ \mathcal{H}_{K,1}(z) & \cdots & \mathcal{H}_{K,U}(z) \end{bmatrix}.$$

To recover the input signals a linear K-input U-output filter $\tilde{\mathcal{W}}(z) = \sum_{m=L_1}^{L_2} \mathbf{W}(m) z^{-m}$ with length $L = L_2 - L_1 + 1$ is applied to the outputs of the sensors. The objective for $\tilde{\mathcal{W}}(z)$ is to achieve distortionless reception.

If we define

$$\tilde{\mathcal{W}}(z) = \begin{bmatrix} \mathcal{W}_{1,1}(z) & \cdots & \mathcal{W}_{1,K}(z) \\ \cdots & \cdots & \cdots \\ \mathcal{W}_{U,1}(z) & \cdots & \mathcal{W}_{U,K}(z) \end{bmatrix},$$

distortionless reception means that

$$\tilde{\mathcal{W}}(z)\,\tilde{\mathcal{H}}(z) = \mathbf{I}_U \tag{6.3}$$

where $\mathbf{I}_U$ is a $U \times U$ identity matrix [13, 14]. The system $\tilde{\mathcal{W}}(z)$ is required to be bounded-input bounded-output (BIBO) stable. The solution (6.3) is achievable only ideally. Since the input signal constellations are symmetric, the statistics of the input signals $x_i(n)$ reflect the same symmetry. Moreover, *signal reconstruction is possible only up to a constant delay, due to the stationarity of the input process.* The recovered signals will be subject to a phase ambiguity,

1. The same model with some marginal changes applies to the fractional sampling case as well. We restrict however the description of the algorithm to the symbol-spaced case for the sake of clarity.

a delay, and a permutation ambiguity. The best possible result for practical distortionless reception by means of a linear filter is

$$\tilde{\mathcal{W}}(z)\,\tilde{\mathcal{H}}(z) = \mathbf{P}\mathcal{D}(z) \tag{6.4}$$

where $\mathbf{P}$ is a permutation matrix and

$$\mathcal{D}(z) = \text{diag}\{e^{j\phi_1}z^{-n_1}, e^{j\phi_2}z^{-n_2}, ..., e^{j\phi_U}z^{-n_U}\}$$

where $\phi_i \in [-\pi, \pi]$, n_i is an integer for $i = 1, 2..., U$ [15–17]. We say that $\tilde{\mathcal{H}}(z)$ satisfies the distortionless reception condition if there exists a BIBO stable distortionless reception filter $\tilde{\mathcal{W}}(z)$. A system $\tilde{\mathcal{H}}(z)$ satisfies the distortionless reception condition if and only if [16, 17]

$$\det\left(\tilde{\mathcal{H}}^H(e^{j\omega})\tilde{\mathcal{H}}(e^{j\omega})\right) \neq 0, \quad \text{for all } \omega \in [-\pi, \pi]. \tag{6.5}$$

In the time-domain the linear filter can be written

$$z_i(n) = \sum_{j=1}^{K}\sum_{m} w_{i,j}(m)y_j(n-m) \qquad i = 1, 2, ..., U \tag{6.6}$$

where $w_{i,j}(m)$ is the filter corresponding to the polynomial $\mathcal{W}_{i,j}(z)$ and $z_i(n)$ is the ith output of the deconvolution filter $\tilde{\mathcal{W}}(z)$. The overall impulse response is characterized by the input/output relation

$$z_i(n) = \sum_{j=1}^{U}\sum_{m} s_{i,j}(m)x_j(n-m) \qquad i = 1, 2, ..., U. \tag{6.7}$$

The impulse response of the two cascaded filters is given by the multivariate convolution

$$\begin{aligned} s_{i,j_1}(m_2) &= \sum_{j=1}^{K}\sum_{m} w_{i,j}(m)h_{j,j_1}(m_2-m) \\ &= \left[\sum_{j=1}^{K}\mathbf{H}_{j,j_1}\mathbf{w}_{i,j}\right]_{m_2} \qquad i, j_1 = 1, 2, ..., U, \end{aligned}$$

where

$$[\mathbf{H}_{i,j}]_{m,n} = h_{i,j}(m-n), \; -\infty \leq m \leq +\infty, \; L_1 \leq n \leq L_2,$$

and $\mathbf{w}_{i,j} = [w_{i,j}(L_1), w_{i,j}(L_1+1), ..., w_{i,j}(L_2)]^T$. In a vector form we can write

$$\tilde{\mathbf{s}}_i = \tilde{\mathbf{H}}\tilde{\mathbf{w}}_i, \quad i = 1, 2, ..., U, \tag{6.8}$$

where $\tilde{\mathbf{H}}$, $\tilde{\mathbf{s}}_i$, $\tilde{\mathbf{w}}_i$ are defined as

$$\begin{aligned}
\tilde{\mathbf{H}} &= \begin{bmatrix} \mathbf{H}_{1,1} & \mathbf{H}_{2,1} & \cdots & \mathbf{H}_{K,1} \\ \mathbf{H}_{1,2} & \mathbf{H}_{2,2} & \cdots & \mathbf{H}_{K,2} \\ \vdots & \vdots & \cdots & \vdots \\ \mathbf{H}_{1,U} & \mathbf{H}_{2,U} & \cdots & \mathbf{H}_{K,U} \end{bmatrix}, \\
\tilde{\mathbf{w}}_i &= \left[\mathbf{w}_{i,1}{}^T, \mathbf{w}_{i,2}{}^T, ..., \mathbf{w}_{i,K}{}^T\right]^T, \\
\tilde{\mathbf{s}}_i &= \left[\mathbf{s}_{i,1}^T, \mathbf{s}_{i,2}^T, \mathbf{s}_{i,3}^T, ..., \mathbf{s}_{i,U}^T\right]^T, \\
\mathbf{s}_{i,j} &= [..., s_{i,j}(-1), s_{i,j}(0), s_{i,j}(1), ...]^T
\end{aligned}$$

The desired response $\tilde{\mathbf{s}}_i$ that completely restores the information signal of the ith transmitter up to the delay n_i, can be expressed as

$$\tilde{\boldsymbol{\delta}}_i = \left[\boldsymbol{\delta}_{i,1}^T, \boldsymbol{\delta}_{i,2}^T, \boldsymbol{\delta}_{i,3}^T, ..., \boldsymbol{\delta}_{i,U}^T\right]^T, \tag{6.9}$$

where

$$\boldsymbol{\delta}_{i,j} = \begin{cases} [..., 0, 0, 0, ...] & i \neq j \\ \hat{\boldsymbol{\delta}}_i & i = j \end{cases}. \tag{6.10}$$

The generic mth element of the vector $\hat{\boldsymbol{\delta}}_i$ is $\delta(m - n_i)$, if we neglect the phase shift and we force the solution not to permute the inputs ($\mathbf{P} = \mathbf{I}_U$).[2] This generalizes the one-dimensional case [7]. It is possible to solve this deconvolution problem by solving the minimization problem

$$\min_{\tilde{\mathbf{w}}_i} \left\|\tilde{\mathbf{H}}\tilde{\mathbf{w}}_i - \tilde{\boldsymbol{\delta}}_i\right\|^2 \qquad i = 1, 2, ..., U. \tag{6.11}$$

2. In practice the permutation ambiguity is not a problem because it can be solved by a user identification matching procedure. Specifically in [18] the coded digital verification color code (CDVCC) section of the frame is intended to be used for this purpose.

6.2 The Super-Exponential Method

Shalvi and Weinstein [1] showed that for the following scalar model

$$y(n) = \sum_{m} h(m)x(n-m) + \eta(n) \qquad (6.12)$$

where $x(n)$ is an i.i.d. non-Gaussian white process with variance σ_x^2 and zero-lag fourth-order cumulant γ_{4x}, an $L = L_2 - L_1 + 1$-long linear filter

$$z(n) = \sum_{m=L_1}^{L_2} w(m)y(n-m) \qquad (6.13)$$

can be found such that $z(n) \simeq x(n-d)e^{j\phi}$, from the super-exponential iterative equations[3]

$$\tilde{\mathbf{w}}^{[1]} = \tilde{\mathbf{\Phi}}^{\dagger}\ \mathbf{D} \Rightarrow \tilde{\mathbf{w}}^{[2]} = \frac{\tilde{\mathbf{w}}^{[1]}}{\sqrt{\tilde{\mathbf{w}}^{[1]^H}\ \tilde{\mathbf{\Phi}}\ \tilde{\mathbf{w}}^{[1]}}} \qquad (6.14)$$

where the generic (n, m)th element of the matrix $\tilde{\mathbf{\Phi}}$ with dimensions $L \times L$, is given by

$$[\mathbf{\Phi}]_{n,m} = \frac{cum\left[y(k-m), y^*(k-n)\right]}{\sigma_x^2} \qquad (6.15)$$

and the nth element of the $L \times 1$ vector $\mathbf{D}$ of fourth-order cumulants is given by

$$[\mathbf{D}]_n = \frac{cum\left[z(k), z(k), z^*(k), y^*(k-n)\right]}{\gamma_{4x}} \qquad (6.16)$$

and evidently $\tilde{\mathbf{w}}^{[1]}$, $\tilde{\mathbf{w}}^{[2]}$, contain the taps $w(m)$ properly organized after the first and the second steps of the generic iteration. The generalization of the approach to MIMO systems was introduced by Martone in [6, 19]. The problem of finding the filter $\tilde{\mathbf{w}}_i$ for $i = 1, 2, ..., U$ that restores the ith transmitted digital stream is addressed next.

3. $\mathbf{A}^{\dagger}$ means pseudoinverse for the matrix $\mathbf{A}$.

6.2.1 The MIMO Super-Exponential Approach

The fundamental assumptions necessary to develop the algorithm are listed as follows:

- **AS1**: $\tilde{\mathcal{H}}(z)$ is *irreducible* and $\mathbf{H}(m) \neq \mathbf{0}$ only for $m \in [J_1, J_2]$ $(J = J_2 - J_1 + 1)$ with $\mathbf{H}(J_2)$ full rank.
- **AS2**: The complex sequence $\{x_i(n)\}$ is constituted by random variables identically non-Gaussian distributed and statistically independent, and the cumulants of $\{x_i(n)\}$ satisfy
  - — $E\{x_i(n)\} = E\{x_i^3(n)\} = 0$
  - — $cum[x_i(n), x_j^*(n)] = \sigma_x^2 > 0$, only for $i = j$.
  - — $E\{x_i(n)x_j(n)\} = E\left\{x_i^*(n)x_j^*(n)\right\} = 0$, for any i, j,
  - — $cum[x_{j_1}(n), x_{j_2}(n), x_{j_3}^*(n), x_{j_4}^*(n)] = \gamma_{4x} \neq 0$, only for $j_1 = j_2 = j_3 = j_4$.

Assumption **AS1** is required to insure the distortionless reception condition for $\tilde{\mathcal{H}}(z)$. In fact under **AS1** is possible to design a deconvolution filter with length $L = L_2 - L_1 + 1 \geq \lceil \frac{(J-1)U}{K-U} \rceil$ such that the *generalized Sylvester matrix* $H_{L,J}$ of the MIMO system is full column rank [13, 14]. This implies that a distortionless reception linear filter exists [i.e., $\tilde{\mathcal{H}}(z)$ satisfies the distortionless reception condition] and that (6.5) is verified. Assumption **AS2** requires the same zero-lag fourth-order cumulant and variance for the U modulated signal sources $x_i(n)$. Since the U signals are assumed to use the same modulation scheme, this appears reasonable. However, a generalization of the algorithm to the case of different zero-lag fourth-order cumulants and variances is possible. In practical applications when $\tilde{\mathcal{H}}(z)$ is a time-varying response, it is almost impossible to guarantee at any time instant the existence of the full-rank least squares solution of (6.11).[4] In fact, even if some propagation parameters may be considered almost time-invariant (for example, delay spread or maximum Doppler shift), the characteristics of the different multipath channels are independently changing in time and

4. We are trying to point out that the Hermitian matrix $\tilde{\mathbf{H}}^H\tilde{\mathbf{H}}$ may be not only singular but also *close* to singular, that happens when $\tilde{\mathbf{H}}$ is nearly rank-deficient; in other words, the least squares problem (6.11) is *ill-conditioned.*

it is in general quite unrealistic to make such a strong assumption. The well-known remedy to rank-deficient (or ill-conditioned) least squares problems is the use of a minimum-norm solution of (6.11) based on a low-rank approximation of the Hermitian matrix $\tilde{\mathbf{H}}^H\tilde{\mathbf{H}}$ which is obtained using a dominant eigendecomposition.[5] This idea will be later developed in an adaptive algorithm based on a subspace tracking concept. The low-rank approximation can also be motivated by computational efficiency. The eigenvalue decomposition of $\tilde{\mathbf{H}}^H\tilde{\mathbf{H}}$ is $\tilde{\mathbf{H}}^H\tilde{\mathbf{H}} = \tilde{\mathbf{V}}\tilde{\mathbf{\Lambda}}\tilde{\mathbf{V}}^H$ where $\tilde{\mathbf{V}} = \begin{bmatrix} \tilde{\mathbf{v}}_1 & \tilde{\mathbf{v}}_2 & \cdots & \tilde{\mathbf{v}}_{KL} \end{bmatrix}$ is the $KL \times KL$ orthonormal matrix of eigenvectors, and

$$\tilde{\mathbf{\Lambda}} = \text{diag}(\tilde{\lambda}_1, \tilde{\lambda}_2, ..., \tilde{\lambda}_{KL})$$

is the diagonal matrix of eigenvalues. The r dominant eigenvalues are those eigenvalues that exceed a certain predetermined tolerance level *tol*. Hence, using the notation $\mathbf{V}_n = \begin{bmatrix} \mathbf{v}_1 & \mathbf{v}_2 & \cdots & \mathbf{v}_n \end{bmatrix}$, $\mathbf{\Lambda}_n = \text{diag}(\lambda_1, \lambda_2, ..., \lambda_n)$, we can write $\tilde{\mathbf{V}} = \begin{bmatrix} \tilde{\mathbf{V}}_r & \tilde{\mathbf{V}}_{KL-r} \end{bmatrix}$ and $\tilde{\mathbf{\Lambda}} = \begin{bmatrix} \tilde{\mathbf{\Lambda}}_r & \mathbf{0} \\ \mathbf{0} & \tilde{\mathbf{\Lambda}}_{KL-r} \end{bmatrix}$. In practice one has to find a sufficiently large integer r such that $\tilde{\mathbf{\Lambda}}_r^{-1}$ exists. Next we define the Moore-Penrose inverse $\left(\tilde{\mathbf{H}}^H\tilde{\mathbf{H}}\right)^\dagger = \tilde{\mathbf{V}}_r\tilde{\mathbf{\Lambda}}_r^{-1}\tilde{\mathbf{V}}_r^H$, so that we can solve the least squares problem (6.11) as

$$\tilde{\mathbf{w}}_i = (\tilde{\mathbf{H}}^H\tilde{\mathbf{H}})^\dagger\tilde{\mathbf{H}}^H\tilde{\boldsymbol{\delta}}_i = \tilde{\mathbf{V}}_r\tilde{\mathbf{\Lambda}}_r^{-1}\tilde{\mathbf{V}}_r^H\tilde{\mathbf{H}}^H\tilde{\boldsymbol{\delta}}_i. \tag{6.17}$$

The following two-step iterative procedure defines a class of algorithms for different values of p and q [6, 7]:

$$s_{i,j}(k)^{[1]} = (s_{i,j}(k))^p(s_{i,j}(k)^*)^q \tag{6.18}$$

5. It is interesting to observe that this modification to the Super-Exponential Algorithm was proposed independently by Ding in [20] where it was shown that the length and zero condition in addition to the low-rank modification results in a globally convergent algorithm. Observe that the case treated in [20] is the SIMO case, and that we are instead treating the more general MIMO case; in fact the length and zero condition translates into **AS1**. It is also important to note that [20] does not address the problem of extreme practical interest of designing an adaptive algorithm to implement the batch low-rank super-exponential procedure.

$$s_{i,j}(k)^{[2]} = s_{i,j}(k)^{[1]} \frac{1}{\sqrt{\sum_{j=1}^{K} \sum_{l} \left| s_{i,j}(l)^{[1]} \right|^2}} \tag{6.19}$$

where $(.)^{[1]}$, $(.)^{[2]}$ stand for the result of the first and second step, respectively. This iterative algorithm converges at a super-exponential rate to the desired solution [7] for $p+q \geq 2$. In this work we choose $p = 2, \quad q = 1$, which gives a solution in terms of fourth-order cumulants. Since obviously $\tilde{\mathbf{s}}_i$ is not available (because $\tilde{\mathbf{H}}$ is not known), we derive a procedure [6] in terms of $\tilde{\mathbf{w}}_i$. If we define $\tilde{\mathbf{g}}_i = \left[\mathbf{g}_{i,1}^T, \mathbf{g}_{i,2}^T, \mathbf{g}_{i,3}^T, ..., \mathbf{g}_{i,U}^T\right]^T$, with $\mathbf{g}_{i,j} = [..., g_{i,j}(-1), g_{i,j}(0), g_{i,j}(1), ...]^T$, as the impulse response vector obtained by $g_{i,j}(k) = s_{i,j}^2(k) s_{i,j}(k)^*$, we can state the least squares minimization problem

$$\min_{\tilde{\mathbf{w}}_i} \left\| \tilde{\mathbf{H}} \tilde{\mathbf{w}}_i^{[1]} - \tilde{\mathbf{g}}_i \right\|^2 \tag{6.20}$$

whose solution is

$$\tilde{\mathbf{w}}_i^{[1]} = (\tilde{\mathbf{H}}^H \tilde{\mathbf{H}})^{\dagger} \tilde{\mathbf{H}}^H \tilde{\mathbf{g}}_i. \tag{6.21}$$

To obtain normalization (6A.3), the second step is

$$\tilde{\mathbf{w}}_i^{[2]} = \frac{\tilde{\mathbf{w}}_i^{[1]}}{\sqrt{\tilde{\mathbf{w}}_i^{[1]^H} (\tilde{\mathbf{H}}^H \tilde{\mathbf{H}}) \tilde{\mathbf{w}}_i^{[1]}}}. \tag{6.22}$$

It should be evident that the solution of the least squares problem (6.20) [given by (6.21)] and subsequent normalization as in (6.22) is equivalent to one step of the iterative procedure (6A.2) and (6A.3). Since (6A.2) and (6A.3) converge to the desired solution (6.9) and (6.10) expressed in terms of $\tilde{\mathbf{w}}_i$ in (6.11), it is also true that at the point of convergence the solution of (6.20) is equivalent to the solution of (6.11). A discussion on the convergence of (6.21) and (6.22), related to the convergence of (6.18) and (6.19) is given in Appendix 6A. The procedure (6.21)–(6.22) can be expressed in terms of the cumulants of the outputs of the sensors. From $y_i(k-n) = \sum_{j=1}^{K} \sum_m h_{i,j}(m-n) x_j(k-m)$, we exploit **AS2** and the properties of the cumulants of linear stationary processes (see property *LIN*) so that we can write

$$cum\left[y_{i_1}(k-n), y_{i_2}{}^*(k-m)\right]$$

$$= cum\left[\sum_{j_1=1}^{U}\sum_{m_1} h_{i_1,j_1}(m_1-n)x_{j_1}(k-m_1),\right.$$
$$\left.\sum_{j_2=1}^{U}\sum_{m_2} h^*_{i_2,j_2}(m_2-m)x^*_{j_2}(k-m_2)\right]$$
$$=\sum_{j=1}^{U}\sum_{k} h_{i_1,j}(k-n)h^*_{i_2,j}(k-m)\sigma_x^2$$
$$=\sigma_x^2\sum_{j=1}^{U}\left[\mathbf{H}_{i_2,j}{}^H\mathbf{H}_{i_1,j}\right]_{m,n} \tag{6.23}$$

due to
$cum\left[x_{j_1}(k-m_1), x^*_{j_2}(k-m_2)\right] =$
$$=\begin{cases} cum\left[x_j(n), x^*_j(n)\right]=\sigma_x^2 & j_1=j_2;\ m_1=m_2 \\ 0 & \text{otherwise}\end{cases}.$$

To derive the second key expression related to the solution (6.21), let us consider

$$cum\left[z_i(k), z_i(k), z_i^*(k), y_{i_3}{}^*(k-n)\right]$$
$$=\sum_{j=1}^{U}\sum_{m} h^*_{i_3,j}(m-n)cum\left[z_i(k), z_i(k), z_i^*(k), x_j^*(k-m)\right]$$

and[6]

$$cum\left[z_i(k), z_i(k), z_i^*(k), x_j^*(k-m)\right]$$
$$=\sum_{j_1=1}^{U}\sum_{j_2=1}^{U}\sum_{j_3=1}^{U}\sum_{l_1}\sum_{m_1}\sum_{n_1} s_{i,j_1}(l_1)s_{i,j_2}(m_1)s^*_{i,j_3}(n_1)$$
$$\times cum\left[x_{j_1}(k-l_1), x_{j_2}(k-m_1), x^*_{j_3}(k-n_1), x_j^*(k-m)\right]$$
$$=\gamma_{4x}s^2_{i,j}(m)s^*_{i,j}(m).$$

6. Due to **AS2**

$$cum\left[x_{j_1}(n-m_1), x_{j_2}(n-m_2), x^*_{j_3}(n-m_3), x_j^*(n-k)\right]$$
$$=\begin{cases} \gamma_{4x} & j_1=j_2=j_3=j;\ m_1=m_2=m_3=k \\ 0 & \text{otherwise}\end{cases}.$$

So we can write

$$cum\left[z_i(k), z_i(k), z_i^*(k), y_{i_3}{}^*(k-n)\right]$$

$$= \sum_{j=1}^{U} \sum_{m} h_{i_3,j}{}^*(m-n) g_{i,j}(m) \gamma_{4x}$$

$$= \gamma_{4x} \sum_{j=1}^{U} \left[\mathbf{H}_{i_3,j}^H \mathbf{g}_{i,j} \right]_n . \tag{6.24}$$

Expressions (6.23) and (6.24) can be substituted in the least squares solution (6.21) and the iterative algorithm (6.25) is obtained

$$\begin{aligned} \tilde{\mathbf{w}}_i^{[1]} &= \tilde{\boldsymbol{\Phi}}^{\dagger} \ \mathbf{D}_i \\ \tilde{\mathbf{w}}_i^{[2]} &= \frac{\tilde{\mathbf{w}}_i^{[1]}}{\sqrt{\tilde{\mathbf{w}}_i^{[1]^H} \ \tilde{\boldsymbol{\Phi}} \ \tilde{\mathbf{w}}_i^{[1]}}} \end{aligned} \tag{6.25}$$

where the matrix $\tilde{\boldsymbol{\Phi}}$ with dimensions $KL \times KL$, is given by

$$\tilde{\boldsymbol{\Phi}} = \begin{bmatrix} \boldsymbol{\Phi}_{1,1} & \boldsymbol{\Phi}_{1,2} & \cdots & \boldsymbol{\Phi}_{1,K} \\ \boldsymbol{\Phi}_{2,1} & \boldsymbol{\Phi}_{2,2} & \cdots & \boldsymbol{\Phi}_{2,K} \\ \vdots & \vdots & \vdots & \vdots \\ \boldsymbol{\Phi}_{K,1} & \boldsymbol{\Phi}_{K,2} & \cdots & \boldsymbol{\Phi}_{K,K} \end{bmatrix}, \tag{6.26}$$

$$\left[\boldsymbol{\Phi}_{i,j}\right]_{n,m} = \frac{cum\left[y_j(k-m), y_i^*(k-n)\right]}{\sigma_x^2} \tag{6.27}$$

and the $KL \times 1$ vector $\mathbf{D}_i$ of fourth-order cumulants is given by:

$$\mathbf{D}_i = \left[\mathbf{d}_{i,1}^T, \mathbf{d}_{i,2}^T, ..., \mathbf{d}_{i,K}^T\right]^T, \tag{6.28}$$

$$\left[\mathbf{d}_{i,m}\right]_n = \frac{cum\left[z_i(k), z_i(k), z_i^*(k), y_m^*(k-n)\right]}{\gamma_{4x}}. \tag{6.29}$$

6.2.2 Adaptive Low-Rank Processing

In this section we derive an adaptive algorithm for on-line computation of the multichannel deconvolution filter parameters. The cumulants of interest in the algorithm can be estimated as

$$\overline{cum}[y(k-n), y^*(k-m)] = \frac{1}{N} \sum_{k=1}^{N} y(k-n)\, y^*(k-m) \tag{6.30}$$

and

$$\overline{cum}\,[z(k), z(k), z^*(k), y^*(k-n)] = \frac{1}{N}\sum_{k=1}^{N} |z(k)|^2 \; y^*(k-n)\, z(k)$$
$$-2\,\frac{1}{N}\sum_{k=1}^{N} |z(k)|^2 \; \frac{1}{N}\sum_{k=1}^{N} z(k)\, y^*(k-n)$$
$$-\frac{1}{N}\sum_{k=1}^{N} z(k)^2 \; \frac{1}{N}\sum_{k=1}^{N} z^*(k)\, y^*(k-n) \tag{6.31}$$

where we have neglected indices for $y(k)$, $z(k)$ for simplicity, and we have indicated the estimated cumulant as $\overline{cum}\,(..)$. At the end of the convergence process

$$\mathbf{D}_i - \tilde{\mathbf{\Phi}}\tilde{\mathbf{w}}_i = \mathbf{0}$$

must be satisfied. We assume that the power constraint (6.22) is always satisfied using an automatic gain control. Clearly at each $n+1$ stage we wish to solve the problem

$$\min_{\tilde{\mathbf{w}}_i} \left\| \begin{bmatrix} \lambda\tilde{\mathbf{Y}}^{(n)} \\ \tilde{\mathbf{y}}(n+1)^T \end{bmatrix} \tilde{\mathbf{w}}_i - \begin{bmatrix} \lambda\tilde{\mathbf{Z}}_i^{(n)} \\ \tilde{z}_i(n+1) \end{bmatrix} \right\|^2 \tag{6.32}$$

with

$$\tilde{\mathbf{Y}}^{(n)} = \begin{bmatrix} \lambda\tilde{\mathbf{Y}}^{(n-1)} \\ \tilde{\mathbf{y}}(n)^T \end{bmatrix},$$

$$\tilde{\mathbf{Z}}_i^{(n)} = \begin{bmatrix} \lambda\tilde{\mathbf{Z}}_i^{(n-1)} \\ \tilde{z}_i(n) \end{bmatrix},$$

$$\tilde{\mathbf{y}}(n) = \left[\breve{\mathbf{y}}_1(n)^T, \breve{\mathbf{y}}_2(n)^T, ..., \breve{\mathbf{y}}_K(n)^T\right]^T,$$

$$\breve{\mathbf{y}}_i(n) = [y_i(n-L_1), y_i(n-L_1-1), ..., y_i(n-L_2)]^T$$

and

$$\tilde{z}_i(n) = \left(|z_i(n-1)|^2 - 2\,\sigma_x^2\right) \; z_i(n-1).$$

In fact, the *normal equations* define the desired minimizer as

$$\tilde{\mathbf{Y}}^{(n+1)^H}\;\tilde{\mathbf{Y}}^{(n+1)}\;\tilde{\mathbf{w}}_i = \tilde{\mathbf{Y}}^{(n+1)^H}\;\tilde{\mathbf{Z}}_i^{(n+1)}$$

which is equivalent to $\tilde{\boldsymbol{\Phi}}\tilde{\mathbf{w}}_i = \mathbf{D}_i$ for $n \to \infty$ and if we employ sample statistic estimators for the cumulants and the covariance matrix. The expression for $\tilde{z}_i(n)$ can be justified by considering the estimation of fourth-order cumulants based on sample average given by (6.31) and **AS2** (the third term on the right-hand side of (6.31) needs not to be estimated since $E\left\{z_i^2(n)\right\} = \sum_j \sum_k s_{i,j}(k)^2 E\left\{x_i(n)\, x_i(n)\right\} = 0$, see **AS2**). Moreover, due to the power normalization (6.22),

$$\begin{aligned} E\left\{|z_i(n)|^2\right\} &= E\left\{z_i(n)\, z_i^*(n)\right\} \\ &= \sum_j \sum_k |s_{i,j}(k)|^2 E\left\{x_i(n)\, x_i^*(n)\right\} = \sigma_x^2. \end{aligned}$$

Observe that the expression (6.32) reveals similarity with the recursive least squares problem where the process $\tilde{z}_i(n)$ is the desired signal. The adaptive solution of the problem could be derived using the RLS algorithm or its variations [9, 10]. One of the important drawbacks of full-rank RLS is that this approach requires at every step, explicitly or implicitly, the existence of the inverse $\left(\tilde{\mathbf{Y}}^{(n+1)^H}\, \tilde{\mathbf{Y}}^{(n+1)}\right)^{-1}$. This cannot realistically be guaranteed in any practical situation and in particular in space/time applications. In fact, the multipath channels follow independent time-varying characteristics so that the data matrix $\tilde{\mathbf{Y}}^{(n)}$ may become ill-conditioned due to statistical fluctuation of the received signals. The most immediate consequence is that rank degeneracy appears in the least squares problem (6.32) which makes the results of any traditional numerical algorithm based on full-rank processing meaningless. In this case one should detect the rank degeneracy by looking at the singular values of $\tilde{\mathbf{Y}}^{(n)}$ and decide that certain columns of this matrix can be ignored. In particular, the $KL - r$ eigenvectors of $\tilde{\mathbf{Y}}^{(n)^H}\, \tilde{\mathbf{Y}}^{(n)}$ [the left singular vectors of $\tilde{\mathbf{Y}}^{(n)}$] corresponding to the small singular values can be ignored in a minimum norm solution of the least-squares problem (6.32).

The eigenvalue decomposition of $\tilde{\mathbf{Y}}^{(n)^H}\, \tilde{\mathbf{Y}}^{(n)}$ is

$$\tilde{\mathbf{Y}}^{(n)^H}\, \tilde{\mathbf{Y}}^{(n)} = \mathbf{V}(n)\boldsymbol{\Lambda}(n)\mathbf{V}(n)^H \tag{6.33}$$

where $\mathbf{V}(n) = \begin{bmatrix} \mathbf{v}_1(n) & \mathbf{v}_2(n) & \cdots & \mathbf{v}_{KL}(n) \end{bmatrix}$ is the $KL \times KL$ orthonormal matrix of eigenvectors, and

$$\boldsymbol{\Lambda}(n) = \operatorname{diag}(\lambda_1(n), \lambda_2(n), ..., \lambda_{KL}(n))$$

is the diagonal matrix of eigenvalues. The r dominant eigenvalues are those eigenvalues that exceed a certain predetermined tolerance level related to the numerical accuracy of the computing device we are using and to the additive noise power. We use the notation

$$\mathbf{V}(n) = \left[\begin{array}{cc} \mathbf{V}_r(n) & \mathbf{V}_{KL-r}(n) \end{array} \right],$$

$$\mathbf{\Lambda}(n) = \left[\begin{array}{cc} \mathbf{\Lambda}_r(n) & \mathbf{0} \\ \mathbf{0} & \mathbf{\Lambda}_{KL-r}(n) \end{array} \right],$$

so that we can define the generalized inverse

$$\left(\tilde{\mathbf{Y}}^{(n)^H} \tilde{\mathbf{Y}}^{(n)} \right)^{\dagger} = \mathbf{V}_r(n)\mathbf{\Lambda}_r(n)^{-1}\mathbf{V}_r(n)^H \tag{6.34}$$

and the low-rank solution of (6.32)

$$\tilde{\mathbf{w}}_i(n) = \mathbf{V}_r(n)\mathbf{\Lambda}_r(n)^{-1}\mathbf{V}_r(n)^H\tilde{\mathbf{Y}}^{(n)^H} \tilde{\mathbf{Z}}_i^{(n)}. \tag{6.35}$$

The *low-rank in-space vector* and *residual vector* are defined respectively as

$$\mathbf{Z}_i^{(n)} = \tilde{\mathbf{Y}}^{(n)}\tilde{\mathbf{w}}_i(n), \tag{6.36}$$

$$\mathbf{u}_i(n) = \tilde{\mathbf{Z}}_i^{(n)} - \mathbf{Z}_i^{(n)}. \tag{6.37}$$

It is possible to show [21, 22] that if we define $\tilde{\mathbf{Y}}^{(n)^H} \tilde{\mathbf{Y}}^{(n)} = \tilde{\mathbf{\Phi}}(n)$ then an iterative procedure to find an *approximate* eigendecomposition [9] of $\tilde{\mathbf{\Phi}}(n)$ (considered as a time-invariant matrix) is

- $\mathbf{A}(l) = \tilde{\mathbf{\Phi}}(n)\mathbf{Q}(l-1)$
- $\mathbf{A}(l) = \mathbf{Q}(l)\tilde{\mathbf{R}}(l)$:
  Square-root decomposition ($\mathbf{Q}(l)$ is constituted by orthonormal columns, $\tilde{\mathbf{R}}(l)$ is upper triangular).

that is

$$\lim_{l\to\infty} \mathbf{Q}(l)\tilde{\mathbf{R}}(l)\mathbf{Q}(l-1)^H \simeq \mathbf{V}_r(n)\mathbf{\Lambda}_r(n)\mathbf{V}_r(n)^H.$$

This is known as simultaneous orthogonal iteration [23]. Strobach [22] and Owsley [24] observed that this iteration can be applied also when $\tilde{\mathbf{\Phi}}(n)$ is slowly varying. The recursive estimation of $\tilde{\mathbf{\Phi}}(n)$, $\mathbf{D}_i(n)$ is

$$\tilde{\mathbf{\Phi}}(n) = \lambda\tilde{\mathbf{\Phi}}(n-1) + (1-\lambda)\tilde{\mathbf{y}}^*(n)\,\tilde{\mathbf{y}}(n)^T \quad (6.38)$$
$$\mathbf{D}_i(n) = \lambda\mathbf{D}_i(n-1) + (1-\lambda)\tilde{\mathbf{y}}^*(n)\tilde{z}_i(n) \quad (6.39)$$

where λ is a parameter close to one that accounts for some exponential weighting [9]. In addition, using the simultaneous orthogonal iteration method and substituting the iteration index l with the time step index n, it is possible to approximate (6.34) as

$$\tilde{\mathbf{\Phi}}(n)^{\dagger} = \mathbf{Q}(n-1)\tilde{\mathbf{R}}(n)^{-1}\mathbf{Q}(n)^H. \quad (6.40)$$

The estimate of the filter weights at time step n is so obtained as

$$\begin{aligned}\tilde{\mathbf{w}}_i(n) &= \mathbf{Q}(n-1)\tilde{\mathbf{R}}(n)^{-1}\mathbf{Q}(n)^H\tilde{\mathbf{Y}}^{(n)^H}\,\tilde{\mathbf{Z}}_i^{(n)} \\ &= \mathbf{Q}(n-1)\tilde{\mathbf{R}}(n)^{-1}\mathbf{Q}(n)^H\mathbf{D}_i(n). \end{aligned} \quad (6.41)$$

It was shown in [22] that the update on matrix $\mathbf{A}(n)$ can be performed as

$$\begin{aligned}\mathbf{A}(n) = \lambda\mathbf{A}(n-1)\mathbf{Q}(n-2)^H\mathbf{Q}(n-1) \\ +(1-\lambda)\tilde{\mathbf{y}}(n)\tilde{\mathbf{y}}(n)^H\mathbf{Q}(n-1). \end{aligned} \quad (6.42)$$

The in-space component, obtained pinning the in-space vector, is

$$z_i(n) = \tilde{\mathbf{y}}(n)^T\mathbf{Q}(n-1)\tilde{\mathbf{R}}(n)^{-1}\mathbf{Q}(n)^H\mathbf{D}_i(n) \quad (6.43)$$

while the residual component, obtained pinning the residual vector, is

$$u_i(n) = z_i(n) - \tilde{z}_i(n).$$

Expressions (6.41)–(6.43) are at the basis of LORAF1 of [22] as modified for complex signals and applied to our blind deconvolution problem. The algorithm (defined LORAF1-HOS) is reported in the following.

1. $\tilde{z}_i(n) = (|z_i(n-1)|^2 - 2\,\sigma_x^2)\,z_i(n-1)$, $\tilde{\mathbf{y}}(n)$, r, λ;
2. $\mathbf{A}(n) = \lambda\mathbf{A}(n-1)\mathbf{Q}(n-2)^H\mathbf{Q}(n-1) + (1-\lambda)\tilde{\mathbf{y}}(n)\tilde{\mathbf{y}}(n)^H\mathbf{Q}(n-1)$;
3. $\mathbf{A}(n) \Rightarrow \mathbf{Q}(n)\tilde{\mathbf{R}}(n)$: QR decomposition;
4. $\mathbf{D}_i(n) = \lambda\mathbf{D}_i(n-1) + (1-\lambda)\tilde{\mathbf{y}}^*(n)\tilde{z}_i(n)$;
5. Solve by back substitution $\tilde{\mathbf{R}}(n)\mathbf{b}(n) = \mathbf{Q}(n)^H\mathbf{D}_i(n)$ for $\mathbf{b}(n)$;

6. $z_i(n) = \mathbf{b}(n)^T \mathbf{Q}(n-1)^H \tilde{\mathbf{y}}(n)$.

The initialization is

$$\mathbf{Q}(n-2) = \mathbf{Q}(n-1) = \begin{bmatrix} \mathbf{I}_r \\ \mathbf{0} \end{bmatrix},$$

$$\mathbf{A}(n-1) = \mathbf{Q}(n-2)\tilde{\mathbf{y}}(n-1)\tilde{\mathbf{y}}(n-1)^H,$$

$$\mathbf{D}_i(n-1) = \tilde{\mathbf{y}}^*(n-1)\tilde{z}_i(n-1).$$

The QR decomposition required at every step can be efficiently performed using Jacobi rotations [9]. Observe that

$$\tilde{\mathbf{w}}_i(n) = \mathbf{Q}(n-1)\mathbf{b}(n)$$

is not explicitly computed. Some important refinements of this basic algorithm can be made to improve computational efficiency. Particularly, the explicit updating of the QR decomposition of $\mathbf{A}(n)$ can be eliminated in a scheme which directly tracks the Q and the R factors as new data snapshots are received. This method requires $\frac{r(r+1)(2r+1)}{6}$ Givens rotations per update and is formalized in the algorithm LORAF2 (see [22] for details). Further simplifications can be used to reduce the number of Givens rotations to $2r-1$ and obtain the ultrafast subspace tracking scheme defined LORAF3 in [22]. The algorithm modified for our problem is reported in Table 6.1. The matrix $\mathbf{G}(n)$ is a Givens rotation matrix [9] that sweeps the last row of the matrix it premultiplies. The structure of (6.32) may even suggest the use of a square-root type algorithm for subspace tracking: This approach would have the immediate advantage of reducing the required dynamic range of the triangular factor $\tilde{\mathbf{R}}(n)$.

Adaptive dominant space order estimation. The order r of the dominant space can be estimated using a threshold level *tol*, which, added to the estimated noise floor (of the AWGN), will help to discriminate the relevant eigenvectors. The subspace section of the algorithm is operated with the order r equal to $r_{\max}$. This maximum subspace order $r_{\max}$ is experimentally determined so that the rank of the autocovariance matrix statistically never exceeds this parameter. A possible adaptive noise power estimator is

$$\sigma_n^2(n) = KL\, p_y(n) - \sigma_x^2,$$

Table 6.1
The Algorithm LORAF3-HOS. The Matrix $\mathbf{G}(n)$ is a Givens Rotation Matrix.

INPUT	
--	$\tilde{z}_i(n) = \left(\lvert z_i(n-1)\rvert^2 - 2\,\sigma_x^2\right)\, z_i(n-1),\ \tilde{\mathbf{y}}(n),\ r,\ \lambda$
TRACKING	
Step 1	$\mathbf{h}(n) = \mathbf{Q}(n-1)^H \mathbf{y}(n)$
Step 2	$\mathbf{y}_\perp(n) = \tilde{\mathbf{y}}(n) - \mathbf{Q}(n-1)\mathbf{h}(n)$
Step 3	$\overline{\mathbf{y}}_\perp(n) = \frac{\mathbf{y}_\perp(n)}{\lvert\lvert \mathbf{y}_\perp(n)\rvert\rvert}$
Step 4	$\begin{bmatrix} \tilde{\mathbf{R}}(n) \\ \mathbf{0}^T \end{bmatrix} \Leftarrow \mathbf{G}(n) \begin{bmatrix} \lambda\tilde{\mathbf{R}}(n-1) + (1-\lambda)\mathbf{h}(n)\mathbf{h}(n)^H \\ (1-\lambda)\mathbf{h}(n)^H \lvert\lvert \mathbf{y}_\perp(n)\rvert\rvert \end{bmatrix}$
Step 5	$\begin{bmatrix} \mathbf{Q}(n) & \mathbf{q}(n) \end{bmatrix} \Leftarrow \begin{bmatrix} \mathbf{Q}(n-1) & \overline{\mathbf{y}}_\perp(n) \end{bmatrix} \mathbf{G}(n)^H$
FILTERING	
Step 1	$\mathbf{D}_i(n) = \lambda \mathbf{D}_i(n-1) + (1-\lambda)\tilde{\mathbf{y}}^*(n)\tilde{z}_i(n)$
Step 2	$\mathbf{f}_i(n) = \lambda \mathbf{f}_i^+(n-1) + (1-\lambda)\mathbf{h}(n)\tilde{z}_i(n)$
Step 3	$\begin{bmatrix} \mathbf{f}_i^+(n) \\ \times \end{bmatrix} = \mathbf{G}(n) \begin{bmatrix} \mathbf{f}_i(n) \\ \overline{\mathbf{y}}_\perp(n)^H \mathbf{D}_i(n) \end{bmatrix}$
Step 4	Solve by back substitution $\tilde{\mathbf{R}}(n)\mathbf{b}(n) = \mathbf{f}_i^+(n)$ for $\mathbf{b}(n)$
Step 5	$z_i(n) = \mathbf{b}(n)^T \mathbf{h}(n)$
Init.	
Step 1	$\mathbf{Q}(n-1) = \begin{bmatrix} \mathbf{I}_r \\ \mathbf{0} \end{bmatrix}$
Step 2	$\mathbf{f}_i^+(n-1) = \mathbf{0}$
Step 3	$\tilde{\mathbf{R}}(n-1) = \mathbf{0}$
Step 4	$\mathbf{D}_i(n-1) = \tilde{\mathbf{y}}^*(n-1)\tilde{z}_i(n-1)$

with

$$p_y(n) = \lambda p_y(n-1) + \frac{1-\lambda}{KL}\, tr[\tilde{\mathbf{y}}(n)\tilde{\mathbf{y}}(n)^H].$$

Only the eigenvectors corresponding to the eigenvalues that exceed $\sigma_n^2(n) + tol$ are used in the computation of the filter $\tilde{\mathbf{w}}_i(n)$ and as a consequence in signal reconstruction. Observe that the eigenvalues (actually their approximation) are obtained by extracting the diagonal elements of the upper triangular matrix $\tilde{\mathbf{R}}(n)$ (which is an *almost diagonal* matrix). The experimental results of Section 6.2.3 do not make use, however, of the adaptive space order estimation.[7] The order $r = r_{\max}$ is actually experimentally determined off-line for a certain propagation scenario as a trade-off between error rate performance and computational complexity.

The simulations in Section 6.2.3 are performed using a word length size of 24 bits, using fixed-point arithmetic. The computational complexity in terms of multiplications and elementary complex 2×2 Givens rotation per update was calculated and compared to the complexity of the adaptive QR-RLS [9]. The approximate number of computations per iteration using filters length equal to L and K sensors is given for QR-RLS by the following:

- Number of real multiplications: $10KL + 30(KL)^2$;
- Number of Givens rotations: KL.

For LORAF3-HOS we have the following:

- Number of real multiplications: $8KLr + 106r^2 + 124r + 20KL - 58$;
- Number of Givens rotations: $2r - 1$.

It is evident that the LORAF3-HOS algorithm here proposed becomes convenient in terms of computational complexity with respect to the QR-RLS scheme if the order r of the dominant space

7. There is marginal performance improvement if real-time estimation and tracking of r is performed in the fixed number of mobile transmitters scenario, which is the case we simulated. Adaptive order estimation becomes indispensable when the number of transmitters U varies dynamically: The rank properties of the space-time autocorrelation matrix may change dramatically in this case.

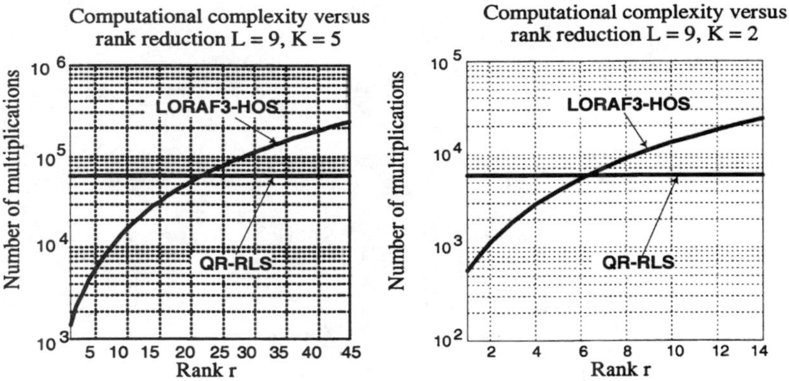

Figure 6.1 Computational complexity of LORAF3-HOS and QR-RLS.

remains below a certain value: In Figure 6.1 for $K = 5$ and $K = 2$ and $L = 9$ the number of multiplications required by the two algorithms is computed and plotted versus the order r. LORAF3-HOS is convenient for $r \leq 21$ for $K = 5$ and $r \leq 6$ for $K = 2$.

6.2.3 Results of Experiments

A TDMA system for cellular communications has been simulated according to [18]. In addition, we present the results of some lab experiments performed using the WJ Communications wideband dual-mode (AMPS and IS-136) base-station system $Base_2$. The main purpose of experimenting the algorithm with data samples collected from hardware equipment is to demonstrate practical applicability of the ideas presented. A block diagram of the receiver section of the base station is shown in Figure 6.2.

The tuner module performs a standard single conversion scheme. The A/D is a high-speed bandpass sampler, while the conversion at baseband is operated by digital downconverters (wideband processing). The slots are 162 symbols long and 14 symbols at the beginning of each slot are known by the receiver.[8] It is obvious that no blind algorithm can achieve convergence in 14 symbols. The scheme we propose is a hybrid scheme that during training performs low-rank least squares filtering, while

8. This corresponds to a frame format similar to the digital traffic channel (DTC) of [18].

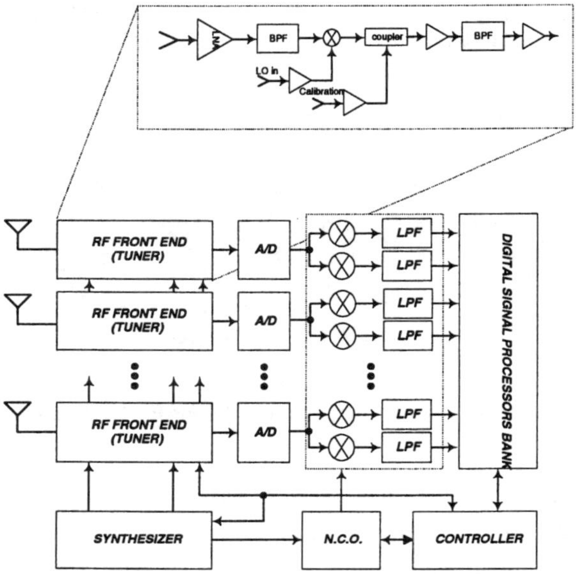

Figure 6.2 Block diagram of the receiver.

after convergence runs in blind mode as described previously. The difference is only in the desired signal that is fed back to the adaptive algorithm. During training the desired signal is $x_i^{\text{train}}(n)$ instead of $\tilde{z}_i(n)$, where $x_i^{\text{train}}(n)$ is the training sequence, that is

$$\tilde{z}_i(n) = \begin{cases} x_i^{\text{train}}(n) & \text{training} \\ \left(|z_i(n-1)|^2 - 2\,\sigma_x^2\right) z_i(n-1) & \text{blind} \end{cases} . \tag{6.44}$$

The MSE is defined as the average of the squared error obtained over $\tilde{M} = 100$ Monte Carlo runs and is given by $\text{MSE}_i = \frac{1}{\tilde{M}}\sum_{\nu=1}^{\tilde{M}} |\epsilon_i^{(\nu)}(n)|^2$. The error obtained at the mth run is $\epsilon_i^{(\nu)}(n) = z_i^{(\nu)}(n) - x_i(n - n_i)$ where n_i is the delay introduced by the filters and $z_i^{(\nu)}(n)$ is the output of the combined filters relative to the ith transmitter obtained at the νth run. Observe that the initial training with known symbols associated with the user ID matching procedure solves the sources permutation ambiguity and the phase uncertainty. Of particular importance is the fact that we always compare the proposed approach LORAF3-HOS with a more traditional QR-RLS approach, explicitly a second-order statistics-based method. The QR-RLS uses the synchronization sequences while in training and past decisions while in decision directed mode.

Computer simulations results. In the simulations we assumed a sensor spacing of $\frac{\tilde{\lambda}}{2}$. One of the transmitters is at array broadside and is the signal of interest. We assume a two-path model and each path impulse response is modeled as a two-ray Rayleigh fading channel ($P_1 = P_2 = 2$) [18], in particular, $h_{k,l}(m)$, the T-sampled impulse response of the lth transmitter and k receiver, is

$$h_{k,l}(m) = \sum_{\kappa=1}^{N_l} \sum_{\nu=1}^{P_\kappa^{(l)}} \rho_{\kappa,\nu}^{(l)} r_p \left(t - \tau_{\kappa,\nu}^{(l)}\right) e^{j\phi_{\kappa,\nu}^{(l)}} e^{-j2\pi kd \frac{\sin \theta_\kappa^{(l)}}{\tilde{\lambda}}} \Bigg|_{t=mT} \tag{6.45}$$

where $r_p(t)$ is the raised cosine function with excess bandwidth 0.35 [21] obtained because we assume that the receiver filters $p_{rx}(t)$ at each antenna element are square root raised cosine filters perfectly matched to the transmitter filters $p_{tx}(t)$, $\tau_{\kappa,\nu}^{(l)}$, $\rho_{\kappa,\nu}^{(l)}$, $\phi_{\kappa,\nu}^{(l)}$ are delay, amplitude, phase of the νth delayed signal in the κth path relative to the lth transmitter. The arrival angles of the paths are spread around 0° with a cluster width of two degrees. The interfering signals are generated with the same parameters but with DOAs clustered around θ_2 and θ_3. In Figure 6.3 the equivalent baseband discrete-time model is shown relative to the ith mobile transmitter. The Doppler frequency usually describes the second order statistics of channel variations. Doppler frequency is related through wavelength $\tilde{\lambda}$ to the ith mobile transmitter velocity V_i expressed in km/hr. The model used in this case is based on the wide sense stationary uncorrelated scattering (WSSUS) assumption [25]. The complex weights are generated as filtered Gaussian processes fully specified by the scattering function. In particular each process has a frequency response equal to the square root of the Doppler power density spectrum.[9] Figure 6.4 shows an experiment where the filters are running on experimental data (not formatted as in the IS-136 TDMA system) in blind adaption using LORAF1-HOS mode for the all

9. The Doppler spectrum is approximated by rational filtered processes. The filters are described by their 3-dB bandwidth which is called the normalized Doppler frequency. The additional assumption is that all channels and complex weights have the same Doppler spectrum.

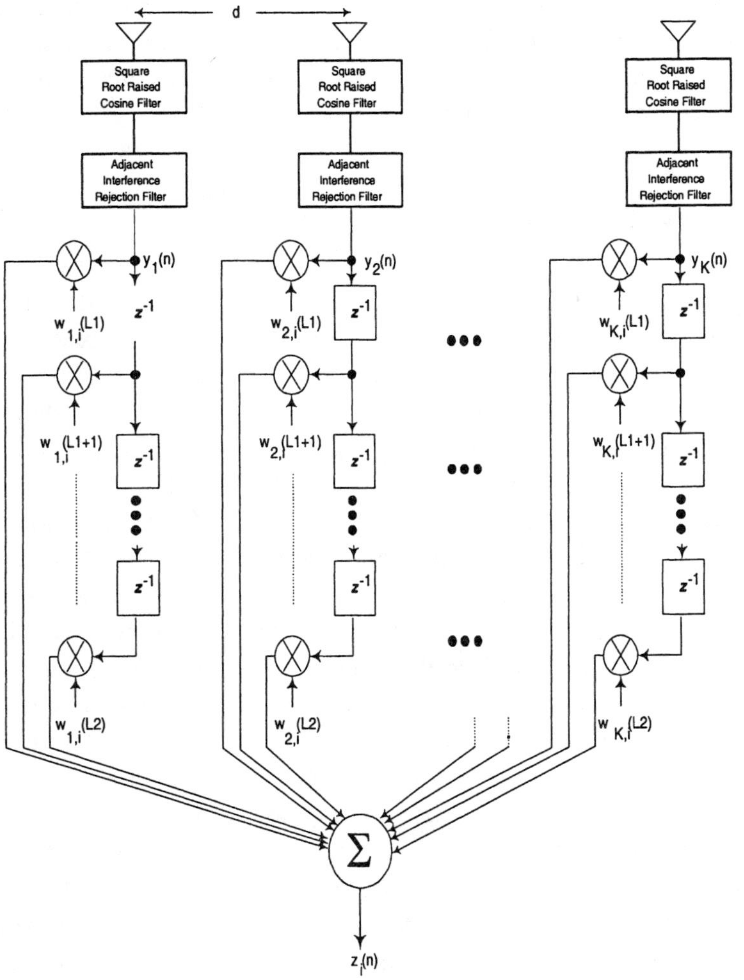

Figure 6.3 Discrete-time model of the ith filter.

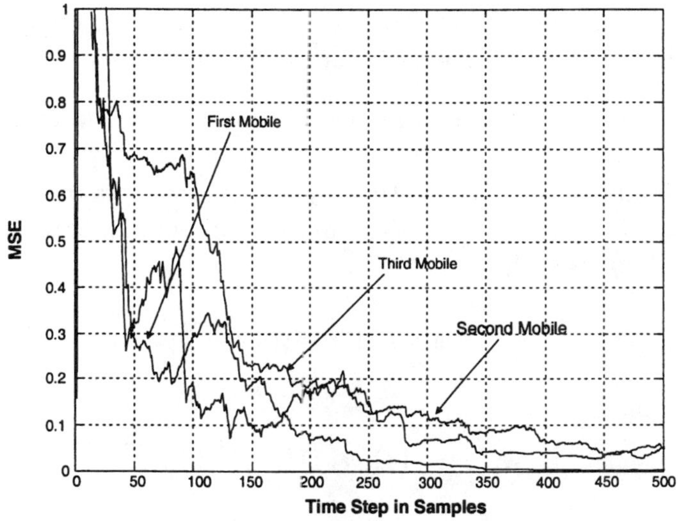

Figure 6.4 MSE versus time step for the three transmitters with blind adaptation (using LORAF1-HOS) during the entire length of the transmitted sequence.

sequence. In Figure 6.4 we have $K = 5, L = 9, r = 22, U = 3, \theta_1 = 0°$, $\theta_2 = 30°$, $\theta_3 = -70°$, $\tau_{1,1}^{(1)} = \tau_{2,1}^{(1)} = \tau_{1,1}^{(2)} = \tau_{2,1}^{(2)} = \tau_{1,1}^{(3)} = \tau_{2,1}^{(3)} = 0$, $\tau_{1,2}^{(1)} = 41.2\ \mu\text{sec}$, $\tau_{2,2}^{(1)} = 41.2\ \mu\text{sec}$, $\tau_{1,2}^{(2)} = 10.3\ \mu\text{sec}$, $\tau_{2,2}^{(2)} = 10.3\ \mu\text{sec}$, $\tau_{1,2}^{(3)} = 5.15\ \mu\text{sec}$, $\tau_{2,2}^{(3)} = 5.15\ \mu\text{sec}$. Velocity of the three mobiles is $V_1 = V_2 = V_3 = 8$ km/hr.

The speed is relatively low but large delay spread is present on the channel, while the cochannel interference suppression capability of the method is evident (the three signals are received with the same average power). It is however also important to note that the experiment shows the weakness of the algorithm: slow start-up convergence. Unfortunately the filter reaches satisfactory MSE only after hundreds of symbols. The solution to the slow start-up problem is, as already mentioned, training, according to the scheme (6.44). The following figures contain plots of the trajectory of the real part of the center tap [10] as compared to the center tap of the

10. The *center* tap is the tap with maximum energy.

ideal distortionless reception filter (constrained to have finite length) computed assuming perfect instantaneous knowledge of $\mathcal{H}(z)$ and shown as the dashed trace. The procedure we used to compute the ideal optimum setting for $\tilde{\mathbf{w}}_i(n)$ was the following:

1. *Freeze* time evolution of the multipath channels parameters at time step nT;
2. Compute the discrete-time symbol-spaced channels $h_{l,k}(m)$ for $m = J_1, J_1 + 1 \ldots J_2$ and compact into matrix $\tilde{\mathbf{H}}$ as in (6.11), say, $\tilde{\mathbf{H}}(n)$;
3. Solve

$$\min_{\tilde{\mathbf{w}}_i(n)} \left\| \tilde{\mathbf{H}}(n)\tilde{\mathbf{w}}_i(n) - \tilde{\boldsymbol{\delta}}_i \right\|^2$$

using

$$\tilde{\mathbf{w}}_i(n) = \left(\tilde{\mathbf{H}}(n)^H \tilde{\mathbf{H}}(n)\right)^{\dagger} \tilde{\mathbf{H}}(n)^H \tilde{\boldsymbol{\delta}}_i.$$

Figure 6.5 shows ideal experiments with one single transmitter and periodic retraining with randomly generated inputs. The main parameters are $K = 2, L = 5, r = 6, U = 1, \theta_1 = 0°, \tau_{1,1}^{(1)} = \tau_{2,1}^{(1)} = 0$, $\tau_{1,2}^{(1)} = 41.2$ μsec, $\tau_{2,2}^{(1)} = 0$ μsec. Velocity is $V_1 = 100$ km/hr. Each case depicts: (a) the Rayleigh fading amplitude evolution in time of the two-ray Rayleigh fading channel, (b) the real part of the center tap and the theoretical behavior of the ideal distortionless reception filter, and (c) the magnitude of the error $|\epsilon(n)|$ for the single transmitter. Several important comments are in order. Observe that fades of the first ray below -10 dB dramatically impact the performance of the adaptive algorithm. In some works it was emphasized the concept of traditional equalizers (based on second-order statistics) not able to identify nonminimum phase channels. Our results indicate that the use of HOS cannot solve totally the problem because of the fast variations of the parameters. The algorithm is extremely sensitive to fades occurring on the first ray, or if the amplitude of the second ray becomes larger than the first one. This problem can be explained by the fact that as the first ray decreases in amplitude, the center tap should move to an adjacent tap, which is a configuration of the taps considerably different from the original one (observe, in fact, the rapid changes in the real part of the center tap in Figure 6.5 in

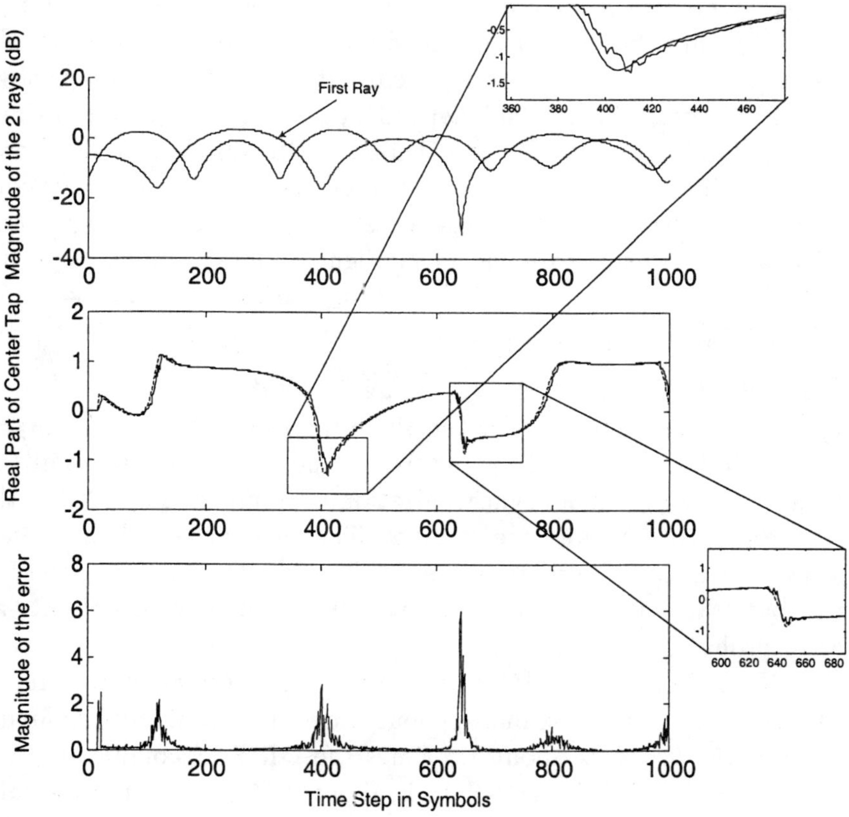

Figure 6.5 For one single transmitter, plots of the two-ray Rayleigh fading channel amplitudes, the real part of the center tap of the filter for the LORAF3-HOS versus the optimum Wiener solution (dashed), and the magnitude of the error. The filter is periodically retrained as the MSE reaches a threshold. Evidently fading events of the first ray dramatically affect performance.

correspondence of a fade of the first ray). The filter has to deconvolve a channel whose spectral response has dramatically changed, and it should be *repositioned in time* with respect to the maximum energy point captured by the synchronizer. Since the synchronization point is not changed dynamically within one slot, the filter is not able to satisfactorily deconvolve the channel. The degradation could probably be reduced increasing the time span of the filter (larger L) or by introducing a feedback filtering section. We reserve the investigation of this important topic to future work. Figure 6.6 shows the MSE averaged over 100 independent computer runs versus time-step in symbols for $\frac{E_b}{N_0} = 21$ dB for real IS-136 slots. It is shown how the blind mode is more robust under fast fading than decision-directed mode of traditional schemes such as the QR-RLS algorithm. BER analysis results are shown in Figure 6.7 (at different speeds) with $K = 5, L = 3:7, r = 15$, $U = 3$, $\theta_1 = 0°$, $\theta_2 = 10°$, $\theta_3 = -45°$, $\tau_{1,1}^{(1)} = \tau_{2,1}^{(1)} = \tau_{1,1}^{(2)} = \tau_{2,1}^{(2)} = \tau_{1,1}^{(3)} = \tau_{2,1}^{(3)} = 0$, $\tau_{1,2}^{(1)} = 41.2$ μsec, $\tau_{2,2}^{(1)} = 0.0$ μsec, $\tau_{1,2}^{(2)} = 41.2$ μsec, $\tau_{2,2}^{(2)} = 0.0$ μsec, $\tau_{1,2}^{(3)} = 41.2$ μsec, $\tau_{2,2}^{(3)} = 0.0$ μsec. Velocity for the three mobiles is $V_1 = 100$ km/hr, $V_2 = V_3 = 50$ km/hr, BER is relative to the first mobile transmitter. Ideal frame and symbol synchronization is assumed. The SNR is the same on each discrete-time channel. The two cochannel interferers are received at an average power 5 dB below the average power of the first transmitter. A sample size of 10^{k+4} was used to estimate an error probability of 10^{-k}.

Note that for $r = 15$, $L = 3$ the computational complexity of LORAF3-HOS is approximately four times the QR-RLS complexity, for $r = 15$, $L = 5$ is about 1.5 times the QR-RLS complexity, and for $r = 15$, $L = 7$ LORAF3-HOS is about 20% less computationally intensive. While we obtain improvement in all cases, the remarkable result is that for $L = 7$ LORAF3-HOS gives for BER=10^{-2} an SNR per bit improvement of about 6 dB.

A simpler and indeed more realistic scenario ($K = 2$, $U = 1$)[11] was studied using data collected from the DSP receiver section of $Base_2$, the dual-mode wideband base station implemented at WJ Communications. A hardware multipath fading simulator is connected to the two antenna ports of the base station as described

11. This testing environment is also specified in [18].

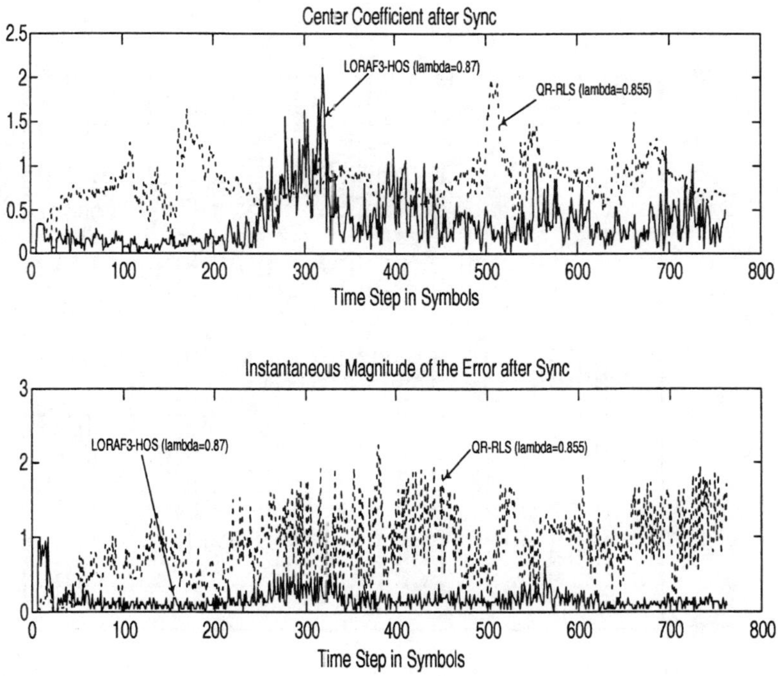

Figure 6.6 Stability experiments for the two algorithms at SNR = 21 dB: QR-RLS and LORAF3-HOS at 100 km/hr. The slot is 750 symbols (for testing purposes only). $K = 2, L = 5, r = 6, U = 1$, $\theta_1 = 0°, \tau_{1,1}^{(1)} = \tau_{2,1}^{(1)} = 0, \tau_{1,2}^{(1)} = 41.2$ μsec, $\tau_{2,2}^{(1)} = 0$ μsec. Velocity is 100 km/hr.

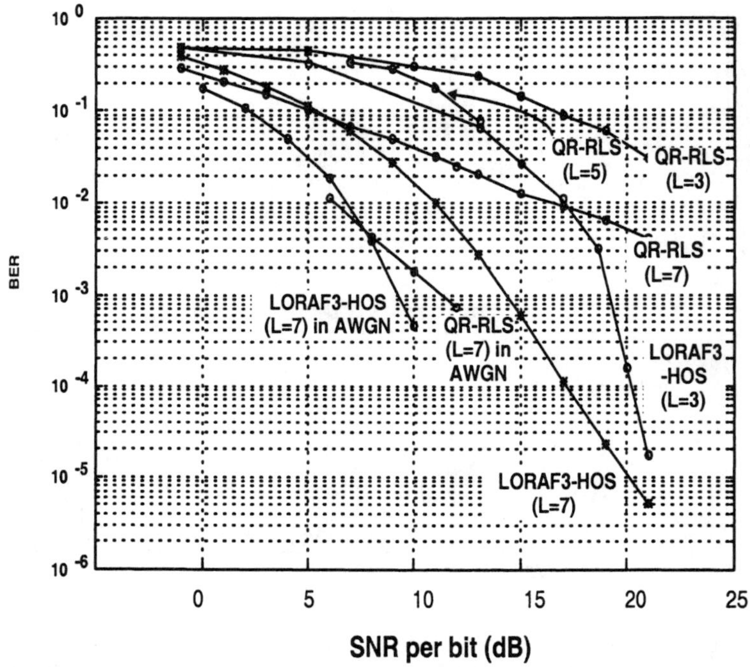

Figure 6.7 BER of the first mobile transmitter in a three-mobile environment. Results for the no-fading case are also shown.

in Chapter 4. The IS-136 signal generator simulates transmission of DTC frames coming from three different mobiles. Additive Gaussian noise is injected on both diversity channels. Observe that the DSP modem receives a sampling rate of 80 kHz (not an integer of the symbol rate). The word length used is 24 and the algorithm has been implemented using simulated fixed-point arithmetic.[12] A polyphase raised-cosine filter concatenated with an adjacent interference rejection filter transforms the rate to $4/T = 97.2$ kHz. Then the two-channel filter works at $2/T$-rate. Automatic gain control is operated on a slot by slot basis. While perfect synchronization was assumed in the previous simulations in the hardware experiments there is an open-loop synchronizer choosing the optimum positioning of the filters at $4/T$ rate.[13] In addition, as required by the specification [18], there is carrier frequency offset between the local oscillator and the transmitted carrier frequency of about 213 Hz. The adaptive filter is not able to track this large frequency offset. Indeed frequency offset is more conveniently estimated using a second-order PLL that compensates for the frequency drift. The results of extensive BER measurements are summarized in Figure 6.8 for $K = 2, L = 9, r = 12$, $U = 1$, $\theta_1 = 0°$, $\tau_{1,1}^{(1)} = \tau_{2,1}^{(1)} = 0$, $\tau_{1,2}^{(1)} = 41.2$ μsec, and $\tau_{2,2}^{(1)} = 0.0$ μsec. The improvement with respect to the QR-RLS is not as impressive as in the previous experiments where multiple interference sources and relatively large number of sensors made more difficult to track the optimum MMSE solution. It is exactly in these circumstances that LORAF3-HOS outperforms known solutions in terms of performance and computational complexity. It is important to mention however that substantial improvement in the two-antenna case can be achieved using QR-based decision-feedback schemes and similar extensions of the cumulant-based algorithm.

12. A simulation analysis of the dynamic range required not to degrade the performance of the algorithm as opposed to the floating point implementation should be carried out, although the description of such analysis is beyond the scope of this book. Different variables of the algorithm required different numbers of fractional bits to avoid overflow. In particular the matrix $\tilde{\mathbf{R}}(n)$ required double-precision representation.

13. The basic strategy is to cross-correlate the incoming signal with the synchronization sequences adequately interpolated.

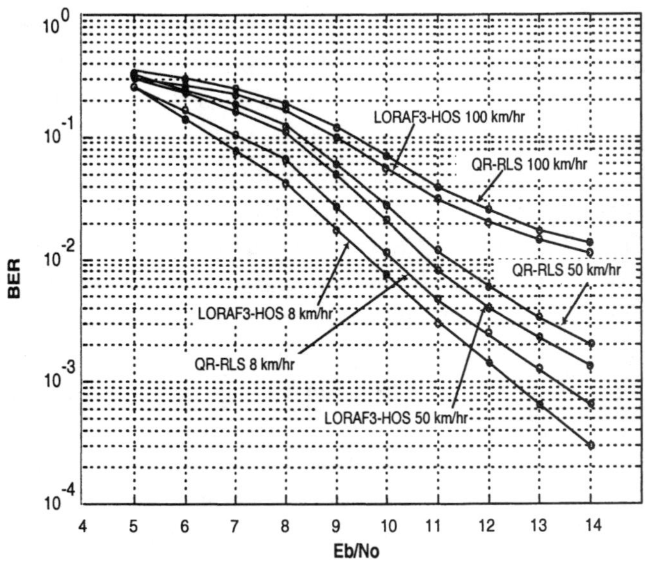

Figure 6.8 BER for the hardware experiments.

6.3 A Nonblind Algorithm

One of the possible solutions to the problem of space-time equalization (as we know from the previous chapters) is the application of the MMSE principle [26], which usually results in multichannel generalizations of single-channel adaptive linear filtering algorithms. Unfortunately the MMSE method has limitations, despite its simplicity, because SOSs alone are processed, which implies the assumption that the underlying processes are all Gaussian processes. Since in a real-world scenario this assumption is rarely satisfied, the use of SOS results in suboptimum performance. The use of HOS has generated high interest in the signal processing community over the last 10 years mainly because of the celebrated theoretical Gaussian rejection property and enhanced identification capability. Most of the works describing HOS-based algorithms for identification/deconvolution are blind and try not to exploit any input-signal knowledge. Many successful applications of HOS were recently proposed in a multichannel setup [19, 27, 28]. However, since the estimation of HOS requires considerably larger sample size than SOS, these blind algorithms result in an intolerably slow convergence behavior that prevented their applicability in practical systems. In TDMA systems (for example, the IS-136 standard for U.S. cellular communications [18]), the bitstream (control data and voice data) is organized into frames. Some known sequences are periodically transmitted in every slot, which allows the receiver to perform frame synchronization, symbol timing recovery, and equalizer training. The length of a slot including the training sequence is usually in the range of 100–200 symbols, depending on the type of frame, while the length of the known sequence is in the range of 14–20 symbols. The use of the known sequence is indispensable to fast start-up equalization. The method proposed in this work is based on the same idea introduced in [11, 12] where high-order cumulants were used to derive MMSE methods for system identification. Here we modify the basic intuition so that it can be applied to the deconvolution of complex vector signals. Moreover, we show that it is indeed possible to derive a fully adaptive implementation based on a recursive Jacobi-type algorithm.

Assume that the generation of the symbols is modeled as

$$a_l(n) = \sum_m b_l(m) x_l(n-m),$$

where $b_l(m)$ is a finite impulse response filter modeling the autocorrelation of the complex sequence $a_l(k)$. The discrete-time model is then

$$y_k(n) = \sum_{l=1}^{U} \sum_m h_{k,l}(m) a_l(n-m) + \eta_k(n) \qquad k = 1, 2, ..., K \tag{6.46}$$

Discrete-time convolution is indicated as $*$. To indicate statistics estimated in a noise-free environment, we will use the notation $\langle ... \rangle_{\mathrm{SNR}=\infty}$. For example, if $r(t) = s(t) + n(t)$ where $n(t)$ is noise and $s(t)$ is signal of interest, we have $\langle E\{r(t)r^*(t+\tau)\}\rangle_{\mathrm{SNR}=\infty} = E\{s(t)s^*(t+\tau)\}$.

Neglecting the additive noise terms, the overall impulse response is characterized by the input/output relation

$$z_i(n) = \sum_{l=1}^{U} \sum_m s_{i,l}(m) a_l(n-m) \qquad i = 1, 2, ..., U. \tag{6.47}$$

where

$$s_{i,l_1}(m_2) = \sum_{l=1}^{K} \sum_{m=L_1}^{L_2} w_{i,l}(m) h_{l,l_1}(m_2 - m).$$

The fundamental assumptions necessary to develop the algorithm are **AS1**, **AS2** of Section 6.2.1 and **AS3'**.

- **AS3'**: The complex sequences $\{a_i(n)\}$ are mutually independent and are all generated by linear stable systems

$$a_i(n) = b_i(n) * x_i(n) \qquad i = 1, 2, ..., U$$

with finite impulse responses $b_i(k)$ $(M_1 \leq k \leq M_2)$ of length $M = M_2 - M_1 + 1$ with $b_i(M_1) \neq 0, b_i(M_2) \neq 0$ for any $i = 1, 2, ..., U$ and no zero on the unit circle. The cumulants of $\{x_i(n)\}$, i.i.d. non-Gaussian processes, satisfy
  - $E\{x_i(n)\} = E\{x_i^3(n)\} = 0$;
  - $cum[x_i(n), x_l^*(n)] = \sigma_x^2 > 0$, only for $i = l$;
  - $E\{x_i(n)x_l(n)\} = E\{x_i^*(n)x_l^*(n)\} = 0$ for any i, l;
  - $cum[x_{l_1}(n), x_{l_2}(n), x_{l_3}^*(n), x_{l_4}^*(n)] = \gamma_{4x} \neq 0$, only for $l_1 = l_2 = l_3 = l_4$.

Assumption **AS3'** generalizes the algorithm to handle colored input processes [29] and attempts to model the fact that the source signals in practice do not satisfy the usual i.i.d. assumption. This is true not only in training but also in decision-directed mode. For example, in [18], 14 symbols at the beginning of any slot are dedicated to synchronization and eventual training. These symbols constitute complex *sounding* sequences that satisfy particular autocorrelation properties. They only approximate the correlation shape of optimal *sounding* sequences. Moreover, in decision-directed mode, symbols coming from the digital traffic channel [18] do not satisfy the ideal i.i.d. assumption. It is, however, extremely important to emphasize the following:

1. The algorithm we propose does not require the impulse responses $b_i(n)$ to be known.
2. It applies with no change (only marginal modifications in the derivation are necessary) also to the case when $a_l(n)$ can be considered i.i.d. processes, that is, when $a_l(n) = x_l(n)$.

6.3.1 Derivation of the Algorithm

The restoration error for the ith channel is defined as

$$\begin{aligned}\epsilon_i(n) &= a_i(n) - \sum_{l=1}^{K}\sum_{m} w_{i,l}(m) y_l(n-m) \\ &= a_i(n) - \sum_{l=1}^{U} s_{i,l}(n) * a_l(n), \end{aligned} \quad (6.48)$$

where we have neglected the contribution of the additive noise. The usual approach is to find the weights $w_{i,l}(m)$ for $l = 1, 2, ..., K$, $m = L_1, L_1+1, ..., L_2$ such that the MSE is minimized. The noise-free MSE in our model can be expressed as

$$\begin{aligned}&\left\langle E\left\{|\epsilon_i(n)|^2\right\}\right\rangle_{\mathrm{SNR}=\infty} \\ &= E\left\{\left|a_i(n) - \sum_{l=1}^{U} s_{i,l}(n) * a_l(n)\right|^2\right\} \\ &= E\left\{\left|\sum_{l=1}^{U}\sum_{k} \tilde{s}_{i,l}(k) x_l(n-k)\right|^2\right\} \\ &= \sigma_x^2 \sum_{l=1}^{U}\sum_{k} |\tilde{s}_{i,l}(k)|^2 \end{aligned} \quad (6.49)$$

where $\tilde{s}_{i,l}(k) = \sum_m [\delta_{i,l}(m) - s_{i,l}(m)]\, b_l(k-m)$, and we have defined $\boldsymbol{\delta}_{i,l} = [..., \delta_{i,l}(-1), \delta_{i,l}(0), \delta_{i,l}(1), ...]^T$, so that $a_i(n)$ can be written as $a_i(n) = \sum_l \sum_m \delta_{i,l}(m) a_i(n-m)$. The MSE as expressed in (6.49) shows that it is possible to obtain some cumulant-based criteria for deconvolution whose objective is the minimization of a cost function proportional to (6.49). If we organize the weights $w_{i,l}(m)$ in a vector

$$\tilde{\mathbf{w}}_i = \left[\mathbf{w}_{i,1}{}^T, \mathbf{w}_{i,2}{}^T, ..., \mathbf{w}_{i,K}{}^T\right]^T,$$

$$\mathbf{w}_{i,l} = [w_{i,l}(L_1), w_{i,l}(L_1+1), ..., w_{i,l}(L_2)]^T,$$

the following theorem can be proved.

Theorem 1: The minimization of the cost function

$$J^{(i)}(\tilde{\mathbf{w}}_i)$$

$$= \left| \sum_{l=1}^{K} \sum_{k=-\infty}^{+\infty} cum \left[\epsilon_i(n), \epsilon_i^*(n), y_l(n-k), y_l^*(n-k) \right] \right|^2$$

with respect to $w_{i,l}(k)$, $l = 1, 2, ..., K$, $k = L_1, ..., L_2$ is equivalent to the minimization of the noise-free MSE $\langle E \{ |\epsilon_i(n)|^2 \} \rangle_{\text{SNR}=\infty}$ up to a scalar factor if

$$\sum_{l_1=1}^{K} \sum_{k} |g_{l_1,i_1}(k)|^2 = \sum_{l_2=1}^{K} \sum_{k} |g_{l_2,i_2}(k)|^2 = \frac{\mathcal{K}}{\gamma_{4x}} \neq 0$$

for any $i_1, i_2 \in [1, 2, ..., U]$ with $g_{i,l}(k) = h_{i,l}(k) * b_l(k)$.

Proof: Using the fact that $\epsilon_i(n) = \sum_{l=1}^{U} \sum_k \tilde{s}_{i,l}(k) x_l(n-k) +$ *noise terms* [see (6.49)], and the noise-rejection properties of fourth-order cumulants, the cost function $J^{(i)}(\tilde{\mathbf{w}}_i)$ can be expressed as

$$\begin{aligned} & J^{(i)}(\tilde{\mathbf{w}}_i) \\ &= \left| \gamma_{4x} \sum_{l_1=1}^{K} \sum_{k} \sum_{l_2=1}^{U} \sum_{m} |\tilde{s}_{i,l_2}(m)|^2 |g_{l_1,l_2}(m-k)|^2 \right|^2 \\ &= \left| \gamma_{4x} \sum_{l_2=1}^{U} \sum_{m} |\tilde{s}_{i,l_2}(m)|^2 \sum_{l_1=1}^{K} \sum_{k} |g_{l_1,r}(k)|^2 \right|^2 \\ &= \left| \sigma_x^2 \sum_{l=1}^{U} \sum_{m} |\tilde{s}_{i,l}(m)|^2 \right|^2 \left| \frac{\mathcal{K}}{\sigma_x^2} \right|^2 \end{aligned} \qquad (6.50)$$

where $\mathcal{K} = \gamma_{4x} \sum_{l=1}^{K} \sum_k |g_{l,r}(k)|^2$. Now it is evident from (6.50) as compared to (6.49) that, as long as $\mathcal{K}$ is different from zero, minimizing $J^{(i)}(\tilde{\mathbf{w}}_i)$ with respect to $w_{i,l}(k)$, $l = 1, 2, ..., K$, $k = L_1, ..., L_2$ is equivalent to minimize $\langle E \{ |\epsilon_i(n)|^2 \} \rangle_{\text{SNR}=\infty}$.

Observe that the condition $\frac{\mathcal{K}}{\gamma_{4x}} \neq 0$ of Theorem 1 holds due to **AS2-AS3**, while $\sum_{l_1=1}^{K} \sum_k |g_{l_1,i_1}(k)|^2 = \sum_{l_2=1}^{K} \sum_k |g_{l_2,i_2}(k)|^2$ for any $i_1, i_2 \in [1, 2, ..., U]$ can be forced in practice using independent AGC circuits.

Theorem 2: The optimum deconvolution filter associated with the minimum of the cost function $J^{(i)}(\tilde{\mathbf{w}}_i)$ satisfies

$$\sum_{l_1} \sum_{k} cum[\epsilon_i(n), y_{l_2}^*(n-m), y_{l_1}(n-k), y_{l_1}^*(n-k)] = 0 \qquad (6.51)$$

and

$$\sum_{l_1}\sum_{k} cum[\epsilon_i^*(n), y_{l_2}(n-m), y_{l_1}(n-k), y_{l_1}^*(n-k)] = 0 \tag{6.52}$$

for $l_2 = 1, 2, ..., K,\ \ m = L_1, L_1+1, ..., L_2,\ \ i = 1, 2, ..., U.$

The proof of this theorem is reported in Appendix 6A.

Using (6.51) it is then true that

$$\begin{aligned}&\sum_{l_1}\sum_{k} cum[a_i(n), y_{l_2}^*(n-m), y_{l_1}(n-k), y_{l_1}^*(n-k)]\\ &\quad -\sum_{l_3}\sum_{m_1} w_{i,l_3}(m_1)\sum_{l_1}\sum_{k}\\ &\times cum[y_{l_3}(n-m_1), y_{l_2}^*(n-m), y_{l_1}(n-k), y_{l_1}(n-k)]\\ &\qquad = 0\end{aligned}$$

and we can write the following equations

$$c_{a,y}^{(i_1,l_1)}(m) = \sum_{l_2=1}^{K}\sum_{m_1=L_1}^{L_2} w_{i_1,l_2}(m_1) c_{y,y}^{(l_2,l_1)}(m-m_1) \tag{6.53}$$

for $i_1 = 1, 2, ..., U$ $l_1 = 1, 2, ..., K$ $m = L_1, L_1+1, ..., L_2$, where

$$\begin{aligned}&c_{a,y}^{(i_1,i_2)}(m)\\ &= \sum_{l=1}^{K}\sum_{k=D_1^{(m)}}^{D_2^{(m)}} cum[a_{i_1}(n), y_{i_2}^*(n-m), y_l(n-k), y_l^*(n-k)],\end{aligned}$$

$$\begin{aligned}&c_{y,y}^{(i_1,i_2)}(m)\\ &= \sum_{l=1}^{K}\sum_{k=B_1^{(m)}}^{B_2^{(m)}} cum[y_{i_1}(n), y_{i_2}^*(n-m), y_l(n-k), y_l^*(n-k)].\end{aligned}$$

where $B_1^{(m)}$, $B_2^{(m)}$, $D_1^{(m)}$, $D_2^{(m)}$ are properly defined by the region of support of the cumulants to be estimated (see Appendix 6A).

Now the equations in (6.53) can be collected in matrix form as

$$\mathbf{C}_{y,y}\tilde{\mathbf{w}}_i = \mathbf{c}_{a,y}^{(i)} \tag{6.54}$$

where $\mathbf{C}_{y,y}$ is a $KL \times KL$ matrix given by

$$\mathbf{C}_{y,y} = \begin{bmatrix} \mathbf{C}_{y,y}^{1,1} & \mathbf{C}_{y,y}^{1,2} & \cdots & \mathbf{C}_{y,y}^{1,K} \\ \mathbf{C}_{y,y}^{2,1} & \mathbf{C}_{y,y}^{2,2} & \cdots & \mathbf{C}_{y,y}^{2,K} \\ \cdots & \cdots & \cdots & \cdots \\ \mathbf{C}_{y,y}^{K,1} & \mathbf{C}_{y,y}^{K,2} & \cdots & \mathbf{C}_{y,y}^{K,K} \end{bmatrix}$$

$$\mathbf{C}_{y,y}^{i,l} = \begin{bmatrix} c_{y,y}^{(i,l)}(0) & \cdots & c_{y,y}^{(i,l)}(L_1 - L_2) \\ c_{y,y}^{(i,l)}(1) & \cdots & c_{y,y}^{(i,l)}(L_1 - L_2 + 1) \\ \cdots & \cdots & \cdots \\ c_{y,y}^{(i,l)}(-L_1 + L_2) & \cdots & c_{y,y}^{(i,l)}(0) \end{bmatrix}$$

and $\mathbf{c}_{a,y}^{(i)}$ is a $KL \times 1$ vector given by

$$\mathbf{c}_{a,y}^{(i)} = \left[\mathbf{c}_{a,y}^{(i,1)T} \mathbf{c}_{a,y}^{(i,2)T} \ldots \mathbf{c}_{a,y}^{(i,K)T}\right]^T$$

$$\mathbf{c}_{a,y}^{(i,l)} = \left[c_{a,y}^{(i,l)}(L_1), c_{a,y}^{(i,l)}(L_1 + 1), \ldots, c_{a,y}^{(i,l)}(L_2)\right]^T.$$

The matrix $\mathbf{C}_{y,y}$ is nonsingular due to **AS1-AS2-AS3** and the fact that $\mathcal{K} \neq 0$. Some remarks regarding this statement are reported in Appendix 6A.

6.3.2 Adaptive Implementation

We can consider a recursive solution of the linear system (6.54) if we adopt the cumulant estimation procedure reported in Appendix 6A. Define as $\mathbf{C}_{y,y}(n)$, $\tilde{\mathbf{w}}_i(n)$, $\mathbf{c}_{a,y}^{(i)}(n)$ the estimation of $\mathbf{C}_{y,y}$, $\tilde{\mathbf{w}}_i$, $\mathbf{c}_{a,y}^{(i)}$, respectively, at time instant n. Since the estimation of fourth-order cumulants by sample-statistics generally gives higher error variance than traditional correlations, the least squares solution of the linear system (6.54) may result in an ill-conditioned problem. This motivated the choice of a particularly powerful tool for the solution of ill-conditioned problems such as total least squares (TLS) [30].

The basic idea in TLS is based on the assumption that *both* the matrix $\mathbf{C}_{y,y}(n)$ and the vector $\mathbf{c}_{a,y}^{(i)}(n)$ are subject to errors (perturbations), say $\mathbf{C}_{y,y}(n)^{\epsilon}$ and $\mathbf{c}_{a,y}^{(i)}(n)^{\epsilon}$. The computational problem to solve becomes[14]

$$\min_{\tilde{\mathbf{w}}_i(n)} \left\| \left[\begin{array}{cc} \mathbf{C}_{y,y}(n) & \mathbf{c}_{a,y}^{(i)}(n) \end{array} \right] - \left[\begin{array}{cc} \mathbf{C}_{y,y}(n)^{\epsilon} & \mathbf{c}_{a,y}^{(i)}(n)^{\epsilon} \end{array} \right] \right\|_F^2$$

$$\text{subject to } \mathbf{C}_{y,y}(n)^{\epsilon}\, \tilde{\mathbf{w}}_i(n) = \mathbf{c}_{a,y}^{(i)}(n)^{\epsilon}.$$

One way of solving (6.55) is by singular value decomposition (SVD) [31]:

- Compute the SVD

$$\left[\begin{array}{cc} \mathbf{C}_{y,y}(n) & \mathbf{c}_{a,y}^{(i)}(n) \end{array} \right] = \mathbf{U}\mathbf{\Sigma}\mathbf{V}^H = \sum_{l=1}^{KL+1} \mathbf{u}_l \sigma_l \mathbf{v}_l^H$$

  where $\sigma_1 \geq \sigma_2 \geq \ldots \geq \sigma_{KL+1}$.
- Determine the largest integer p such that
  1. $\sigma_p > \sigma_{p+1}$;
  2. $[\mathbf{V}]_{KL+1,l} \neq 0$ for $l = p+1, p+2, \ldots, KL+1$.
- Partition $\mathbf{V}$ as $\mathbf{V} = \begin{bmatrix} \mathbf{V}_{11} & \mathbf{V}_{12} \\ \mathbf{V}_{21} & \mathbf{V}_{22} \end{bmatrix}$ where $\mathbf{V}_{11}$ is $KL \times p$, $\mathbf{V}_{12}$ is $KL \times KL - p + 1$, $\mathbf{V}_{21}$ is $1 \times p$, $\mathbf{V}_{22}$ is $1 \times KL - p + 1$.
- The minimum-norm TLS solution is

$$\tilde{\mathbf{w}}_i(n) = -\mathbf{V}_{12}\mathbf{V}_{22}^{\dagger} = -\mathbf{V}_{12}\mathbf{V}_{22}^H(\mathbf{V}_{22}\mathbf{V}_{22}^H)^{-1}. \qquad (6.55)$$

This algorithm includes the extensions from [30].[15] The required

---

14. We use the notation $||\mathbf{v}|| = \sqrt{\sum_{i=1}^{M} |v_i|^2}$ for the 2-norm of the complex M-vector $\mathbf{v} = [v_1, \ldots, v_M]^T$, and $||\mathbf{M}||_F = \sqrt{\sum_{i=1}^{M}\sum_{j=1}^{N} |m_{i,j}|^2}$ for the Frobenius norm of the $M \times N$ complex matrix $\mathbf{M}$ whose generic i, j element is $m_{i,j}$.
15. The conditions in step 2 ensure that the algorithm computes the unique minimum-norm solution and that such a solution exists. Observe also that

$$||\tilde{\mathbf{w}}_i(n)|| = \sqrt{||\mathbf{V}_{22}||^{-2} - 1}$$

and

$$\left\| \left[\begin{array}{cc} \mathbf{C}_{y,y}(n) & \mathbf{c}_{a,y}^{(i)}(n) \end{array} \right] - \left[\begin{array}{cc} \mathbf{C}_{y,y}(n)^{\epsilon} & \mathbf{c}_{a,y}^{(i)}(n)^{\epsilon} \end{array} \right] \right\|_F = \sqrt{\sum_{l=p+1}^{KL+1} \sigma_l^2}.$$

Whenever $p = KL$ the full TLS solution is obtained. If $p < KL$, the so-called truncated TLS (T-TLS) solution is obtained.

SVD in the TLS solution makes the solution prohibitive in terms of computational complexity in real-time applications.

However, SVD can be approximated using a QR factorization followed by a Jacobi-type diagonalization procedure based on an SVD-update scheme inspired to [32]. Denote $[\mathbf{C}_{y,y}(n)]_l$ as the lth row of $\mathbf{C}_{y,y}(n)$. Similarly, $\left[\mathbf{c}_{a,y}^{(i)}(n)\right]_l$ is the lth element of $\mathbf{c}_{a,y}^{(i)}(n)$.

The *approximated* SVD (which is actually a URV decomposition) at step n is $\mathbf{U}(n)\mathbf{R}(n)\mathbf{V}^H(n)$, where $\mathbf{R}(n)$ is an upper triangular and *almost* diagonal matrix. After appending a set of new data rows to this factorization, the problem is to obtain the new decomposition $\mathbf{U}(n+1)\mathbf{R}(n+1)\mathbf{V}^H(n+1)$. This can be performed by means of the following steps:

- $\mathbf{R}(n+1) \Leftarrow \lambda\mathbf{R}(n)$
- $\mathbf{V}(n+1) \Leftarrow \mathbf{V}(n)$
- *for* $l = 1, 2, ..., KL$ *do*

— *1.* Incorporate the new cumulants estimates:

$$\left[\tilde{\mathbf{r}}^T \tilde{d}\right] = \left[[\mathbf{C}_{y,y}(n+1)]_l \left[\mathbf{c}_{a,y}^{(i)}(n+1)\right]_l\right] \mathbf{V}(n+1),$$

— *2.* Update the QR factors:

$$\begin{bmatrix} \hat{\mathbf{R}}(n+1) \\ [\mathbf{0}^T\ 0] \end{bmatrix} \Leftarrow \mathbf{Q}(n+1)^H \begin{bmatrix} \mathbf{R}(n+1) \\ \left[\tilde{\mathbf{r}}^T\ \tilde{d}\right] \end{bmatrix}$$

— *3.* Diagonalize by $\mathcal{N}$-step iterative plane rotations: *for* $\nu = 1, ..., \mathcal{N}$ *do*

$$\begin{aligned} \mathbf{R}(n+1) &\Leftarrow \mathbf{P}_{1,2}\mathbf{\Theta}_{1,n,l}^H\hat{\mathbf{R}}(n+1)\mathbf{\Phi}_{1,n,l}\mathbf{P}_{1,2} \\ \mathbf{V}(n+1) &\Leftarrow \mathbf{V}(n)\mathbf{\Phi}_{1,n,l}\mathbf{P}_{1,2} \end{aligned}$$

for $\kappa = 2, ..., KL$ *do*

$$\begin{aligned} \mathbf{R}(n+1) &\Leftarrow \mathbf{P}_{\kappa,\kappa+1}\mathbf{\Theta}_{\kappa,n,l}^H\mathbf{R}(n+1)\mathbf{\Phi}_{\kappa,n,l}\mathbf{P}_{\kappa,\kappa+1} \\ \mathbf{V}(n+1) &\Leftarrow \mathbf{V}(n+1)\mathbf{\Phi}_{\kappa,n,l}\mathbf{P}_{\kappa,\kappa+1} \end{aligned}$$

end

end

- *end*.

Initialization is $\mathbf{V}(0) \Leftarrow \mathbf{I}_{KL}$, and $\mathbf{R}(0) \Leftarrow \mathbf{0}_{KL}$, where $\mathbf{I}_{KL}$ is a $KL \times KL$ identity matrix and $\mathbf{0}_{KL}$ is a $KL \times KL$ matrix whose entries are all zero. At this point partitioning $\mathbf{V}(n+1)$ as $\mathbf{V}(n+1) = \begin{bmatrix} \mathbf{V}_{11}(n+1) & \mathbf{V}_{12}(n+1) \\ \mathbf{V}_{21}(n+1) & \mathbf{V}_{22}(n+1) \end{bmatrix}$, it is possible to obtain the minimum-norm TLS solution for $\tilde{\mathbf{w}}_i(n+1)$ using (6.55). Observe that the **U**-factor of the factorization is not propagated. In Step 2 the QR factorization update [31] is performed by means of a sweep of the new input data. The plane rotations $\mathbf{\Phi}_{\kappa,n,l}$ and $\mathbf{\Theta}_{\kappa,n,l}$ of Step 3 are unitary transformations chosen as to annihilate the $(\kappa, \kappa+1)$ element of $\mathbf{R}(n+1)$ [32, 33], while leaving it upper triangular.[16] The matrices $\mathbf{P}_{\kappa,\kappa+1}$ are permutation matrices in the $(\kappa, \kappa+1)$ plane

$$\mathbf{P}_{\kappa,\kappa+1} = \begin{bmatrix} \mathbf{I}_{\kappa-1} & & & \\ & 0 & 1 & \\ & 1 & 0 & \\ & & & \mathbf{I}_{KL-\kappa} \end{bmatrix}.$$

Stewart [33] was the first one to observe that two-sided orthogonal transformations of this type iteratively applied to a matrix will reduce the R-factor to almost diagonal and the diagonal elements will approximate the singular values [32]. The number of iterations one has to perform to obtain the desired degree of accuracy is indicated as $\mathcal{N}$. In applications where computational efficiency is of concern $\mathcal{N}$ must be as low as 1. In an extremely fast-fading environment where tracking capability of the algorithm is top-priority the value of $\mathcal{N}$ should be in the range 3-5. In any case the higher the value of $\mathcal{N}$, the closer the SVD is approximated at every step. It is important to

16. The plane rotations $\mathbf{\Phi}_{\kappa,n,l}$ and $\mathbf{\Theta}_{\kappa,n,l}$ of Step 3 have the following structure

$$\mathbf{\Theta}_{\kappa,n,l} = \begin{bmatrix} \mathbf{I}_{\kappa-1} & & & \\ & \cos\theta_{\kappa,n,l} & r_{\kappa,n,l}\sin\theta_{\kappa,n,l} & \\ & -r^*_{\kappa,n,l}\sin\theta_{\kappa,n,l} & \cos\theta_{\kappa,n,l} & \\ & & & \mathbf{I}_{KL-\kappa} \end{bmatrix},$$

$$\mathbf{\Phi}_{\kappa,n,l} = \begin{bmatrix} \mathbf{I}_{\kappa-1} & & & \\ & \cos\psi_{\kappa,n,l} & s_{\kappa,n,l}\sin\psi_{\kappa,n,l} & \\ & -s^*_{\kappa,n,l}\sin\psi_{\kappa,n,l} & \cos\psi_{\kappa,n,l} & \\ & & & \mathbf{I}_{KL-\kappa} \end{bmatrix}$$

where $\mathbf{I}_n$ is a $n \times n$ identity matrix and $r_{\kappa,n,l}$, $s_{\kappa,n,l}$, $\theta_{\kappa,n,l}$, $\psi_{\kappa,n,l}$ are obtained from $\mathbf{R}(n+1)$ (see [9, 31, 33] for more details).

observe that several successful refinements of this elegant algorithm were proposed by the authors of [32]. To detail these refinements is out of the scope of this work. One of the problems revealed in [32] is the fact that the matrix $\mathbf{V}(n)$ as computed by accumulation of Jacobi rotation matrices gradually loses orthogonality. Our application of the method, however, never justified (in terms of performance improvement) the increased computational complexity subsequent to the orthogonalization procedure proposed in [32]. One of the reasons is the bursty nature of the TDMA system: The receiver never processes more than 500–1,000 samples (depending on the sampling rate) and it appears that the deviation from orthogonality becomes consistent for larger sample sizes. However, the algorithm is very effective in tracking slow variations of the data matrices and, more importantly, is much more stable than a full-RLS solution. The value of the scheme is in the fact that the complexity per time step is an $\mathcal{O}((KL+1)^2)$ if $\mathcal{N} = 1$. Moreover, including orthogonalization of the V-matrix produces an algorithm with high accuracy (similar to the cyclic Jacobi algorithm for SVD [9]).

6.3.3 Results of Experiments

The simulations are performed using a word length size of 16 bits fixed-point arithmetic. The approximate number of instructions per second using filters length equal to $L = 7$ and $K = 2$, including open-loop synchronization, frequency offset compensation, and convolutional decoding of the bit stream is expressed in MIPS per time-slot (full-rate channel)

$$\text{QR-RLS} \Rightarrow 32\ \text{MIPS}$$
$$\text{HOS-TLS}\ (\mathcal{N} = 1) \Rightarrow 48\ \text{MIPS}.$$

Square roots are performed using series approximation; divisions use the Analog Devices engine. It is evident that the HOS algorithm here proposed has considerably higher computational complexity than the traditional QR-RLS scheme, but it is important to emphasize that the significant improvement in performance may justify the choice. In the simulations we assumed a sensor spacing equal $\frac{\lambda}{2}$. The antenna has five elements and there are three transmitting mobile stations. One of the transmitters is at array broadside and is the signal of interest. We assume a two-ray model and each path impulse response is modeled as a two-path Rayleigh fading channel ($P_1^{(1)} = P_2^{(1)} = 2$) [18]. The MSE is defined as the average of the squared error obtained over $\tilde{M} = 100$ Montecarlo runs and is given by $\text{MSE}_1 = \frac{1}{\tilde{M}} \sum_{\nu=1}^{\tilde{M}} |\epsilon_i^{(\nu)}(n)|^2$. The error obtained at the mth run is $\epsilon_i^{(\nu)}(n) = z_i^{(\nu)}(n) - a_i(n - n_i)$ where n_i is the delay introduced by the filters and $z_i^{(\nu)}(n)$ is the output of the combined filters obtained at the νth run relative to the ith transmitter. Observe that the proposed algorithm uses the known synchronization sequence while in *training mode* and past decisions while in decision-directed mode. The slots are 162 symbols long and 14 symbols at the beginning of each slot are known at the receiver. [17] The SNR for each discrete-time channel impulse response is defined as in [26]. In Figure 6.9 the traces of the filtering stage relative to the first user are shown using particular test-slots for $\text{SNR}_{bit} = 21$ dB. The parameters are $K = 5, L = 7, r = 15, U = 3$, $\theta_1 = 0°$, $\theta_2 = 10°$, $\theta_3 = -45°$, $\tau_{1,1}^{(1)} = \tau_{2,1}^{(1)} = \tau_{1,1}^{(2)} = \tau_{2,1}^{(2)} = \tau_{1,1}^{(3)} = \tau_{2,1}^{(3)} = 0$, $\tau_{1,2}^{(1)} = 41.2$ μsec, $\tau_{2,2}^{(1)} = 0.0$ μsec, $\tau_{1,2}^{(2)} = 41.2$ μsec, $\tau_{2,2}^{(2)} = 0.0$ μsec, $\tau_{1,2}^{(3)} = 41.2$ μsec, $\tau_{2,2}^{(3)} = 0.0$ μsec. Velocity for the three mobiles is $V_1 = 100$ km/hr, $V_2 = V_3 = 50$ km/hr. Observe that the filter after the first 14 symbols is running in decision-directed mode. The QR-RLS algorithm on the left is compared with the proposed approach. The plots represent the traces of the four largest magnitudes among the taps of the three five-input, one-input filters of length $L = 7$. Figure 6.10 shows an experiment where the synchronization sequences are assumed to be infinitely long (that is, the receiver has perfect knowledge of the transmitted information) with $K = 5, L = 3, r = 15, U = 3$, $\theta_1 = 0°$, $\theta_2 = 10°$, $\theta_3 = -45°$, $\tau_{1,1}^{(1)} = \tau_{2,1}^{(1)} = \tau_{1,1}^{(2)} = \tau_{2,1}^{(2)} = \tau_{1,1}^{(3)} = \tau_{2,1}^{(3)} = 0$, $\tau_{1,2}^{(1)} = 41.2$ μsec, $\tau_{2,2}^{(1)} = 0.0$ μsec, $\tau_{1,2}^{(2)} = 41.2$ μsec, $\tau_{2,2}^{(2)} = 0.0$ μsec, $\tau_{1,2}^{(3)} = 41.2$ μsec, $\tau_{2,2}^{(3)} = 0.0$ μsec. Velocity for the three mobiles is $V_1 = 100$ km/hr, $V_2 = V_3 = 20$ km/hr. The plots show the trajectory of the magnitude of the weights as compared to the optimum Wiener filter (constrained to have finite length) computed assuming perfect instantaneous knowledge of $\tilde{\mathcal{H}}(z)$. The forgetting factor are $\lambda = 0.97, 0.9, 0.85$, respectively, in the plots and $\text{SNR}_{bit} = 30$ dB.

---

17. This corresponds to a frame format similar to the data traffic channel (DTC) of [18].

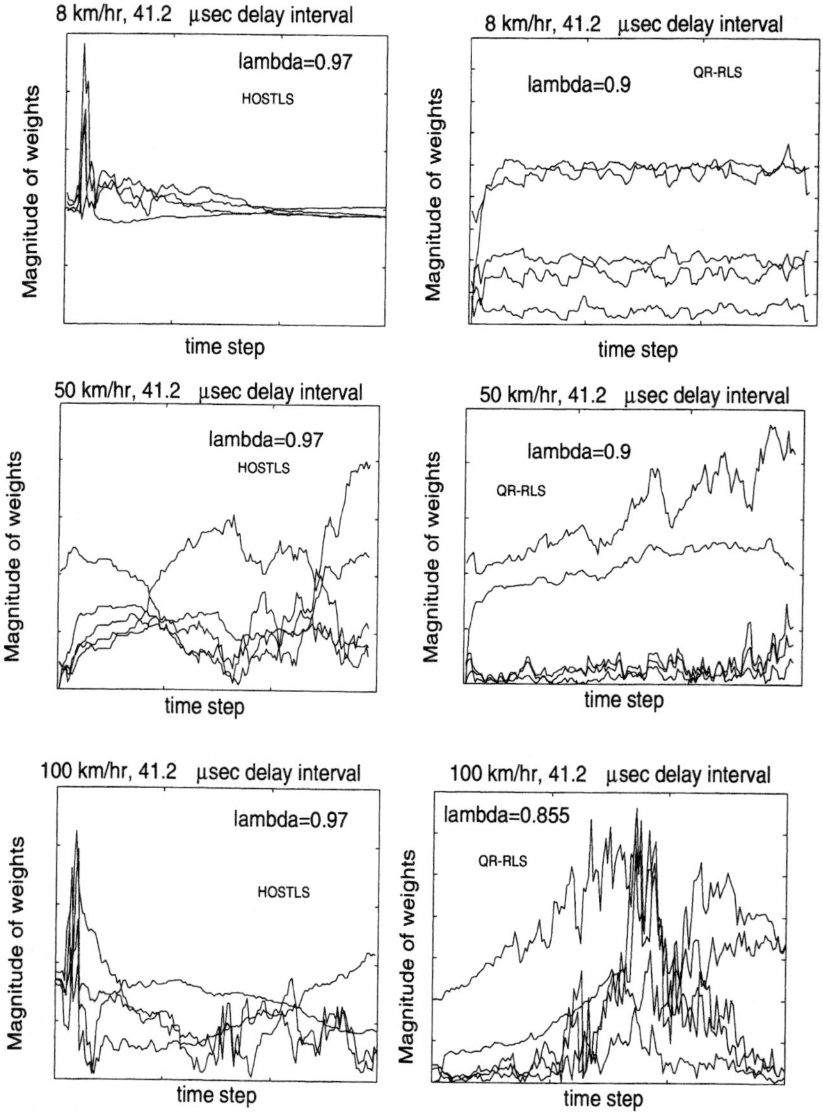

Figure 6.9 Trajectories of the magnitudes of the filter weights corresponding to the first mobile for different speeds.

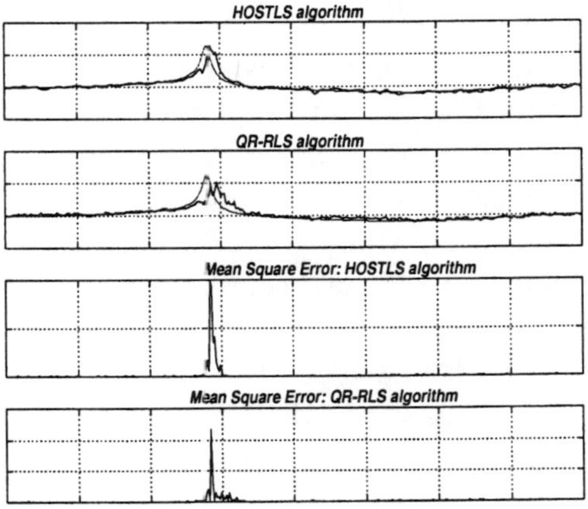

Figure 6.10 An experiment that shows the tracking capability of the algorithm. The dashed trace is the optimum Wiener filter computed assuming perfect knowledge of the discrete-time of the channel impulse response.

6.4 Key Points of This Chapter

We have studied two solutions to the array processing problem based on higher than second order statistics. The first method is based on a vector generalization [6] of the idea presented in [7] and the low-rank adaptive processing concept of [22]. The algorithm is sufficiently fast to track channel variations caused by moving transmitters, while at the same time being highly attractive from the computational point of view, proving that the use of HOS does not necessarily imply slow convergence and hence extremely large sample size. Fast start-up convergence is achieved by the use of training sequences. The experimental investigation is of particular interest because the applicability of the new method was verified in a realistic environment specified by the current U.S. standard for digital cellular communications [18]. BER results were shown using also data collected from hardware.

The algorithm LORAF3-HOS is perfectly adequate to track channel variations due to Doppler shifts larger than 80 Hz more efficiently than traditional decision-directed based methods especially in multiple cochannel interference and fast fading scenarios when using relatively large number of sensors. In these cases tracking the optimum MMSE solution using traditional methods such as RLS or its variations is not only computationally intensive but also not satisfactory in terms of performance, due to the rank degeneracy problem of the space-time correlation matrix. The best known method for RLS filtering in terms of numerical properties, the QR-based RLS (QR-RLS), was compared to the proposed algorithm, LORAF3-HOS. The remarkable result is that at BER $= 10^{-2}$ the improvement of LORAF3-HOS in SNR per bit with respect to the traditional decision-directed QR-RLS is about 6 dB using the same length of the filters and same number of sensors. Moreover, only 80% of the computational complexity per update of the full-rank QR-RLS is required to obtain such improvement. The computational complexity reduction that one can obtain with the low-rank tracking idea is very attractive; in fact, it is possible to obtain satisfactory BER results (compliant with [18]) using less than 60% of the QR-RLS computational complexity.

The second method uses HOS but it also exploits the knowledge of the training/synchronization sequences. The MSE of traditional second-order statistics algorithms has been shown to be proportional to a cost function involving cumulants of the restoration error. From this basic observation we derived a set of linear equations relating the separating filters to input-output fourth-order cumulants. The well-known high-variance exhibited by short data record estimates of cumulants is mitigated by the use in the adaptive implementation of a recursive total least squares approach.

References

[1] O. Shalvi and E. Weinstein, "Universal Methods for Blind Deconvolution," in *Blind Deconvolution*, S. Haykin, Ed., Englewood Cliffs, NJ: Prentice Hall 1994.

[2] Porat Boaz, *Digital Processing of Random Signals,* Englewood Cliffs, NJ: Prentice-Hall, 1994.

[3] J. M. Mendel, "Tutorial on Higher Order Statistics (Spectra) in Signal Processing and System Theory: Theoretical Results and Some Applications," *Proc. IEEE*, Vol. 79, pp. 278–305, March 1991.

[4] D. Hatzinakos and C. L. Nikias, "Blind Equalization Using a Tricepstrum Based Algorithm," *IEEE Trans. Commun.*, Vol. 39, pp. 669–682, Aug. 1991.

[5] T. Ohgane, et al., "An Implementation of a CMA Adaptive Array for High Speed GMSK Transmission in Mobile Communications," *IEEE Trans. Vehic. Tech.*, Vol. 42, pp. 282–288, Aug. 1993.

[6] M. Martone, "Non-Gaussian Multivariate Adaptive AR Estimation Using the Super Exponential Algorithm," *IEEE Trans. Sig. Proc.*, Vol. 44, pp. 2640–2644, Oct. 1996.

[7] O. Shalvi and E. Weinstein, "Super Exponential Methods for Blind Deconvolution," *IEEE Trans. Information Theory*, Vol. 39, pp. 504–519, March 1993.

[8] L. L. Scharf, *Statistical Signal Processing*, Reading, MA: Addison Wesley, 1991.

[9] S. Haykin, *Adaptive Filter Theory*, Englewood Cliffs, NJ: Prentice Hall, 1986.

[10] J. M. Cioffi and T. Kailath, "Fast Recursive Least Squares Transversal Filters for Adaptive Filtering," *IEEE Trans. on Acou. Sp. Sig. Proc.*, Vol. 32, pp. 302–337, April 1984.

[11] C. Feng and C. Chi, "Design of Wiener Filters Using a Cumulant-Based MSE Criterion," *Signal Processing*, Elsevier Vol. 54, pp. 23–48, 1996.

[12] A. N. Delopoulos and G. B. Giannakis, "Strongly Consistent Identification Algorithms and Noise Insensitive MSE Criteria," *IEEE Trans. Sig. Proc.*, Vol. 30, pp. 1955–1970, Aug. 1992.

[13] T. Kailath, *Linear Systems*, Englewood Cliffs, NJ: Prentice Hall, 1980.

[14] R.R. Bitmead, et al., "Greatest Common Divisors Via Generalized Sylvester and Bezout Matrices," *IEEE Trans. Aut. Contr.*, Vol. 23, pp. 1043–1047, Dec. 1978.

[15] A. Swami, G. Giannakis, and S. Shamshunder, "Multichannel ARMA processes," *IEEE Trans. Sig. Proc.*, Vol. 42, pp. 898–913, Apr. 1994.

[16] Y. Li and K. J. Ray Liu, *Blind MIMO FIR Channel Identification Based on Second-Order Statistics with Multiple Signal Recovery,* ISR TR 96–7 Technical Report, University of Maryland, College Park.

[17] Y. Li and K. J. Ray Liu, "On Blind Equalization of MIMO Channels," *ICC 96, Proceedings*, pp. 1000–1004.

[18] TIA/EIA/IS–136.1-A, "TDMA Cellular/PCS–Radio Interface–Mobile Station–Base Station Compatibility - Digital Control Channel," Oct. 1996.

[19] M. Martone, "An Adaptive Algorithm for Adaptive Antenna Array Low-Rank Processing in Cellular TDMA Base Stations," *IEEE Trans. Commun.*, Vol. 46, pp. 627–643, May 1998.

[20] Z. Ding, "On Convergence Analysis of Fractionally Spaced Adaptive Blind Equalizers," *IEEE Trans. Sig. Proc.*, Vol. 45, pp. 650–657, March 1997.

[21] J. G. Proakis, *Digital Communications*, New York: McGraw-Hill, 1989.

[22] P. Strobach, "Low-Rank Adaptive Filters," *IEEE Trans. Sig. Proc.*, Vol. 44, pp. 2932–2947, Dec. 1996.

[23] G. W. Stewart, "Methods of Simultaneous Iteration for Calculating Eigenvectors of Matrices," in *Topics in Numerical Analysis II*, pp. 169–185, J. H. Miller, Ed., New York: Academic, 1975.

[24] N. L. Owsley, "Adaptive Data Orthogonalization," in *Proc. IEEE ICASSP*, pp. 109–112, 1978.

[25] P. A. Bello, "Characterization of Randomly Time Variant Linear Channels," *IEEE Trans. Comm. Syst. Tech.*, Vol. 11, pp. 360–393, Dec. 1963.

[26] P. Balaban and J. Salz, "Optimum Diversity Combining and Equalization in Digital Data Transmission with Applications to Cellular Mobile Radio: Part I and II," *IEEE Trans. Comm.*, Vol. 40, pp. 805–907, May 1992.

[27] M. Martone, "On-Board Regeneration of Uplink Signals Using a Blind Adaptive Multichannel Estimator," *IEEE Trans. Aerosp. Electronic Sys.*, Vol. 34, pp. 49–62, Jan. 1998.

[28] M. Martone, "Cumulant-Based Adaptive Multichannel Filtering for Wireless Communication Systems with Multipath RF Propagation Using Antenna Arrays," *IEEE Trans. Vehicular Technology*, Vol. 47, pp. 377–391, May 1998.

[29] Y. Inouye and K. Hirano, "Cumulant-Based Blind Identification of Linear Multi-Input Multi-Output Systems Driven by Colored Inputs," *IEEE Trans. Sig. Proc.*, Vol. 45, pp. 1543–1552, June 1997.

[30] S. Van Huffel and J. Vandewalle, *The Total Least Squares Problem: Computational Aspects and Analysis,* Philadelphia: SIAM, 1991.

[31] G. H. Golub and C. F. Van Loan, *Matrix Computation,* North Oxford Academic Publishing Co., Johns Hopkins Press, 1988.

[32] M. Moonen, P. Van Dooren, and J. Vanderwalle, "An SVD Updating Algorithm for Subspace Tracking," *SIAM J. Matrix Anal. Appl.*, Vol. 13, No. 4, 1992.

[33] G. W. Stewart, "A Jacobi-Like Algorithm for Computing the Schur Decomposition of a Nonhermitian Matrix," *SIAM J. Sci. Stat. Comp.,* Vol. 6, No. 4, pp. 853-863, 1985.

Appendix 6A: Essential Higher-Order Statistics

6A.1 Cumulants of Stationary Processes

If $\mathbf{x} = [x_1, x_2, ..., x_K]^T$ is a collection of random variables, and $\mathbf{v} = [v_1, v_2, ..., v_K]^T$ is a collection of deterministic variables, then the Kth order cumulant is defined as the Kth coefficient in the MacLaurin series expansion of the cumulant generating function

$$\mathcal{K}_x(\mathbf{v}) = lnE\left\{e^{j\,\mathbf{v}^T\mathbf{x}}\right\}. \tag{6A.1}$$

Alternatively [3] one can define Kth order cumulants as combinations of joint moments of orders up to K, particularly for zero mean random variables and $K = 2, 3, 4$

$$\begin{aligned}
cum\left[x_1^{(*)}, x_2^{(*)}\right] &= E\{x_1^{(*)}x_2^{(*)}\},\\
cum\left[x_1^{(*)}, x_2^{(*)}, x_3^{(*)}\right] &= E\{x_1^{(*)}x_2^{(*)}x_3^{(*)}\},\\
cum\left[x_1^{(*)}, x_2^{(*)}, x_3^{(*)}, x_4^{(*)}\right] &= E\{x_1^{(*)}x_2^{(*)}x_3^{(*)}x_4^{(*)}\}\\
&\quad -E\{x_1^{(*)}x_2^{(*)}\}E\{x_3^{(*)}x_4^{(*)}\}\\
&\quad -E\{x_1^{(*)}x_3^{(*)}\}E\{x_2^{(*)}x_4^{(*)}\}\\
&\quad -E\{x_1^{(*)}x_4^{(*)}\}E\{x_2^{(*)}x_3^{(*)}\},
\end{aligned}$$

where $x_l^{(*)}$ is an eventual complex conjugation for x_l. In this work the random variables are samples collected at time nT_s from the the sensor outputs of an array. These collections of samples are modeled as stationary random processes. The fundamental properties of cumulants of random zero mean stationary processes are listed as follows:

- *LIN:* $cum\left[\sum_i f_i\, x_i(n), ...\right] = \sum_i f_i\, cum\left[\sum_i x_i(n), ...\right]$.
- *STATIND:* If the samples of a process can be divided into two (or more) statistically independent subsets, then their joint cumulants are zero.
- *GAUSS:* If the samples of a process are jointly Gaussian, then their joint K-cumulant is zero for K greater than two.

When the samples are well separated in time, and if the cumulants are absolutely summable, then the theoretical cumulants are consistently estimated from a data record of N samples and ensemble averages can be approximated by empirical averages, simply exploiting the cumulant to moment equations.

6A.2 On the Convergence of the MIMO Super-Exponential Algorithm

The iterative procedure

$$s_{i,j}(k)^{[1]} = (s_{i,j}(k))^p (s_{i,j}(k)^*)^q \tag{6A.2}$$

$$s_{i,j}(k)^{[2]} = s_{i,j}(k)^{[1]} \frac{1}{\sqrt{\sum_{j=1}^{K} \sum_l \left|s_{i,j}(l)^{[1]}\right|^2}} \tag{6A.3}$$

[where $(.)^{[1]}$, $(.)^{[2]}$ stand for the result of the first and second step, respectively], applied to the vector $\tilde{\mathbf{s}}_i$ maintains the index of the tap with largest magnitude which was called in [7] the *leading tap* (see, for details, [7]). From $\tilde{\mathbf{s}}_i^0$ with elements $s_{i,j}^0(k)$, we want to achieve convergence after l iterations to $\tilde{\mathbf{s}}_i^l$ with elements $s_{i,j}^l(k)$ that are nonzero only for a specific $j = j_0$ and $k = k_0$. While there is no specific requirement on k_0 (which will result in an arbitrary delay), it is evident that it must hold $j_0 = i$ to restore the information of the ith transmitter and not an arbitrary j_0th transmitter. Although this problem in our algorithm has been solved by use of training sequences, the convergence to the ith transmitter can be insured if the leading tap of $\tilde{\mathbf{s}}_i^0$ is contained in the vector $\mathbf{s}_{i,i}^0$, that is, if $|s_{i,i}^0(k_0)| > |s_{i,j}^0(k)|$ is verified for some k_0, all $j \neq i$ and any k. This situation is always verified as long as one initializes $\tilde{\mathbf{w}}_i$ with $\mathbf{w}_{i,j} = \mathbf{0}$ for all $j \neq i$. The initialization of $\mathbf{w}_{i,i}$ is arbitrary. Another important aspect is the global convergence of the algorithm in the $\tilde{\mathbf{w}}_i$ domain to the same solution of the algorithm in the $\tilde{\mathbf{s}}_i$ domain: This cannot be generally guaranteed when $\tilde{\mathbf{w}}_i$ has finite length. The problem of the convergence of a certain function $\Psi_{\tilde{\mathbf{w}}_i}(\tilde{\mathbf{w}}_i) = \Psi(\tilde{\mathbf{H}}\tilde{\mathbf{w}}_i)$ (when $\tilde{\mathbf{s}}_i = \tilde{\mathbf{H}}\tilde{\mathbf{w}}_i$) to the same solution of the function $\Psi(\tilde{\mathbf{s}}_i)$ is a well known and investigated problem in the blind deconvolution literature [1]. This convergence can not always be guaranteed for both functions. First, observe that the procedure (6A.2)–(6A.3) is a gradient-based search to solve the maximization problem

$$\max_{\tilde{\mathbf{s}}_i} \Psi(\tilde{\mathbf{s}}_i) \tag{6A.4}$$

subject to the constraint $||\tilde{\mathbf{s}}_i||^2 = 1$. In fact, the two steps in (6A.2)–(6A.3) are equivalent to the gradient-based iteration

$$\tilde{\mathbf{s}}_i = \tilde{\mathbf{s}}_i + \mu \frac{\partial \Psi(\tilde{\mathbf{s}}_i)}{\partial \tilde{\mathbf{s}}_i} \tag{6A.5}$$

with $\Psi(\tilde{\mathbf{s}}_i) = \sum_j \sum_k s_{i,j}(k)^2 s_{i,j}(k)^{*2}$, $\frac{\partial f(\mathbf{v})}{\partial \mathbf{v}}$ indicates the gradient of the vector $\mathbf{v}$ and we choose a *very large* step size μ. However, we have translated the maximization procedure for $\Psi(\tilde{\mathbf{s}}_i)$ over $\tilde{\mathbf{s}}_i$ into a maximization for a certain $\Psi_{\tilde{\mathbf{w}}_i}(\tilde{\mathbf{w}}_i)$ over $\tilde{\mathbf{w}}_i$ where

$$\Psi_{\tilde{\mathbf{w}}_i}(\tilde{\mathbf{w}}_i) = \Psi(\tilde{\mathbf{H}}\tilde{\mathbf{w}}_i) = \Psi(\tilde{\mathbf{s}}_i)$$

and we have used $\tilde{\mathbf{s}}_i = \tilde{\mathbf{H}}\tilde{\mathbf{w}}_i$.

Let us assume that $\tilde{\mathbf{s}}_i^E$ is an extremum for $\Psi(\tilde{\mathbf{s}}_i)$, that is,

$$\left.\frac{\partial \Psi(\tilde{\mathbf{s}}_i)}{\partial \tilde{\mathbf{s}}_i}\right|_{\tilde{\mathbf{s}}_i = \tilde{\mathbf{s}}_i^E} = \mathbf{0}.$$

It is then obviously true that $\tilde{\mathbf{w}}_i^E$ such that

$$\tilde{\mathbf{s}}_i^E = \tilde{\mathbf{H}}\tilde{\mathbf{w}}_i^E$$

is also an extremum for $\Psi_{\tilde{\mathbf{w}}_i}(\tilde{\mathbf{w}}_i)$ because

$$\left.\frac{\partial \Psi_{\tilde{\mathbf{w}}_i}(\tilde{\mathbf{w}}_i)}{\partial \tilde{\mathbf{w}}_i}\right|_{\tilde{\mathbf{w}}_i = \tilde{\mathbf{w}}_i^E} = \mathbf{0}.$$

The converse may not be true. That is, if we assume that $\overline{\tilde{\mathbf{w}}}_i^E$ is an extremum for $\Psi_{\tilde{\mathbf{w}}_i}(\tilde{\mathbf{w}}_i)$, it may not be true that $\overline{\tilde{\mathbf{s}}}_i^E = \tilde{\mathbf{H}}\overline{\tilde{\mathbf{w}}}_i^E$ is an extremum for $\Psi(\tilde{\mathbf{s}}_i)$. In fact we may have $\frac{\partial\Psi(\tilde{\mathbf{H}}\tilde{\mathbf{w}}_i)}{\partial\tilde{\mathbf{w}}_i}\Big|_{\tilde{\mathbf{w}}_i=\overline{\tilde{\mathbf{w}}}_i^E} = \mathbf{0}$ if $\frac{\partial\Psi(\tilde{\mathbf{s}}_i)}{\partial\tilde{\mathbf{s}}_i}\Big|_{\overline{\tilde{\mathbf{s}}}_i^E=\tilde{\mathbf{H}}\overline{\tilde{\mathbf{w}}}_i^E} \neq \mathbf{0}$ belongs to the kernel of $\tilde{\mathbf{H}}$ which is orthogonal to the subspace spanned by $\tilde{\mathbf{H}}$, which means that $\overline{\tilde{\mathbf{w}}}_i^E$ can be far from the desired solution.

We investigate this last important issue. By the chain rule we have

$$\frac{\partial\Psi(\tilde{\mathbf{H}}\tilde{\mathbf{w}}_i)}{\partial w_{i,n}(m)} = \sum_{k=1}^{U}\sum_{l} h_{n,k}^*(m-l) \cdot \frac{\partial\Psi(\tilde{\mathbf{s}}_i)}{\partial s_{i,k}(l)}\Bigg|_{s_{i,k}(l)=\sum_{k_1=1}^{K}\sum_{m_1} w_{i,k_1}(m_1)h_{k_1,k}(l-m_1)} \tag{6A.6}$$

or in vector form

$$\frac{\partial\Psi(\tilde{\mathbf{H}}\tilde{\mathbf{w}}_i)}{\partial\tilde{\mathbf{w}}_i} = \tilde{\mathbf{H}}^H\frac{\partial\Psi(\tilde{\mathbf{s}}_i)}{\partial\tilde{\mathbf{s}}_i}\Bigg|_{\tilde{\mathbf{s}}_i=\tilde{\mathbf{H}}\tilde{\mathbf{w}}_i}. \tag{6A.7}$$

Now multiply both sides of (6A.7) by $\tilde{\mathbf{H}}\left(\tilde{\mathbf{H}}^H\tilde{\mathbf{H}}\right)^\dagger$ to obtain

$$\tilde{\mathbf{H}}\left(\tilde{\mathbf{H}}^H\tilde{\mathbf{H}}\right)^\dagger\frac{\partial\Psi(\tilde{\mathbf{H}}\tilde{\mathbf{w}}_i)}{\partial\tilde{\mathbf{w}}_i} \simeq \mathbf{I}_U\frac{\partial\Psi(\tilde{\mathbf{s}}_i)}{\partial\tilde{\mathbf{s}}_i}\Bigg|_{\tilde{\mathbf{s}}_i=\tilde{\mathbf{H}}\tilde{\mathbf{w}}_i} = \frac{\partial\Psi(\tilde{\mathbf{s}}_i)}{\partial\tilde{\mathbf{s}}_i}\Bigg|_{\tilde{\mathbf{s}}_i=\tilde{\mathbf{H}}\tilde{\mathbf{w}}_i} \tag{6A.8}$$

because $\tilde{\mathbf{H}}\left(\tilde{\mathbf{H}}^H\tilde{\mathbf{H}}\right)^\dagger\tilde{\mathbf{H}}^H \simeq \mathbf{I}_U$, the $U\times U$ identity matrix. It is then evident from (6A.8) that if $\overline{\tilde{\mathbf{w}}}_i^E$ is an extremum for $\Psi_{\tilde{\mathbf{w}}_i}(\tilde{\mathbf{w}}_i)$, that is, if $\frac{\partial\Psi(\tilde{\mathbf{H}}\tilde{\mathbf{w}}_i)}{\partial\tilde{\mathbf{w}}_i}\Big|_{\tilde{\mathbf{w}}_i=\overline{\tilde{\mathbf{w}}}_i^E} = \mathbf{0}$ then $\frac{\partial\Psi(\tilde{\mathbf{s}}_i)}{\partial\tilde{\mathbf{s}}_i}\Big|_{\overline{\tilde{\mathbf{s}}}_i^E=\tilde{\mathbf{H}}\overline{\tilde{\mathbf{w}}}_i^E} \simeq \mathbf{0}$. Since $\left\|\frac{\partial\Psi(\tilde{\mathbf{s}}_i)}{\partial\tilde{\mathbf{s}}_i}\right\|$ is continuous with respect to $w_{i,n}(m)$, and $h_{k,n}(m)$, then we only need to require that for a sufficiently small $\Delta > 0$ it is satisfied

$$\left\|\tilde{\mathbf{H}}\left(\tilde{\mathbf{H}}^H\tilde{\mathbf{H}}\right)^\dagger\tilde{\mathbf{H}}^H - \mathbf{I}_U\right\|_F < \Delta \tag{6A.9}$$

to guarantee that $\left\|\frac{\partial\Psi(\tilde{\mathbf{s}}_i)}{\partial\tilde{\mathbf{s}}_i}\right\|$ is arbitrarily small, which implies that there exists an extremum $\tilde{\mathbf{s}}_i^E$ such that $\left\|\bar{\tilde{\mathbf{s}}}_i^E - \tilde{\mathbf{s}}_i^E\right\|$ is arbitrarily small. In other words, the extremum for $\Psi_{\tilde{\mathbf{w}}_i}(\tilde{\mathbf{w}}_i)$ is arbitrarily close to the extremum for $\Psi(\tilde{\mathbf{s}}_i)$ if there exists a sufficiently small Δ such that (6A.9) is verified. Condition (6A.9) emphasizes the importance of selecting the rank-order of the low-rank filter $r = r_{\max}$ carefully. The selection of a low value for r will alleviate the computational effort of the algorithm but will make the algorithm prone to misconvergence, due to (6A.9).

6A.3 Time-Varying Cumulants

The $\mathcal{K}$th–order time-varying cumulant for the process

$$y_k(n) = \sum_{l=1}^{U}\sum_{m} h_{k,l}(m)x_l(n-m) + \eta_k(n)$$

$$k = 1, 2, ..., K \qquad (6A.10)$$

(in this book $\mathcal{K} = 2$ or $\mathcal{K} = 4$) at lag $m_0, m_1, ..., m_{\mathcal{K}-1}$ is indicated as

$$c_{y^{(*)}_{i_1,l_1},...,y^{(*)}_{i_\mathcal{K},l_\mathcal{K}}}(n; m_0, ..., m_{\mathcal{K}-1})$$
$$= cum\left[y^{(*)}_{i_1,l_1}(n+m_0), ..., y^{(*)}_{i_\mathcal{K},l_\mathcal{K}}(n+m_{\mathcal{K}-1})\right] \qquad (6A.11)$$

where $x^{(*)}$ denotes an optional complex conjugation of x. It is easily shown that a generalization of the equations shown in [19] (valid for time-invariant channels) can be obtained from the time-varying Bartlett equation (see [2]) (valid for $\mathcal{K} = 2$ only in the absence of noise)

$$c_{y^{(*)}_{i_1,l_1},...,y^{(*)}_{i_\mathcal{K},l_\mathcal{K}}}(n; m_0, ..., m_{\mathcal{K}-1})$$
$$= \gamma_{\mathcal{K}a}\sum_{k=1}^{K}\sum_{m}\prod_{p=1}^{\mathcal{K}} h^{(*)}_{i_p,k,l_p}(n, m+m_{i-1}), \qquad (6A.12)$$

for $i_1, ..., i_\mathcal{K}$ chosen in the set $[1, 2, ..., K]$, $l_1, ..., l_\mathcal{K}$ chosen in the set $[1, 2, ..., Q]$. We do not report the detailed derivation of the time-varying super-exponential equations because they can be easily derived from equations already available in [19, 27]. In fact evidently in the absence of noise

$$c_{y_{i_1,l_1},y^*_{i_2,l_2}}(n;m_0,m_1)$$
$$= \sigma_a^2 \sum_{k=1}^{K} \sum_{m} h_{i_1,k,l_1}(n,m+m_0) h^*_{i_2,k,l_2}(n,m+m_1)$$

and

$$c_{y^*_{i_1,l_1},z_i,z^*_i,z_i}(n;m_0,0,0,0)$$
$$= \gamma_{4a} \sum_{k=1}^{K} \sum_{m} h^*_{i_1,k,l_1}(n,m+m_0)|s_{i,k}(n,m)|^2 s_{i,k}(n,m)$$

are equations analogous to (20) and (21) of [19] respectively, with

$$s_{i,j_1}(n,m_2) = \sum_{\kappa=1}^{A} \sum_{j=1}^{K} \sum_{l=1}^{Q} \sum_{m} w^{\kappa}_{i,j,l}(n,m) h^{\kappa}_{j,j_1,l}(n,m_2-m)$$

for $i, j_1 = 1, 2, ..., K$.

6A.4 Proof of Theorem 2

In Chapter 5 we have used the following theorem.

Theorem 2: The optimum deconvolution filter associated with the minimum of the cost function $J^{(i)}(\tilde{\mathbf{w}}_i)$ satisfies

$$\sum_{l_1} \sum_{k} \text{cum}[\epsilon_i(n), y^*_{l_2}(n-m), y_{l_1}(n-k), y^*_{l_1}(n-k)] = 0 \tag{6A.13}$$

and

$$\sum_{l_1} \sum_{k} \text{cum}[\epsilon^*_i(n), y_{l_2}(n-m), y_{l_1}(n-k), y^*_{l_1}(n-k)] = 0 \tag{6A.14}$$

for $l_2 = 1, 2, ..., K,\ \ m = L_1, L_1+1, ..., L_2,\ \ i = 1, 2, ..., U$.

Suppose $w^+_{i,l}(n)$ are the filter weights organized into $\tilde{\mathbf{w}}^+_i$ satisfying (6A.13) and (6A.14), and minimizing $J^{(i)}(\tilde{\mathbf{w}}^+_i)$ while

$w_{i,l}(n)$ organized into $\tilde{\mathbf{w}}_i$ are the weights of an other arbitrary filter. Define

$$\epsilon_i^+(n) = a_i(n) - \sum_l w_{i,l}^+(n) * y_l(n)$$

and observe that it is possible to express $\epsilon_i(n)$ as

$$\begin{aligned} \epsilon_i(n) &= \left[a_i(n) - \tilde{\mathbf{w}}_i^{+^T}\tilde{\mathbf{y}}(n)\right] + \left(\tilde{\mathbf{w}}_i^+ - \tilde{\mathbf{w}}_i\right)^T \tilde{\mathbf{y}}(n) \\ &= \epsilon_i^+(n) + \left(\tilde{\mathbf{w}}_i^+ - \tilde{\mathbf{w}}_i\right)^T \tilde{\mathbf{y}}(n) \qquad (6A.15) \end{aligned}$$

where

$$\tilde{\mathbf{y}}(n) = \left[\tilde{\mathbf{y}}_1^T(n), ..., \tilde{\mathbf{y}}_K^T(n)\right]^T,$$

$$\tilde{\mathbf{y}}_l(n) = [y_l(n - L_1), ..., y_l(n - L_2)]^T.$$

Substituting (6A.15) into the expression for $J^{(i)}(\tilde{\mathbf{w}}_i)$ and using (6A.13) and (6A.14), it is possible to write

$$\begin{aligned} &J^{(i)}(\tilde{\mathbf{w}}_i) \\ &= \Bigg| \sum_{l_1}\sum_k \text{cum}[\epsilon_i^+(n), \epsilon_i^{+^*}(n), y_{l_1}(n-k), y_{l_1}^*(n-k)] \\ &+ \sum_{l_1}\sum_{l_2}\sum_{m_1}\sum_{m_2} \left[w_{i,l_1}^+(m_1) - w_{i,l_1}(m_1)\right] \\ &\qquad \left[w_{i,l_2}^+(m_2) - w_{i,l_2}(m_2)\right]^* \\ &\qquad \times \mathcal{K} \sum_{l_3}\sum_n g_{l_3,l_1}(n-m_1) g_{l_3,l_2}^*(n-m_2) \Bigg|^2 \\ &= \Bigg| \frac{\mathcal{K}}{\sigma_x^2} \left\langle E\left\{|\epsilon_i^+(n)|^2\right\}\right\rangle_{\text{SNR}=\infty} \\ &\qquad + \mathcal{K} \sum_l \sum_n \left|\left(\tilde{\mathbf{w}}_i^+ - \tilde{\mathbf{w}}_i\right) \mathbf{g}_l(n)\right|^2 \Bigg|^2 \\ &= \left|\frac{\mathcal{K}}{\sigma_x^2}\right|^2 \Bigg| \left\langle E\left\{|\epsilon_i^+(n)|^2\right\}\right\rangle_{\text{SNR}=\infty} \\ &\qquad + \sigma_x^2 \sum_l \sum_n \left|\left(\tilde{\mathbf{w}}_i^+ - \tilde{\mathbf{w}}_i\right) \mathbf{g}_l(n)\right|^2 \Bigg|^2 \qquad (6A.16) \end{aligned}$$

where

$$\mathbf{g}_i(n) = \left[\mathbf{g}_{i,1}^T(n), \mathbf{g}_{i,2}^T(n), ..., \mathbf{g}_{i,K}^T(n)\right]^T,$$

$$\mathbf{g}_{i,l}(n) = [g_{i,l}(n-L_1), ..., g_{i,l}(n-L_2)]^T.$$

Of course (6A.16) is minimized when $\tilde{\mathbf{w}}_i^+ = \tilde{\mathbf{w}}_i$.

6A.4.1 On the Choice of $D_1^{(m)}$, $B_1^{(m)}$, $D_2^{(m)}$, $B_2^{(m)}$

Assume that $g_{i,l}(n)$ are FIR of length $r = r_2 - r_1 + 1$ and observe that

$$\begin{aligned}
&\text{cum}[y_i(n), y_{i_1}^*(n-m), y_l(n-k), y_l^*(n-k)] \\
&= \text{cum}[y_i(n+k), y_{i_1}^*(n+k-m), y_l(n), y_l^*(n)] \\
&= \gamma_{4x} \sum_{l_1} \sum_{n} g_{i,l_1}(n+k) g_{i_1,l_1}^*(n+k-m) |g_{l,l_1}(n)|^2
\end{aligned}$$

is different from zero only for

$$\max\{-r+1, -r+1+m\} \leq k \leq \min\{r-1, r-1+m\}$$

so that $B_1^{(m)}$ and $B_2^{(m)}$ can be chosen as

$$B_1^{(m)} \leq \max\{-r+1, -r+1+m\},$$

$$B_2^{(m)} \geq \min\{r-1, r-1+m\}.$$

Remember that $b_i(n)$ (see **AS3**) is of length $M = M_2 - M_1 + 1$ and observe that

$$\begin{aligned}
&\text{cum}[a_i(n), y_{i_1}^*(n-m), y_l(n-k), y_l^*(n-k)] \\
&= \text{cum}[a_i(n+k), y_{i_1}^*(n+k-m), y_l(n), y_l^*(n)] \\
&= \gamma_{4x} \sum_{l_1} \sum_{n} b_i(n+k) g_{i_1,l_1}^*(n+k-m) |g_{l,l_1}(n)|^2
\end{aligned}$$

is different from zero only for $M_1 \leq n+k \leq M_2$, $r_1 \leq n+k-m \leq r_2$, $r_1 \leq n \leq r_2$. This means also that $M_1 - n \leq k \leq M_2 - n$, $r_1 - n + m \leq k \leq r_2 - n + m$ and $r_1 \leq n \leq r_2$,

$$\max\{M_1 - n, r_1 - n + m\} \leq k \leq \min\{M_2 - n, M_2 - n + m\},$$

and $r_1 \leq n \leq r_2$, that is,

$$\max\{M_1 - r_2, r_1 - r_2 + m\} \leq k \leq \min\{M_2 - r_1, r_2 - r_1 + m\}.$$

Hence, $D_1^{(m)}$ and $D_2^{(m)}$ can be chosen as

$$D_1^{(m)} \leq \max\{M_1 - r_2, -M + 1 + m\},$$
$$D_2^{(m)} \geq \min\{M_2 - r_1, r - 1 + m\}.$$

6A.4.2 On the Rank of the Matrix $\mathbf{C}_{y,y}$

Define the $K(M + J + L - 2) \times KL$ matrix

$$\tilde{\mathbf{G}} = \begin{bmatrix} \mathbf{G}_{1,1} & \mathbf{G}_{2,1} & \cdots & \mathbf{G}_{K,1} \\ \mathbf{G}_{1,2} & \mathbf{G}_{2,2} & \cdots & \mathbf{G}_{K,2} \\ \vdots & \vdots & \cdots & \vdots \\ \mathbf{G}_{1,U} & \mathbf{G}_{2,U} & \cdots & \mathbf{G}_{K,U} \end{bmatrix}$$

$$[\mathbf{G}_{i,k}]_{m,n} = \begin{cases} g_{i,k}(m-n) & M_1 + J_1 \leq m - n \leq M_2 + J_2 \\ 0 & \text{elsewhere} \end{cases}$$
$$M_1 + J_1 + L_1 \leq m \leq M_2 + J_2 + L_2, \;\; L_1 \leq n \leq L_2,$$

the $K(J + L - 1) \times KL$ matrix

$$\tilde{\mathbf{H}} = \begin{bmatrix} \mathbf{H}_{1,1} & \mathbf{H}_{2,1} & \cdots & \mathbf{H}_{K,1} \\ \mathbf{H}_{1,2} & \mathbf{H}_{2,2} & \cdots & \mathbf{H}_{K,2} \\ \vdots & \vdots & \cdots & \vdots \\ \mathbf{H}_{1,U} & \mathbf{H}_{2,U} & \cdots & \mathbf{H}_{K,U} \end{bmatrix}$$

$$[\mathbf{H}_{i,k}]_{m,n} = \begin{cases} h_{i,k}(m-n) & J_1 \leq m - n \leq J_2 \\ 0 & \text{elsewhere} \end{cases}$$
$$J_1 + L_1 \leq m \leq J_2 + L_2, \; L_1 \leq n \leq L_2,$$

and the $K(M + J + L - 2) \times K(J + L - 1)$ matrix

$$\tilde{\mathbf{B}} = \begin{bmatrix} \mathbf{B}_1 & \mathbf{0} & \cdots & \mathbf{0} \\ \mathbf{0} & \mathbf{B}_2 & \cdots & \mathbf{0} \\ \vdots & \vdots & \cdots & \vdots \\ \mathbf{0} & \mathbf{0} & \cdots & \mathbf{B}_U \end{bmatrix}$$

$$[\mathbf{B}_i]_{m,n} = \begin{cases} b_i(m-n) & M_1 \le m-n \le M_2 \\ 0 & \text{elsewhere} \end{cases}$$
$$M_1 + J_1 + L_1 \le m \le M_2 + J_2 + L_2,$$
$$J_1 + L_1 \le n \le J_2 + L_2.$$

Now observe that

$$\begin{aligned} c_{y,y}^{(i_1,i_2)}(m_2 - m_1) &= \sum_{l=1}^{K} \sum_{k=-\infty}^{\infty} \\ &\mathrm{cum}[y_{i_1}(n-m_1), y_{i_2}^*(n-m_2), y_l(n-k), y_l^*(n-k)] \\ &= \gamma_{4x} \sum_{l=1}^{K} \sum_{k=-\infty}^{\infty} \sum_{l_1=1}^{U} \sum_{n} g_{i_1,l_1}(n-m_1) \\ &\qquad \times g_{i_2,l_1}^*(n-m_2) |g_{l,l_1}(n-k)|^2 \\ &= \gamma_{4x} \sum_{l=1}^{K} \sum_{k=-\infty}^{\infty} |g_{l,r}(k)|^2 \\ &\qquad \sum_{l_1=1}^{U} \sum_{n} g_{i_1,l_1}(n-m_1) g_{i_2,l_1}^*(n-m_2) \\ &= \mathcal{K} \sum_{l_1=1}^{U} \left[\mathbf{G}_{i_2,l_1}^H \mathbf{G}_{i_1,l_1} \right]_{m_2,m_1}. \end{aligned} \tag{6A.17}$$

The last equality states that

$$\mathbf{C}_{y,y} = \mathcal{K} \tilde{\mathbf{G}}^H \tilde{\mathbf{G}}. \tag{6A.18}$$

The three following facts imply that $\tilde{\mathbf{G}}$ is a full (column) rank matrix and that $\tilde{\mathbf{G}}^H \tilde{\mathbf{G}}$ has rank equal to KL.

- $\tilde{\mathbf{H}}$ is full-rank [16, 17, 19, 29] due to **AS2**.
- $\tilde{\mathbf{B}}$ is full-rank due to **AS3**.
- $\tilde{\mathbf{G}} = \tilde{\mathbf{B}} \tilde{\mathbf{H}}$.

Since $\mathcal{K} \neq 0$, the full-rank property of $\tilde{\mathbf{G}}^H \tilde{\mathbf{G}}$ extends to $\mathbf{C}_{y,y}$ due to **AS1** and (6A.18).

6A.4.3 Cumulants Estimation

Adaptive estimation of cumulants can be implemented by means of the method also used in [4]. We define

$$m^{a,y}_{i_1,i_2,i_3,i_4}(m,k) = E\left\{a_{i_1}(n), y^*_{i_2}(n-m), y_{i_3}(n-k), y^*_{i_4}(n-k)\right\},$$

$$m^{y,y}_{i_1,i_2,i_3,i_4}(m,k) = E\left\{y_{i_1}(n), y^*_{i_2}(n-m), y_{i_3}(n-k), y^*_{i_4}(n-k)\right\},$$

$$m^{a,y}_{i,l}(k) = E\left\{a_i(n), y^*_l(n+k)\right\},$$

$$m^{y,y}_{i,l}(k) = E\left\{y_i(n), y^*_l(n+k)\right\},$$

and the estimates of the respective moments based on sample statistics as $\hat{m}^{a,y}_{i_1,i_2,i_3,i_4}(m,k)^{(n)}$, $\hat{m}^{y,y}_{i_1,i_2,i_3,i_4}(m,k)^{(n)}$, $\hat{m}^{a,y}_{i,l}(k)^{(n)}$, $\hat{m}^{y,y}_{i,l}(k)^{(n)}$ using n samples. Assuming we have available at iteration 0, $y_i(0), ..., y_i(I_{\text{lag}})$ for $i = 1,2,...,K$, at iteration n we can update $\hat{m}^{a,y}_{i_1,i_2,i_3,i_4}(m,k)^{(n)}$, from $\hat{m}^{a,y}_{i_1,i_2,i_3,i_4}(m,k)^{(n-1)}$, as follows

$$\begin{aligned}\hat{m}^{a,y}_{i_1,i_2,i_3,i_4}(m,k)^{(n)} &= (1-\alpha(n))\ \hat{m}^{a,y}_{i_1,i_2,i_3,i_4}(m,k)^{(n-1)} \\ +\alpha(n)\ a_{i_1}(\Gamma^{(n,m,k)})\ y^*_{i_2}(\Gamma^{(n,m,k)}-m)&\ y_{i_3}(\Gamma^{(n,m,k)}-k) \\ &\times y^*_{i_4}(\Gamma^{(n,m,k)}-k)\end{aligned}$$

and $\hat{m}^{y,y}_{i_1,i_2,i_3,i_4}(m,k)^{(n)}$, from $\hat{m}^{y,y}_{i_1,i_2,i_3,i_4}(m,k)^{(n-1)}$, as follows

$$\begin{aligned}\hat{m}^{y,y}_{i_1,i_2,i_3,i_4}(m,k)^{(n)} &= (1-\alpha(n))\ \hat{m}^{y,y}_{i_1,i_2,i_3,i_4}(m,k)^{(n-1)} \\ +\alpha(n)\ y_{i_1}(\Gamma^{(n,m,k)})\ y^*_{i_2}(\Gamma^{(n,m,k)}-m)&\ y_{i_3}(\Gamma^{(n,m,k)}-k) \\ &\times y^*_{i_4}(\Gamma^{(n,m,k)}-k)\end{aligned}$$

where

$$\alpha(n) = \frac{1}{n} + I_{\text{lag}}$$

and

$$\Gamma^{(n,m,k)} = \min\left(n + I_{\text{lag}}, n + I_{\text{lag}} + k, n + I_{\text{lag}} + m\right).$$

Similarly, for the second-order moments, we can write

$$\hat{m}_{i,l}^{a,y}(k)^{(n)} = (1-\alpha(n))\ \hat{m}_{i,l}^{a,y}(k)^{(n-1)} + \alpha(n)\ a_i(\Gamma_2^{(n,k)})\ y_l^*(\Gamma_2^{(n,k)}-k). \tag{6A.19}$$

where $\alpha(n) = \frac{1}{n} + I_{\text{lag}}$ and $\Gamma_2^{(n,k)} = \min\left(n + I_{\text{lag}}, n + I_{\text{lag}} + k\right)$.

Evidently we can obtain the cumulants estimation at point n as

$$\hat{c}_{a,y}^{i_1,i_2}(m)^{(n)} = \sum_{l=1}^{K} \sum_{k=D_1^{(m)}}^{D_2^{(m)}} \left(\hat{m}_{i_1,i_2,l,l}^{a,y}(m,k)^{(n)} - \hat{m}_{i_1,i_2}^{a,y}(m)^{(n)}\ \hat{m}_{l,l}^{y,y}(0)^{(n)} - \hat{m}_{i_1,l}^{a,y}(k)^{(n)}\ \hat{m}_{i_2,l}^{y,y}(m-k)^{(n)}\right)$$

$$\hat{c}_{y,y}^{i_1,i_2}(m)^{(n)} = \sum_{l=1}^{K} \sum_{k=B_1^{(m)}}^{B_2^{(m)}} \left(\hat{m}_{i_1,i_2,l,l}^{y,y}(m,k)^{(n)} - \hat{m}_{i_1,i_2}^{y,y}(m)^{(n)}\ \hat{m}_{l,l}^{y,y}(0)^{(n)} - \hat{m}_{i_1,l}^{y,y}(k)^{(n)}\ \hat{m}_{i_2,l}^{y,y}(m-k)^{(n)}\right)$$

and an obvious organization of these quantities into the matrices of the system

$$\mathbf{C}_{y,y}\tilde{\mathbf{w}}_i = \mathbf{c}_{a,y}^{(i)} \tag{6A.20}$$

gives the estimates $\mathbf{C}_{y,y}(n)$ and $\mathbf{c}_{a,y}(n)$.

7

Advanced MIMO Modems

7.1 Introduction

We have seen in the previous chapters that the MIMO model is the appropriate model for multiantenna systems. A first superficial analysis of the demodulation problem could suggest a straightforward generalization of well-known single-input-single-output (SISO) schemes. However, MIMO detection presents some peculiar problems. In practical long-range microwave applications the channel responses (representing propagation distortions between each antenna at the transmitter and each antenna at the receiver) are afflicted by frequency-selective distortion. The problem of demodulating a wavefield-modulated signal may become a significant challenge. A wireless multipath channel with a delay spread that exceeds a few tens of symbol periods is still today a challenging problem for communication engineers. Actually, it has been stated that delay spread is the ultimate obstacle to mass deployment of reliable high data rate modems for wireless communications. Optimal approaches to detection for known channel equalization are given by MAP or MLSE approaches. These techniques suffer from significant complexity when the channel time duration exceeds a few symbols. The time span of an equalizer can be in practice extended using suboptimum approaches like linear filters and decision feedback linear filters. However, even suboptimum techniques suffer from high complexity when the channel length L exceeds a few tens of symbols. Orthogonal frequency division multiplexing (OFDM) has been advocated by many as a possible remedy to the large time dispersion problem. Traditional multicarrier techniques of the OFDM type perform a frequency domain decomposition of a channel characterized by frequency selective distortion in a plurality of subchannels that are affected by frequency flat distortion. The distortion in each independent subchannel can then be easily compensated by simple gain and phase adjustments. The price to pay in OFDM is that the complexity (and cost) of the transceiver is shifted from the baseband segment of the digital radio to the RF front end. In fact, multicarrier signals are vulnerable to distortions induced by RF electronics. The effects of intermodulation distortions, phase noise, and transmitter power amplifier nonlinearities contribute to a significant loss of mutual information and performance degradations that cannot be recovered by channel coding. An accurate RF design is necessary. In this chapter we detail the baseband processing of MIMO systems and describe the RF specifications required to implement such processing. We finally report the results of some experimental outdoor field trials which demonstrated the hardware feasibility of fixed broadband wireless radio links with spectral efficiencies in excess of 50 bps/Hz in non-LOS environments with frequency selective multipath. We describe a hardware experimental setup that allowed data transmission in the range 60–200 Mbps occupying a bandwidth of 2–4.6 MHz at MMDS frequencies ($\simeq$ 2.5 GHz). Moreover, using calibrated antenna collections and sequential spatio-temporal channel estimates in a static environment, transmission with antenna arrays with up to 40 transmit elements and up to 60 receive elements is evaluated. This last experiment demonstrates the possibility of a 1 Gbps wireless transmission with a raw spectral efficiency of about 208 bps/Hz.

7.2 Spatio-Temporal Wavefield Modeling

We have shown in Chapter 3 that wavefields generated by linear channels can be expressed as

$$s(t, \mathbf{r}) = \int_T \int_\Gamma h(t, \tau, \mathbf{r}, \mathbf{r}') x(\tau, \mathbf{r}') d\mathbf{r}' d\tau \qquad (7.1)$$

where $h(t, \tau, \mathbf{r}, \mathbf{r}')$ is the time-variant/space-variant wavefield response of the channel (assumed space-time selective), T is the time interval of interest and Γ is the spatial volume of interest; $\mathbf{r}$ is a 3-D or 2-D parameterization of the spatial domain and $x(\tau, \mathbf{r})$ is the transmitted spatio-temporal signal. Observe the change in notation for the coordinate vectors: A spatial position is identified in Chapter 3 as $\vec{r}$, while in this chapter we will use the notation $\mathbf{r}$. When the channel is time-invariant, or relatively slowly varying over T, $[h(t, \tau, \mathbf{r}, \mathbf{r}') = h(t - \tau, \mathbf{r}, \mathbf{r}')]$, we obtain a simple model for the wavefield of interest

$$S(\omega, \mathbf{r}) = \int_{\Gamma} H(\omega, \mathbf{r}, \mathbf{r}')X(\omega, \mathbf{r}')d\mathbf{r}', \tag{7.2}$$

where $x(t, \mathbf{r}) = \int d\omega X(\omega, \mathbf{r}, \mathbf{r}')e^{j\omega t}$, $h(t, \mathbf{r}, \mathbf{r}') = \int d\omega H(\omega, \mathbf{r}, \mathbf{r}')e^{j\omega t}$.

7.2.1 On Wavefield Modeling

When we assume a farfield scattering model we can write $H(\omega, \mathbf{r}, \mathbf{r}')$ as a superposition of plane waves

$$H(\omega, \mathbf{r}, \mathbf{r}') = \int_{\Theta} \int_{\Theta} d\gamma d\gamma' \rho_h(\gamma, \gamma', \omega) e^{j\mathbf{r}\cdot\hat{\gamma}\mathcal{K}} e^{j\mathbf{r}'\cdot\hat{\gamma}'\mathcal{K}}$$

where:

- Θ is the set of possible directions of arrival,

$$\Theta = \begin{cases} \phi \in [-\pi, \pi] & \text{if 2-D} \\ \theta \in [0, \pi],\ \phi \in [-\pi, \pi] & \text{if 3-D} \end{cases};$$

- γ and γ' are points in Θ;
- $\hat{\gamma}$ and $\hat{\gamma}'$ are unit vectors pointed in the directions γ and γ';
- $\rho_h(\gamma, \gamma', \omega)$ is a *scattering radiation density.*

$X(\omega, \mathbf{r})$, as generated by N_{tx} antenna elements, is expressed as

$$X(\omega, \mathbf{r}) = \sum_{n=1}^{N_{\text{tx}}} q_n(\omega)\delta(\mathbf{r} - \mathbf{r}_n)$$

where $q_n(\omega)$ are the N_{tx} information-bearing signals transmitted at the N_{tx} elements of the transmit array, and $\mathbf{r}_n$, $n = 1, 2, ..., N_{\text{tx}}$ are the locations of the elements in the transmit array.

Since a plane wave $e^{j\mathbf{r}\cdot\hat{\gamma}\mathcal{K}}$ is solution of the Helmoltz equation, $\rho_h(\gamma, \gamma', \omega)$ is an arbitrary complex function fully specified by the spatio-temporal propagation modes of the channel (see Chapter 2). With the addition of spatially and temporally white noise, we have

$$R(\omega, \mathbf{r}) = S(\omega, \mathbf{r}) + N(\omega, \mathbf{r}). \tag{7.3}$$

We can expand the scattering radiation density with a complete and orthogonal (two-dimensional for 2-D propagation or four-dimensional for 3-D propagation) basis [denoted $f_{n,m}(\gamma, \gamma')$],

$$\rho_h(\gamma, \gamma') = \sum_n \sum_m \psi_{n,m} f_{n,m}(\gamma, \gamma') \tag{7.4}$$

where $\psi_{n,m}$ are coefficients of the expansion given by

$$\psi_{n,m} = \int_\Theta \int_\Theta d\gamma d\gamma' \rho_h(\gamma, \gamma') f^*_{n,m}(\gamma, \gamma').$$

Using (7.4) we obtain

$$H(\mathbf{r}, \mathbf{r}') = \sum_{n,m} \psi_{n,m} \mathcal{H}_{n,m}(\mathbf{r}, \mathbf{r}')$$

where

$$\mathcal{H}_{n,m}(\mathbf{r}, \mathbf{r}') = \int_\Theta \int_\Theta d\gamma d\gamma' f_{n,m}(\gamma, \gamma') e^{j\mathbf{r}\cdot\hat{\gamma}\mathcal{K}} e^{j\mathbf{r}'\cdot\hat{\gamma}'\mathcal{K}}.$$

So we can express

$$S(\mathbf{r}) = \sum_{k=1}^{N_{\text{tx}}} q_k \sum_{n,m} \psi_{n,m} \mathcal{H}_{n,m}(\mathbf{r}, \mathbf{r}_k). \tag{7.5}$$

Now consider the receive array as a *sampling operator*: From $S(\omega, \mathbf{r})$ it returns a vector $\mathbf{s}(\omega)$ of measurements (the array output)

$$\mathbf{s} = A_R \circ S(\mathbf{r}) \tag{7.6}$$

where A_R is the *array sampling operator* (a vector valued linear and continuous functional). Define $\mathbf{a}_{R,n,m,k}$ as the response of the receive array to a wavefield of the form $\mathcal{H}_{n,m}(\mathbf{r}, \mathbf{r}_k)$,

$$\mathbf{a}_{R,n,m,k} = A_R \circ \mathcal{H}_{n,m}(\mathbf{r}, \mathbf{r}_k).$$

From the linearity of the sampling operator, we get

$$\begin{aligned}\mathbf{s} &= A_R \circ S(\mathbf{r}) = \sum_k q_k \sum_{n,m} \psi_{n,m} A_R \circ \mathcal{H}_{n,m}(\mathbf{r}, \mathbf{r}_k) \\ &= \sum_k q_k \sum_{n,m} \psi_{n,m} \mathbf{a}_{R,n,m,k} = \sum_k q_k \mathbf{A}_{R,k} \mathbf{\Psi}, \end{aligned} \tag{7.7}$$

where $\mathbf{\Psi}$ is a vector of coefficients $\psi_{n,m}$ and $\mathbf{A}_{R,k}$ is a matrix of vectors $\mathbf{a}_{R,n,m,k}$ ordered as to properly represent $\sum_{n,m} \psi_{n,m} \mathbf{a}_{R,n,m,k} = \mathbf{A}_{R,k}\mathbf{\Psi}$.

Considering the additive noise, we obtain the following spatio-temporal model

$$\mathbf{r} = \sum_k q_k \mathbf{A}_{R,k} \mathbf{\Psi} + \mathbf{n}. \tag{7.8}$$

The value of (7.7) is that it decouples completely the effect of the spatio-temporal channel and the parameters of the transmit/receive array. The vector $\mathbf{\Psi}$ fully captures the spatio-temporal channel. An array of isotropic elements in a 2-D spatial representation gives

$$[\mathbf{a}_{R,n,m,k}]_l = \mathcal{H}_{n,m}(\mathbf{r}_l = (r_l, \phi_l), \mathbf{r}_k = (r_k, \phi_k))$$

where $\mathbf{r}_l = (r_l, \phi_l)$ is the location of the lth element in the receive array and $\mathbf{r}_k = (r_k, \phi_k)$ is the location of the kth element in the transmit array.

7.2.2 Wavefield Representations

One straightforward representation for the "channel" wavefield can be used employing Fourier bases. Consider the 2-D case with

$$f_{n,m}(\gamma, \gamma') = \frac{1}{2\pi} e^{jn\phi} e^{jm\phi'}. \tag{7.9}$$

The basis set $\mathcal{H}_{n,m}(\mathbf{r}, \mathbf{r}')$ can be written [1]

$$\begin{aligned}\mathcal{H}_{n,m}(\mathbf{r} = (r, \phi), \mathbf{r}' = (r', \phi')) &= \int_0^{2\pi} \int_0^{2\pi} d\theta d\theta' \\ &\quad e^{jKr\cos(\theta-\phi) - jn\theta} e^{jKr\cos(\theta'-\phi') - jm\theta'} \\ &= 2\pi j^{n+m} J_n(Kr) J_m(Kr') e^{-j(n\phi + m\phi')}, \end{aligned} \tag{7.10}$$

where $J_n(\mathcal{K}r)$ is the Bessel function of the first kind. So we obtain

$$[\mathbf{a}_{R,n,m,k}]_l = 2\pi j^{n+m} J_n(\mathcal{K}r_l) J_m(\mathcal{K}r_k) e^{-j(n\phi_l + m\phi_k)}$$
$$l = 1,2,...,N_{\text{rx}}, k = 1,2,...,N_{\text{tx}}. \qquad (7.11)$$

It is worth examining in more detail the relationship of the wavefield model with the traditional model based on manifolds. Define $\mathbf{a}_R(\gamma)$ and $\mathbf{a}_T(\gamma)$ as the manifolds of the receive and transmit array (in other words, $[\mathbf{a}_R(\gamma)]_m$ and $[\mathbf{a}_T(\gamma)]_m$ are the responses of the mth sensor to a plane wave propagating in the direction γ). Since

$$\mathbf{a}_R(\gamma) = A_R \circ e^{j\mathcal{K}\mathbf{r}\cdot\hat{\gamma}}$$

and

$$\mathbf{a}_T(\gamma) = A_T \circ e^{j\mathcal{K}\mathbf{r}'\cdot\hat{\gamma}'}$$

where A_R and A_T are the *array sampling operators* (receiver and transmitter, respectively), we get

$$\begin{aligned}
\mathbf{a}_{R,n,m,k} &= A_R \circ \int_\Theta \int_\Theta d\gamma d\gamma' f_{n,m}(\gamma,\gamma') e^{j\mathbf{r}\cdot\hat{\gamma}\mathcal{K}} e^{j\mathbf{r}_k\cdot\hat{\gamma}'\mathcal{K}} \\
&= \int_\Theta \int_\Theta d\gamma d\gamma' f_{n,m}(\gamma,\gamma') \mathbf{a}_R(\gamma) \left[\mathbf{a}_T(\gamma')\right]_k. \qquad (7.12)
\end{aligned}$$

This says that $[\mathbf{a}_{R,n,m,k}]_l$ is the Fourier decomposition of the product between the manifolds of the transmit and receive arrays, which implies that

$$\mathbf{A}_{R,k}\mathbf{w}_{\gamma,\gamma'} = \mathbf{a}_R(\gamma) \left[\mathbf{a}_T(\gamma')\right]_k$$

where $\mathbf{w}_{\gamma,\gamma'}$ is a vector organization of the basis functions $f^*_{n,m}(\gamma,\gamma')$. For example, we might consider the *uniform linear array* configuration, both at the transmitter and at the receiver. We have

$$[\mathbf{a}_T(\gamma)]_k = e^{j\frac{2\pi}{\lambda}(k-1)d_T \sin\theta} \quad k = 1,2,...,N_{\text{tx}}$$

and

$$[\mathbf{a}_R(\gamma)]_k = e^{j\frac{2\pi}{\lambda}(k-1)d_R \sin\theta} \quad k = 1,2,...,N_{\text{rx}}$$

where d_T and d_R are element separations, respectively, at the transmitter and at the receiver arrays, and λ is the carrier wavelength.

7.2.3 The Connection with MIMO Models

Recall that (7.8) is function of the frequency. The transmitted signals can be expressed in the time domain as

$$\begin{aligned} \tilde{q}_k(t) &= \sum_l a_k(l)p_k(t - lT_s) \\ &= \sum_l a_k(l)p_{k,l}(t) = \int d\omega q_k(\omega)e^{j\omega t}, \end{aligned} \tag{7.13}$$

where $a_k(l)$ is a QAM symbol, T_s is the symbol period and $p_k(t)$ is a pulse-shaping filter. Denoting convolutions as $*$, we have

$$r_i(t) = \sum_{k,l,n,m} a_k(l)\tilde{a}_{R,k,n,m,i}(t) * \tilde{\psi}_{n,m}(t) * p_{k,l}(t) + \tilde{n}_i(t) \tag{7.14}$$

where

$$\tilde{a}_{R,k,n,m,i}(t) = \int d\omega \left[\mathbf{a}_{R,n,m,k}(\omega)\right]_i e^{j\omega t}, \tag{7.15}$$

$r_i(t) = \int d\omega \left[\mathbf{r}(\omega)\right]_i e^{j\omega t}$, $\tilde{n}_i(t) = \int d\omega \left[\mathbf{n}(\omega)\right]_i e^{j\omega t}$ and $\tilde{\psi}_{n,m}(t) = \int d\omega \psi_{n,m}(\omega)e^{j\omega t}$.

The connection with the commonly assumed MIMO model (see for example [2, 3])

$$r_i(t) = \sum_{k=1}^{N_{\text{tx}}} \sum_{l=-\infty}^{+\infty} a_k(l)h_{k,i}(t - lT_s) + \tilde{n}_i(t), \quad i = 1, ..., N_{\text{rx}} \tag{7.16}$$

is evidently obtained as

$$h_{k,i}(t) = \sum_{n,m} \tilde{a}_{R,k,n,m,i}(t) * \tilde{\psi}_{n,m}(t) * p_k(t),$$

or in the frequency domain

$$H_{k,i}(\omega) = \sum_{n,m} a_{R,k,n,m,i}(\omega)\psi_{n,m}(\omega)P_k(\omega),$$

where $p_k(t) = \int d\omega P_k(\omega)e^{j\omega t}$, and $\left[\mathbf{a}_{R,n,m,k}(\omega)\right]_i = a_{R,k,n,m,i}(\omega)$.

7.3 Detection

For simplicity we consider the uniform linear array configuration, both at the transmitter and at the receiver [4, 5]. We then have

$$[\mathbf{a}_T(\gamma)]_k = e^{j\frac{2\pi}{\lambda}(k-1)d_T \sin\theta} \quad k = 1, 2, ..., N_{\text{tx}}$$

and

$$[\mathbf{a}_R(\gamma)]_k = e^{j\frac{2\pi}{\lambda}(k-1)d_R \sin\theta} \quad k = 1, 2, ..., N_{\text{rx}}$$

where d_T and d_R are element separations, respectively, at the transmitter and at the receiver arrays and λ is the carrier wavelength. These expressions can be used in (7.12) and numerical computation of $\mathbf{a}_{R,n,m,k}$ can be performed. The vector $\Psi(\omega)$ can be truncated to $4N+2$ terms $[\psi_{-N,-N}(\omega), ..., \psi_{N,N}(\omega)]^T$. According to the maximum likelihood principle, the optimal estimation of the channel parameters can be obtained maximizing

$$\begin{aligned} L_N(\Psi(\omega)) &= -\sum_{i=1}^{N_{\text{rx}}} \left| [\mathbf{r}(\omega)]_i - \left[\sum_k q_k \mathbf{A}_{R,k} \Psi(\omega) \right]_i \right|^2 \\ &= -||\mathbf{r}(\omega) - \sum_k q_k(\omega) \mathbf{A}_{R,k} \Psi(\omega)||^2 \\ &= -||\mathbf{r}(\omega) - \mathbf{S}(\omega)\Psi(\omega)||^2. \end{aligned} \tag{7.17}$$

The solution of this maximization problem is

$$\hat{\Psi}(\omega) = \left[\mathbf{S}(\omega)^H \mathbf{S}(\omega) \right]^{-1} \mathbf{S}(\omega)^H \mathbf{r}(\omega).$$

The matrix $\mathbf{S}(\omega)$ is formed from appropriate training data.

The transmitted signals can be expressed in the time domain as

$$\begin{aligned} \tilde{q}_k(t) &= \sum_l a_k(l) p_k(t - lT_s) \\ &= \sum_l a_k(l) p_{k,l}(t) = \int d\omega q_k(\omega) e^{j\omega t}, \end{aligned} \tag{7.18}$$

where $a_k(l)$ is a QAM symbol, T_s is the symbol period and $p_k(t)$ is the pulse-shaping filter. Denoting convolutions as $*$, we have

$$r_i(t) = \sum_{k,l} a_k(l) \sum_{n,m} \tilde{a}_{R,k,n,m,i} \tilde{\psi}_{n,m}(t) * p_{k,l}(t) + \tilde{n}_i(t)$$

where

$$\tilde{a}_{R,k,n,m,i} = [\mathbf{a}_{R,n,m,k}]_i$$

$r_i(t) = \int d\omega \, [\mathbf{r}(\omega)]_i \, e^{j\omega t}$, $\tilde{n}_i(t) = \int d\omega \, [\mathbf{n}(\omega)]_i \, e^{j\omega t}$ and $\tilde{\psi}_{n,m}(t) = \int d\omega \psi_{n,m}(\omega) e^{j\omega t}$. At this point we can look for the optimum minimum MSE (MMSE) linear filter $W_{k,i}(\omega)$, whose T_s-spaced sampled output can be expressed as

$$Z_k(\omega) = \sum_{m=-\infty}^{+\infty} \sum_{i=1}^{N_{\mathrm{TX}}} W_{k,i}(\omega - 2\pi m/T_s) R_i(\omega - 2\pi m/T_s).$$

We have considered spectrum folding due symbol-rate sampling and used the notation $R_i(\omega) = [\mathbf{r}(\omega)]_i$. The MSE is written as

$$\mathrm{MSE}_k = T_s \int_{-2\pi/2T_s}^{+2\pi/2T_s} E\left\{ |Z_k(\omega) - A_k(\omega)|^2 \right\} d\omega,$$

where $A_k(\omega)$ is the Fourier domain representation of

$$\sum_{m=-\infty}^{+\infty} a_k(m) \delta(t - mT_s).$$

For simplicity we assume that there is no excess bandwidth, that is that $P_k(\omega)$ is exactly band-limited to the Nyquist bandwidth. Since we have

$$H_{k,i}(\omega) = \mathbf{\Psi}(\omega)^T \mathbf{a}_{R,k,i} P_k(\omega) = \mathbf{a}_{R,k,i}^T \mathbf{\Psi}(\omega) P_k(\omega)$$

we can compact in vector notation

$$\mathbf{H}_k(\omega) = \mathbf{A}_{R,k} \mathbf{\Psi}(\omega) P_k(\omega) \tag{7.19}$$

where

$$\mathbf{H}_k(\omega)^T = [H_{k,1}(\omega), H_{k,2}(\omega), ..., H_{k,N_{\mathrm{TX}}}(\omega)] \, .$$

Basically the expression (7.19) states that $\mathbf{H}_k(\omega)$ is in the subspace spanned by the columns of $\mathbf{A}_{R,k}$ which forces the channel vector to lie in the appropriate a priori known subspace: the subspace of the baseband channels generated by plane waves obeying the Helmoltz equation. Compared to unstructured methods [2, 6, 7] this channel parameterization in practice reduces the MSE of an MMSE detector computed using (7.19). However, the structure of the model is such that no estimation of angle of arrivals or channel gains is necessary.

The optimum linear MMSE filter is obtained from

$$\begin{aligned}\mathbf{W}_k(\omega) &= \left[W_{k,1}(\omega), W_{k,2}(\omega), ..., W_{k,N_{\mathrm{rx}}}(\omega)\right]^T \\ &= \left[\mathbf{R}_H(\omega) + \mathbf{R}_N(\omega)\right]^{-1} \mathbf{H}_k(\omega) \\ &= \mathbf{R}(\omega)^{-1} \mathbf{A}_{R,k} \mathbf{\Psi}(\omega) P_k(\omega) \end{aligned} \quad (7.20)$$

where $[\mathbf{R}_N(\omega)]_{m,n} = E\{N_n(\omega)N_m^*(\omega)\}$ with $[\mathbf{n}(\omega)]_m = N_m(\omega)$ [1] and

$$\begin{aligned}\mathbf{R}_H(\omega) &= \sum_{k=1}^{N_{\mathrm{tx}}} \mathbf{H}_k(\omega)^* \mathbf{H}_k(\omega)^T \\ &= \sum_{k=1}^{N_{\mathrm{tx}}} \mathbf{A}_{R,k}^* \mathbf{\Psi}(\omega)^* \mathbf{\Psi}(\omega)^T \mathbf{A}_{R,k}^T |P_k(\omega)|^2. \end{aligned} \quad (7.21)$$

The evident problem is that the expansion of $\rho_h(\gamma, \gamma', \omega)$ is exact only with an infinite number of components in $\mathbf{\Psi}(\omega)$. Reducing the number of components in $\mathbf{\Psi}(\omega)$, one could trade off bias and variance [8] of the reconstruction error. The MMSE filters can be implemented in the time domain or in the frequency domain. The preferred implementation used in the experimental prototype is a multirate architecture [9]. The length of the expansion in (7.19) used in the hardware experiments is $N = 64$.

7.4 Practical MIMO Detection

Assume that data at every antenna is cyclically prefixed with L symbols and L is such that LT_s exceeds the delay spread of the wireless channel. It is well-known that at the receiver, after removal of L samples in the received data block, we can express the convolutions

1. The modification of these expressions for the excess-bandwidth case can be easily determined considering the folded spectra for the channels $H_{k,i}(\omega - 2\pi m/T_s)$ and noise $N_i(\omega - 2\pi m/T_s)$, for $m = -J, -J+1, ..., J$.

$$\mathbf{H}_k(\omega)^T = \left[H_{k,1}(\omega - 2\pi J/T_s), ..., H_{k,N_{\mathrm{rx}}}(\omega + 2\pi J/T_s)\right],$$

where $2\pi J/T_s$ is large enough to cover for the excess bandwidth of $P_k(\omega)$.

as cyclic convolutions and we can write

$$\mathbf{y}_n = \mathbf{H}\mathbf{a}_n + \mathbf{n}_n \tag{7.22}$$

where

$$\mathbf{y}_n = \left[\mathbf{y}_{n,1}^T \mathbf{y}_{n,2}^T \ldots \mathbf{y}_{n,N_{\mathrm{rx}}}^T\right]^T, \quad \mathbf{n}_n = \left[\mathbf{n}_{n,1}^T \mathbf{n}_{n,2}^T \ldots \mathbf{n}_{n,N_{\mathrm{rx}}}^T\right]^T,$$

$$\mathbf{a}_n = \left[\mathbf{a}_{n,1}^T \mathbf{a}_{n,2}^T \ldots \mathbf{a}_{n,N_{\mathrm{tx}}}^T\right]^T,$$

$$\mathbf{y}_{n,i} = \left[y_i(n), y_i(n-1), \ldots, y_i(n-N+1)\right]^T,$$

$$\mathbf{a}_{n,i} = \left[a_i(n), a_i(n-1), \ldots, a_i(n-N+1)\right]^T,$$

$$y_i(n) \triangleq r_i(nT_s)$$

$$\mathbf{H} = \begin{bmatrix} \mathbf{H}_{1,1} & \mathbf{H}_{1,2} & \ldots & \mathbf{H}_{1,N_{\mathrm{tx}}} \\ \mathbf{H}_{2,1} & \mathbf{H}_{2,2} & \ldots & \mathbf{H}_{2,N_{\mathrm{tx}}} \\ \cdots & \cdots & \cdots & \cdots \\ \mathbf{H}_{N_{\mathrm{rx}},1} & \mathbf{H}_{N_{\mathrm{rx}},2} & \ldots & \mathbf{H}_{N_{\mathrm{rx}},N_{\mathrm{tx}}} \end{bmatrix}.$$

The matrices $\mathbf{H}_{k,i}$ are $N \times N$ cyclic convolution matrices relative to the response

$$h_{k,i}(t) = \sum_{n,m} a_{R,k,n,m,i}\psi_{n,m}(t)p_k(t),$$

sampled at rate $1/T_s$. Since the vectors $\mathbf{a}_{n,i}$ have been cyclically prefixed, a DFT matrix $\mathbf{F}$, will diagonalize[2] $\mathbf{H}_{k,i}$ as

$$\mathbf{F}\mathbf{H}_{k,i}\mathbf{F}^H = \mathbf{\Lambda}_{k,i},$$

and

$$\left[\mathbf{\Lambda}_{k,i}\right]_n = H_{k,i}(\omega)\big|_{\omega=\frac{2\pi n}{T_s}} \triangleq H_{k,i}(n)$$

is the nth point in the N-point DFT of $h_{k,i}(mT_s)$. We then select permutation matrices $\mathbf{P}_i$ $i = 1, 2$ such that

$$\mathbf{P}_1\tilde{\mathbf{F}}_1\mathbf{H}\tilde{\mathbf{F}}_2^H\mathbf{P}_2 = \mathbf{B} \tag{7.23}$$

---

2. The k, l element of the DFT matrix $\mathbf{F}$ is defined as $\frac{1}{\sqrt{N}}e^{\frac{-j2\pi kl}{N}}$.

where

$$\tilde{\mathbf{F}}_i = \begin{bmatrix} \mathbf{F} & \mathbf{0} & \cdots \\ \mathbf{0} & \mathbf{F} & \cdots \\ \mathbf{0} & \cdots & \mathbf{F} \end{bmatrix} \qquad i = 1, 2$$

($\tilde{\mathbf{F}}_1$ is $N_{\text{rx}}N \times N_{\text{rx}}N$ while $\tilde{\mathbf{F}}_2$ is $N_{\text{tx}}N \times N_{\text{tx}}N$) and

$$\mathbf{B} = \begin{bmatrix} \mathbf{L}_1 & \mathbf{0} & \cdots \\ \mathbf{0} & \mathbf{L}_2 & \cdots \\ \mathbf{0} & \cdots & \mathbf{L}_N \end{bmatrix}.$$

The k, i element of the matrices $\mathbf{L}_n$ is $H_{k,i}(n)$. Observe that at the receiver the vector $\mathbf{y}_n$ is processed to obtain $\mathbf{Y}_n$ as in

$$\mathbf{Y}_n = \mathbf{P}_1 \tilde{\mathbf{F}}_1 \mathbf{y}_n,$$

where

$$\mathbf{Y}_n = \Big[\overbrace{Y_{1,1}^{(n)}, \ldots, Y_{N_{\text{rx}},1}^{(n)}}^{\text{freq. bin 1}}, \overbrace{Y_{1,2}^{(n)}, \ldots, Y_{N_{\text{rx}},2}^{(n)}}^{\text{freq. bin 2}}, \ldots \overbrace{Y_{1,N}^{(n)}, \ldots, Y_{N_{\text{rx}},N}^{(n)}}^{\text{freq. bin } N} \Big]^T$$

In other words, $Y_{i,k}^{(n)}$, $k = 1, 2, \ldots, N$ is the N-point FFT of $\mathbf{y}_{n,i}$. Since we can also write

$$\mathbf{a}_n = \tilde{\mathbf{F}}_2^H \mathbf{P}_2 \mathbf{A}_n,$$

where

$$\mathbf{A}_n = \Big[\overbrace{A_{1,1}^{(n)}, \ldots, A_{N_{\text{tx}},1}^{(n)}}^{\text{freq. bin 1}}, \overbrace{A_{1,2}^{(n)}, \ldots, A_{N_{\text{tx}},2}^{(n)}}^{\text{freq. bin 2}}, \ldots \overbrace{A_{1,N}^{(n)}, \ldots, A_{N_{\text{tx}},N}^{(n)}}^{\text{freq. bin } N} \Big]^T,$$

and $A_{i,k}^{(n)}$, $k = 1, 2, \ldots, N$ is the N-point FFT of $\mathbf{a}_{n,i}$, in the absence of noise we can write

$$\mathbf{y}_n = \mathbf{H}\mathbf{a}_n = \mathbf{H}\tilde{\mathbf{F}}_2^H \mathbf{P}_2 \mathbf{A}_n.$$

So $\mathbf{Y}_n$ can be expressed as

$$\mathbf{Y}_n = \mathbf{P}_1 \tilde{\mathbf{F}}_1 \mathbf{H} \tilde{\mathbf{F}}_2^H \mathbf{P}_2 \mathbf{A}_n = \mathbf{B}\mathbf{A}_n.$$

Due to the block diagonal nature of **B**, this problem can be decoupled in

$$\left[Y_{1,k}^{(n)}, Y_{2,k}^{(n)}, ..., Y_{N_{\mathrm{TX}},k}^{(n)}\right]^T = \mathbf{L}_k \left[A_{1,k}^{(n)}, A_{2,k}^{(n)}, ..., A_{N_{\mathrm{tx}},k}^{(n)}\right]^T,$$

or with additive noise

$$\left[Y_{1,k}^{(n)}, Y_{2,k}^{(n)}, ..., Y_{N_{\mathrm{TX}},k}^{(n)}\right]^T = \mathbf{L}_k \left[A_{1,k}^{(n)}, A_{2,k}^{(n)}, ..., A_{N_{\mathrm{tx}},k}^{(n)}\right]^T + \mathbf{N}_{k,n}.$$

The MMSE filter for the recovery of

$$\left[A_{1,k}^{(n)}, A_{2,k}^{(n)}, ..., A_{N_{\mathrm{tx}},k}^{(n)}\right]^T$$

from

$$\left[Y_{1,k}^{(n)}, Y_{2,k}^{(n)}, ..., Y_{N_{\mathrm{TX}},k}^{(n)}\right]^T$$

is

$$\mathbf{W}_k = \mathbf{L}_k^H \left[\mathbf{L}_k \mathbf{L}_k^H + \sigma_n^2 \mathbf{I}\right]^{-1}, \tag{7.24}$$

where $\sigma_n^2 = E\left\{|\mathbf{N}_{k,n}|^2\right\}$ and we have assumed uncorrelated symbols so that $E\left\{A_{l,k}^{(n)} A_{m,k}^{(n)*}\right\} = \delta(l-m)$. Define

$$\tilde{\mathbf{W}} = \begin{bmatrix} \mathbf{W}_1 & \mathbf{0} & \cdots \\ \mathbf{0} & \mathbf{W}_2 & \cdots \\ \mathbf{0} & \cdots & \mathbf{W}_N \end{bmatrix}.$$

Processing $\mathbf{Y}_n$ as

$$\tilde{\mathbf{F}}_2^H \mathbf{P}_2 \tilde{\mathbf{W}} \mathbf{Y}_n,$$

we obtain an estimate of $\mathbf{a}_n$. To see this more clearly just consider that in the noiseless case we have

$$\begin{aligned} \tilde{\mathbf{F}}_2^H \mathbf{P}_2 \tilde{\mathbf{W}} \mathbf{Y}_n &= \tilde{\mathbf{F}}_2^H \mathbf{P}_2 \tilde{\mathbf{W}} \mathbf{P}_1 \tilde{\mathbf{F}}_1 \mathbf{H} \tilde{\mathbf{F}}_2^H \mathbf{P}_2 \mathbf{A}_n = \tilde{\mathbf{F}}_2^H \mathbf{P}_2 \tilde{\mathbf{W}} \mathbf{B} \mathbf{A}_n \\ &= \tilde{\mathbf{F}}_2 \mathbf{P}_2 \mathbf{A}_n = \mathbf{a}_n, \end{aligned}$$

because $\mathbf{P}_1 \tilde{\mathbf{F}}_1 \mathbf{H} \tilde{\mathbf{F}}_2^H \mathbf{P}_2 = \mathbf{B}$ and in the absence of noise $\tilde{\mathbf{W}}\mathbf{B} = \mathbf{I}$. An explanation of the processing executed at baseband on the complex samples is reported in Figure 7.1.

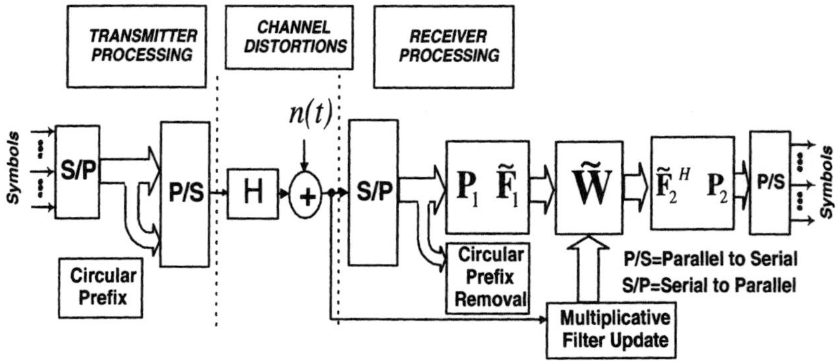

Figure 7.1 Block diagram of the baseband processing chain.

7.4.1 MIMO-OFDM

The basic concept of MIMO-OFDM can be easily described extending the previous derivations. Again the basic idea stems from the fact that

$$\mathbf{P}_1\tilde{\mathbf{F}}_1\mathbf{H}\tilde{\mathbf{F}}_2^H\mathbf{P}_2 = \mathbf{B}$$

with $\mathbf{B}$ block diagonal matrix. The transmitter performs IFFT operations on the blocked data so that the transmitted vectors are $\tilde{\mathbf{F}}_2^H\mathbf{P}_2\mathbf{A}_n$, while at the receiver after proper prefix removal the received time domain vector is transformed back in the frequency domain using FFT operations

$$\mathbf{Y}_n = \mathbf{P}_1\tilde{\mathbf{F}}_1\mathbf{y}_n.$$

This is more easily understood looking at the processing reported in Figure 7.2.

7.4.2 Channel Estimation

We have seen that the problem of compensating for distortions caused by channel time dispersion can be simply decoupled by proper processing to the block channel model

$$\begin{aligned}\left[Y_{1,k}^{(n)}, Y_{2,k}^{(n)}, ..., Y_{N_{\mathrm{rx}},k}^{(n)}\right]^T &= \mathbf{Y}_k^{(n)} \qquad (7.25)\\ &= \mathbf{L}_k\left[A_{1,k}^{(n)}, A_{2,k}^{(n)}, ..., A_{N_{\mathrm{tx}},k}^{(n)}\right]^T + \mathbf{N}_{k,n}\\ &= \mathbf{L}_k\mathbf{A}_k^{(n)} + \mathbf{N}_{k,n}, \quad k = 1, 2, ..., N.\end{aligned}$$

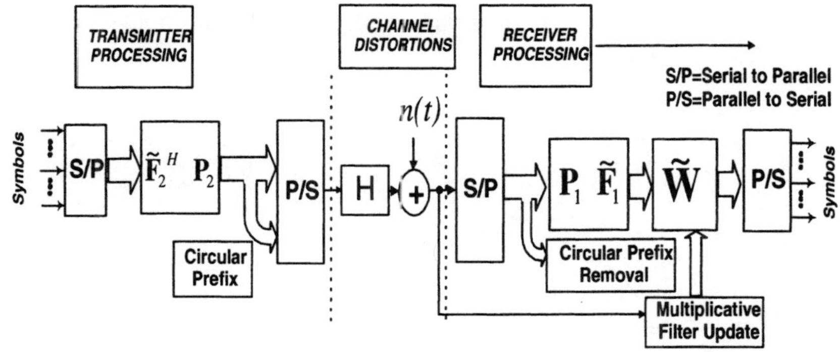

Figure 7.2 Block diagram of the baseband processing chain for MIMO-OFDM.

The channel matrices $\mathbf{L}_k$ are unknown at the receiver and can be estimated using a number of known blocks of symbols $\mathbf{A}_k^{(n)}$ $n = 1, 2, ..., N_S$ periodically inserted in the data stream. Organize the matrices $\mathbf{A}_k^{(n)}$ $n = 1, 2, ..., N_S$ in

$$\overline{\mathbf{A}}_k = \left[\mathbf{A}_k^{(1)}, \mathbf{A}_k^{(2)}, ..., \mathbf{A}_k^{(N_S)}\right],$$

$$\overline{\mathbf{Y}}_k = \left[\mathbf{Y}_k^{(1)}, \mathbf{Y}_k^{(2)}, ..., \mathbf{Y}_k^{(N_S)}\right],$$

$$\overline{\mathbf{N}}_k = \left[\mathbf{N}_{k,1}, \mathbf{N}_{k,2}, ..., \mathbf{N}_{k,N_S}\right].$$

The channel matrices can be estimated as

$$\hat{\mathbf{L}}_k = \overline{\mathbf{A}}_k^H \left[\overline{\mathbf{A}}_k \overline{\mathbf{A}}_k^H\right]^{-1} \overline{\mathbf{Y}}_k.$$

For AWGN this estimator is also the maximum likelihood estimator.

7.5 The Characterization of RF Distortions

The carrier-to-noise ratio (CNR) at the RF front-end output is probably the most important indicator of the capability of the receiver to properly demodulate the signal. Given a particular modulation scheme we typically want to achieve some quality of service (QOS) specified by a target BER. In turn, the BER will specify the CNR required to achieve in an ideal AWGN environment, such as QOS. The usual simplifying assumption is that distortions can be treated as noise in terms of signal quality degradation. The CNR γ is defined as

$$\gamma = \left[\frac{1}{10^{-0.1\gamma_{\mathrm{NF}}} + 10^{-0.1\gamma_{\mathrm{PN}}} + 10^{-0.1\gamma_{\mathrm{IM}}}}\right]_{\mathrm{dB}}$$

where

1. γ_{IM} is the signal-to-interference ratio caused by intermodulation distortions.
2. γ_{NF} is the signal-to-noise ratio caused by the noise figure of the receiver.
3. γ_{PN} is the signal to interference ratio caused by phase noise distortion.

7.5.1 Noise Figure

The electronic circuitry in a receiver always adds a small amount of noise to an incoming signal in excess of that contributed by thermal effects. This circuit noise is normally the limiting factor in detecting low-level signals.

The noise figure of a receiver is a very fundamental parameter and is basically a measure of the noise added by the receiver circuitry. An expression for defining the noise figure of an electronic device in terms of SNR is

$$\mathrm{NF} = 10\log_{10}\left[\frac{S_i/N_i}{S_o/N_o}\right]$$

where

- NF = noise figure in decibels;
- S_i = signal power at device input;
- N_i = noise power at device input;
- S_o = signal power at device output;
- N_o = noise power at device output.

In general, at low input level, γ depends on the noise figure F of the receiver as

$$\gamma_{\mathrm{NF}} = 10\log_{10}\left[\frac{P_i}{KTBF}\right]$$

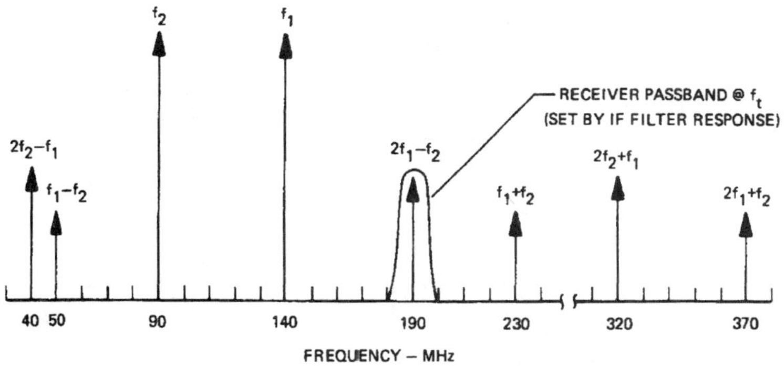

Figure 7.3 Second and third order two-tone intermodulation products for two unwanted input signals at f_1 and f_2 with receiver tuned to f_t.

where K is the Boltzmann constant, F is the noise figure in antilogarithm, T is the absolute temperature[3], and B is the system bandwidth. Since we typically want an SNR per symbol (that is, SNR over the Nyquist bandwidth) greater than some target SNR, say, r, that will provide a target BER $\leq \text{BER}_r$, we need $\gamma_{\text{IM}} >> \gamma_{\text{NF}}$ and $\gamma_{\text{PN}} >> \gamma_{\text{NF}}$, if we want $\gamma_{\text{NF}} \simeq r$. For example, we may say

$$\gamma_{\text{IM}} \geq \gamma_{\text{NF}} + 10 \text{ dB}$$

and

$$\gamma_{\text{PN}} \geq \gamma_{\text{NF}} + 10 \text{ dB}.$$

7.5.2 Intermodulation Distortions

All receivers employ RF-IF signal processing circuitry that is inherently nonlinear. When two sufficiently strong, but unwanted signals are applied to the antenna input of a receiver, they will mix in the RF stages to create spurious signals known as intermodulation products. If the frequency of one of these products is close to the receiver operating frequency, the product will be processed by the RF-IF and detector stages as though it were a real incoming signal of the same frequency. This problem is illustrated in Figure 7.3. Second- and third-order intermodulation distortion are the most common

3. Observe that KTB_1 in dBm is -174 dBm, for $B_1 = 1$ Hz.

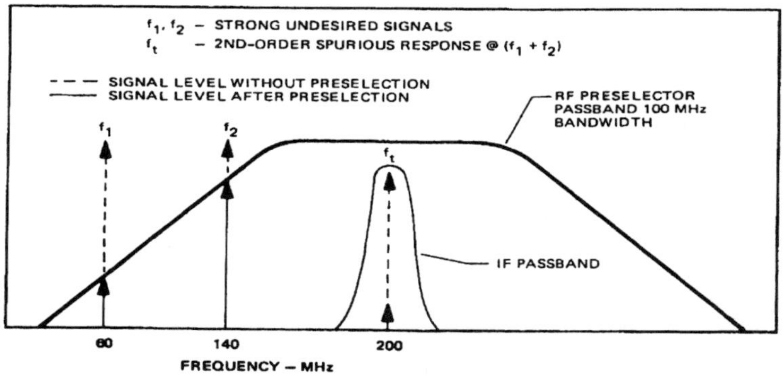

Figure 7.4 Reduction in second-order interference using suboctave preselector.

types encountered, and the frequency relationships involved for these two cases are given by the following:

- $f_1 \pm f_2 = f_c$ second-order intermodulation distortion;
- $2f_1 \pm f_2 = f_c$ third-order intermoduation distortion;

where f_1, f_2 are the frequencies of strong undesired signals and f_c is the frequency of intermodulation product created at the receiver tuned frequency.

Second-order, two-tone intermodulation distortion is not an uncommon problem, especially in a receiver having a broadband RF front end, but it can be minimized by use of a double-balanced mixer in the first converter stage plus use of a push-pull RF preamplifier. Also, with the addition of an RF preselector employing suboctave bandwidth bandpass filters (tunable or fixed), second-order interference can be reduced to an insignificant level. The suboctave preselector filter serves to attenuate strong signals, lying within a range of critical frequencies, which are capable of creating second-order products at the receiver tuned frequency. This reduction in second-order interference by use of RF preselection is illustrated in Figure 7.4.

More difficult to control is third-order, two-tone intermodulation distortion, since RF preselection provides only a partial solution to the problem. This is due to the following distinctive property of

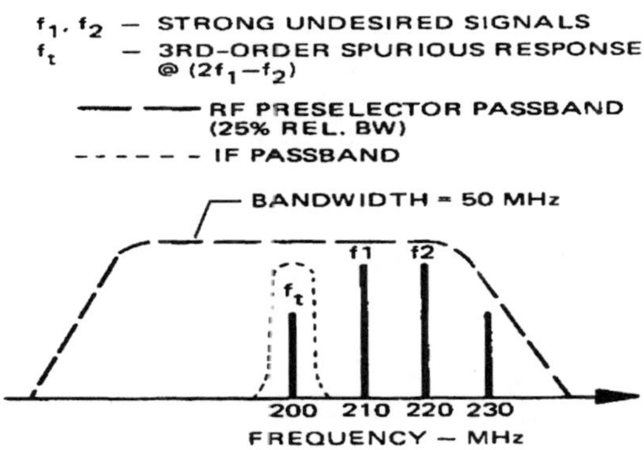

Figure 7.5 Third-order intermodulation interference caused by two strong inband signals.

third-order two-tone interference. Two strong undesired signals both falling within the passband of the preselector will produce the third-order products $(2f_1 - f_2)$ or $(2f_2 - f_1)$ one or both of which may also fall in-band. Decreasing the preselector bandwidth will reduce the frequency range over which the receiver is susceptible to this type of interference. Unfortunately, due to considerations such as size, complexity, and insertion loss, a practical lower limit for the relative bandwidth of preselector filters used in general-coverage receivers is around 20%. Therefore, in a dense signal environment there is always the possibility that two strong signals will fall within the preselector passband and produce an undesired spurious response at the receiver tuned frequency. This situation is illustrated in Figure 7.5. Third-order intermodulation distortion is not limited to the RF front end of a receiver. The above description of the in-band characteristics of third-order interference is also applicable to the receiver IF stages. When the frequency spacing between strong incoming signals is small compared to the bandwidth of the first IF stage, then intermodulation distortion in the IF is likely to occur. Consequently, the problem must be minimized through proper circuit and system design, as well as component selection, for the overall RF-IF receiver chain. Intermodulation performance is commonly tested by applying two signals of equal power to the receiver input and then measuring the equivalent input level of the distortion product created at the

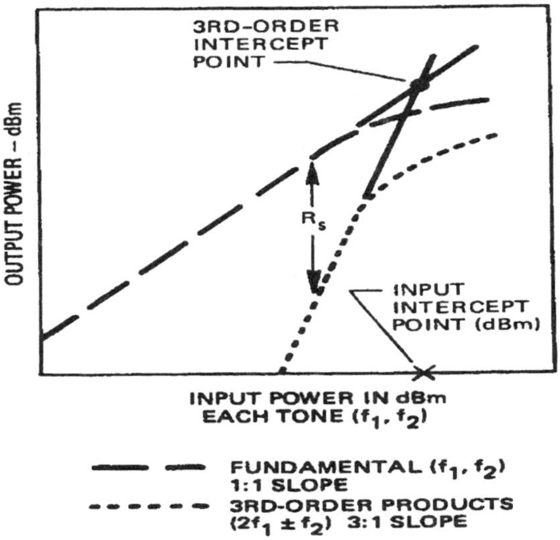

Figure 7.6 Third-order intermodulation distortion.

receiver tuned frequency. The ratio in dB of the distortion product to the input level of the test signals is known as the intermodulation ratio and was frequently used in the past to specify intermodulation performance. In recent years though, the intercept point concept has become a more popular method for characterizing the intermodulation distortion performance of many types of electronic equipment as well as radio receivers. The intercept-point method yields a single number, usually in dBm, which is independent of the input signal levels. The intercept-point concept for two-tone, third-order intermodulation is most easily understood by referring to Figure 7.6. The curves in Figure 7.6 show the typical input-output power relationships for the fundamental and third-order responses in a mildly nonlinear system. The dashed line represents the variation in output power as a function of input power for the two fundamental input signals. Note that for small enough input levels (i.e., below compression) the curve is linear with a slope of unity indicating the output power of the fundamental changes on a decibel-for-decibel basis with the input power. The dotted line depicts the behavior of the output power of the intermodulation products as a function of the fundamental input power. Again the curve is linear for small enough input signals but has a slope of three. This slope indicates that the power in the third-order intermodulation products increases 3 dB for each decibel increase in the input signal levels.

The fictitious extension of the linear portions of these two curves until they intersect establishes the intercept point. The input power level at which this point of intersection occurs is the third-order, two-tone intermodulation input intercept point of the system.

The intercept point of a system cannot be measured directly and is computed from

$$\mathrm{IP} = 1/2R_s + P_i \tag{7.26}$$

where IP is the third-order input intercept point in dBm, R_s is the relative suppression in dB of third-order products, and P_i is the input power level in dBm at which relative suppression is measured.

The above relative suppression term R_s is the amount in dB by which the third-order intermodulation products are suppressed below the fundamental responses when measured at the system output, and is shown graphically in Figure 7.5 as the vertical difference between the two curves. The typical test setup used for evaluating two-tone intermodulation performance and the method for determining R_s is shown in the simplified illustrations of Figure 7.7. Another important relationship, involving intercept point, which can be used to determine the equivalent input level of intermodulation products is

$$\mathrm{IM} = 3(P_i) - 2(\mathrm{IP}) \tag{7.27}$$

where IM is the equivalent input power in dBm of third-order intermodulation products, and IP is the third-order input intercept point in dBm.

It is important to note that the higher the third-order intercept point of a receiver, the less susceptible it will be to spurious responses caused by strong multiple in-band signals. Unfortunately, the desire for a high intercept-point receiver is often not compatible with the simultaneous requirement for low noise figure. Compromises usually must be made when specifying both the receiver noise figure and its third-order intercept point.

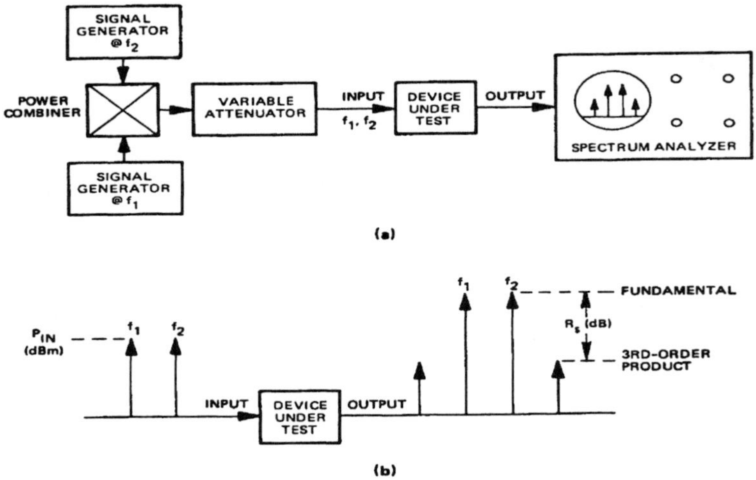

Figure 7.7 (a) Test setup for two-tone intermodulation distortion and (b) measurement of R_s term used in calculation of intercept point.

As a general rule of thumb, a receiver designed for high intercept-point performance will employ RF-IF amplifier stages and double-balanced mixers all having high 1-dB compression points. These power amplifier stages generally have higher noise figures and consume more supply power than small-signal amplifiers. High-power mixers require large local oscillator drive levels, which, in turn, can result in higher LO radiation levels and larger internally generated spurious responses, as well as more power consumption. Thus, the requirement for an extremely high intercept-point receiver can create a multitude of design problems and result in very expensive equipment.

In a two-tone test γ_{IM} is given by the power at the input to the receiver minus the power at the third-order IM component at the output of the receiver,

$$\gamma_{\mathrm{IM3}}^{(2)} = P_i - P_{\mathrm{IM}}.$$

In a multicarrier system with N carriers the two tones are divided in N subcarriers, each one with power $P_i - 10\log_{10} N/2$ so that the average power of the third order IM products created by the two-tone subcarriers is $P_{\mathrm{IM}} - 30\log_{10} N/2$. Of course, one has to find out how many IM products (say, M) fall in a certain subcarrier position. It turns out that the center subcarrier (which is also the position of the RF carrier) is the subcarrier that will receive the worst IM interference. So we can assume that the signal to IM distortion ratio of the receiver is the worst case one and write

$$\begin{aligned}\gamma_{\mathrm{IM}} &= P_i - 10\log_{10} N/2 \\ &\quad - (P_{\mathrm{IM}} - 30\log_{10} N/2 + 10\log_{10} M) \\ &= \gamma_{\mathrm{IM3}}^{(2)} + 20\log_{10} N/2 - 10\log_{10} M. \end{aligned} \qquad (7.28)$$

Since $\gamma_{\mathrm{IM}} \geq \gamma_{\mathrm{NF}} + 10$ dB, we have the following specification

$$\gamma_{\mathrm{IM3}}^{(2)} \geq \gamma_{\mathrm{NF}} + 10\ \mathrm{dB} - 20\log_{10} N/2 + 10\log_{10} M.$$

Spurious-Free Dynamic Range

The expression *spurious-free dynamic range* (SFDR) is typically used to mean that portion of the total dynamic range where there are no third-order spurious responses exceeding the noise floor by 3 dB when two equal-power input signals are applied. The dynamic range of a receiver is the range of input signal levels over which the receiver is usable. Various criteria have been used to define the upper and lower limits for this usable range. One criterion often used to establish the lower limit of the dynamic range is called the minimum detectable signal and is defined as a signal 3 dB greater than the equivalent noise power for a given IF bandwidth. The minimum detectable signal (MDS) is

$$P_L = \mathrm{MDS} = -171\ \mathrm{dBm} + \mathrm{NF} + 10\log_{10} B \qquad (7.29)$$

where P_L is the lower power limit of dynamic range in dBm, MDS is the minimum detectable signal in dBm, NF is the noise figure in dB, and B is the IF bandwidth in Hz. The upper limit for the SFDR is typically set by the level of two equal input signals necessary to create a third-order intermodulation product equivalent to the minimum detectable signal. We have

$$\mathrm{MDS} = 3(P_u) - 2(\mathrm{IP})$$

where P_u is the upper power limit of the SFDR in dBm, and IP is the receiver third-order input intercept point in dBm. The upper power limit is given as

$$\begin{aligned} P_u &= 1/3(\mathrm{MDS} + 2\mathrm{IP}) \\ &= 1/3(-171\ \mathrm{dBm} + \mathrm{NF} + 10\log B) + 2/3(\mathrm{IP}) \end{aligned} \quad (7.30)$$

Using the above expressions for P_L and P_u, the SFDR can now be found from

$$\begin{aligned} \mathrm{SFDR} &= P_u - P_L = 1/3(\mathrm{MDS} + 2\mathrm{IP}) - \mathrm{MDS} \\ &= 2/3(\mathrm{IP} - \mathrm{MDS}) \\ &= 2/3(\mathrm{IP} - \mathrm{NF} - 10\log B + 171\ \mathrm{dBm}) \end{aligned} \quad (7.31)$$

where SFDR is in dBm.

Thus, it is seen that the SFDR is directly proportional to intercept point, but inversely proportional to noise figure and IF bandwidth. In other words, the dynamic range increases with lower noise figures and narrower IF bandwidths as well as for higher intercept points.

As an example of computing the SFDR for a typical high-performance receiver, assume a noise figure of 10 dB, an IF bandwidth of 10 kHz, and an input intercept point of −5 dBm. Substituting these quantities yields:

$$\mathrm{SFDR} = 2/3(-5\ \mathrm{dBm} - 10\ \mathrm{dB} - 40\ \mathrm{dB} + 171\ \mathrm{dBm}) = 77.3\ \mathrm{dB}.$$

7.5.3 Phase Noise and Distortions Associated with the Conversion Scheme

For a good explanation of this type of distortions we need to analyze carefully the RF architecture of interest. This analysis will provide an understanding of the fundamental key elements that are critical in achieving high accuracy of representation of the signal at baseband, in digital form.

IF Versus Low-IF Topologies

IF receivers downconvert the wanted signal from its carrier frequency by mixing with a single sinusoidal signal. The signal can be demodulated at the IF or further filtered and downconverted. The main disadvantage of IF receivers is that apart from the wanted signal, an image signal is downconverted to IF. If we define f_{IF}

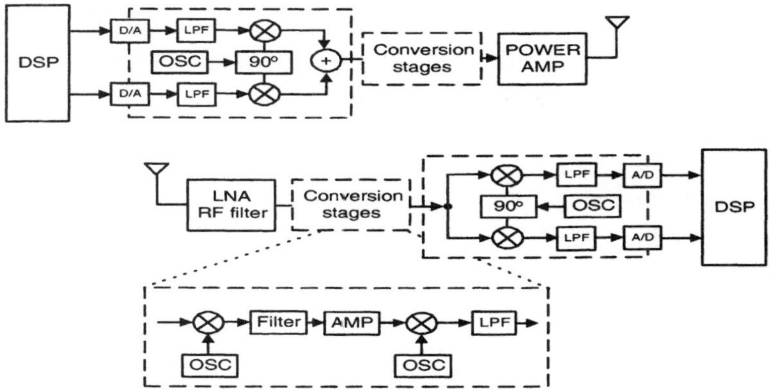

Figure 7.8 A general RF topology for OFDM systems. The analog implementation for the quadrature mod/demod is typically imposed by high data rate applications and/or high level of integration.

as the IF frequency and f_c as the carrier frequency, this image frequency is situated at $f_c - 2f_{\text{IF}}$. Suppose that the wanted signal is $s(t) = \cos(\omega_x t + m(t) + \phi)$. At the output of the mixer we have

$$s(t) \cdot \sin(\omega_c t - \omega_{IF} t + \psi) = \frac{1}{2}\{\sin[\omega_x t + \omega_c t - \omega_{IF} t + \phi + \psi] \\ - \sin[\omega_x t - \omega_c t + \omega_{IF} t + \phi - \psi]\}.$$

The suppression of the image frequency components is typically done by means of a high frequency filter before downconversion. This filter unfortunately has typically very stringent specifications (high Q and high order even for low f_c/f_{IF} ratios) which makes the overall RF front-end expensive and vulnerable (because made of discrete components that have to be tuned in production). Moreover, the integration of these filters into IC is practically impossible. For all these reasons low IF transceivers (or low-IF transceivers) are typically favored in commercial applications. The low IF architecture (shown in Figure 7.8) requires the quadrature modulation and demodulation to be performed in the analog domain. This is typically imposed by high data rate applications that make a digital implementation of the quadrature modem unfeasible with state-of-the-art digital technology. It is then extremely important to include in the analysis of the system performance I/Q amplitude and phase mismatches caused by the analog implementation of the baseband shaping filters and phase error of phase splitters. The main problem afflicting this RF topology is amplitude/phase imbalances and dc offsets. The nonideal behavior of the quadrature modulator can be summarized as

- A dc offset A_I, and A_Q;
- Amplitude gains α_I, and α_Q;
- Phase splitter error with respect to $\pi/2$ of ϕ (we assume symmetrical error).

Typical values for gain imbalance are 2–3% and phase error of 2–3 degrees. Typical dc offsets are 2–3% of full scale. If we represent the I/Q components of $\tilde{s}(t)$ as $\tilde{s}_I(t)$ and $\tilde{s}_Q(t)$ and we define the output of the quad modulator as $v(t)$ we can write

$$\mathbf{v}(t) = \mathbf{M}\mathbf{s}(t) + \mathbf{M}\mathbf{a}$$

where

$$\mathbf{M} = \begin{bmatrix} \alpha_{T,I}\cos(\phi_T/2) & -\alpha_{T,Q}\sin(\phi_T/2) \\ \alpha_{T,Q}\sin(\phi_T/2) & \alpha_{T,I}\cos(\phi_T/2) \end{bmatrix}$$

$$\mathbf{a} = \begin{bmatrix} A_I \\ A_Q \end{bmatrix}, \quad \mathbf{v}(t) = \begin{bmatrix} v_I(t) \\ v_Q(t) \end{bmatrix}, \quad \mathbf{s}(t) = \begin{bmatrix} \tilde{s}_I(t) \\ \tilde{s}_Q(t) \end{bmatrix}.$$

The same distortion is introduced by the quadrature demodulator at the receiver and we have

$$\mathbf{v}_d(t) = \mathbf{D}\mathbf{v}(t) + \mathbf{D}\mathbf{b},$$

where

$$\mathbf{D} = \begin{bmatrix} \alpha_{R,I}\cos(\phi_R/2) & -\alpha_{R,Q}\sin(\phi_R/2) \\ \alpha_{R,Q}\sin(\phi_R/2) & \alpha_{R,I}\cos(\phi_R/2) \end{bmatrix}, \quad \mathbf{b} = \begin{bmatrix} B_I \\ B_Q \end{bmatrix}.$$

The effect of phase noise and/or frequency offset is instead represented by the matrix

$$\mathbf{R}(t) = \begin{bmatrix} \cos(\theta(t)) & -\sin(\theta(t)) \\ \sin(\theta(t)) & \cos(\theta(t)) \end{bmatrix}$$

where $\theta(t)$ is the instantaneous difference in phase between the phase of the oscillator at the transmitter and at the receiver. The overall model is given by

$$\mathbf{v}_d(t) = \mathbf{F}(t)\mathbf{s}(t) + \mathbf{c}(t) + \mathbf{m}(t),$$

where $\mathbf{F}(t) = \mathbf{DR}(t)\mathbf{M}$, $\mathbf{c}(t) = \mathbf{F}(t)\mathbf{a} + \mathbf{b}$ and the noise term is $\mathbf{m}(t) = \mathbf{Dn}(t)$.

Mixing the input signal having the desired information with a local oscillator signal yields upper and lower sidebands, each containing the identical information present in the input frequency. The upper sideband is the sum of the input and the local oscillator frequencies, and the lower sideband is the difference between the input and the local oscillator frequencies. The upper or lower sideband, whichever is selected for use, is called the intermediate frequency (IF). In most receiving systems, the lower sideband (the downconverted product) is used, whereas in transmitting systems the upper sideband (the upconverted product) is used. Changing the frequency of a signal without altering the information it carries is necessary because signal processing components, such as amplifiers, are much less expensive and perform better when designed to operate at lower frequencies. Since it is much less expensive to amplify a signal in the MHz range than in the GHz range, the incoming microwave signal is first downconverted in frequency and then processed. Likewise, in a transmitter it is less expensive to generate, modulate, and amplify a signal in the MHz range and then upconvert it in frequency into the GHz range. Figure 7.9 shows the placement of a mixer in a receiver front end with the schematic symbol most commonly used for mixers. Sometimes X is used instead of I to denote the I-port. For testing purposes, attenuators are placed on all three ports for better matching and to dampen intermodulation products exiting the mixer so that they are minimized in power level before the system reflects them back into the mixer to remix and cause further intermodulation products. When matching and intermodulation products are a problem in a system, isolators are used instead of attenuators on the R- and I-ports so that system sensitivity is not degraded, and on the L-port if LO power is limited.

Since a mixer converts modulated power from one frequency to another, it is sometimes called a frequency converter, but the

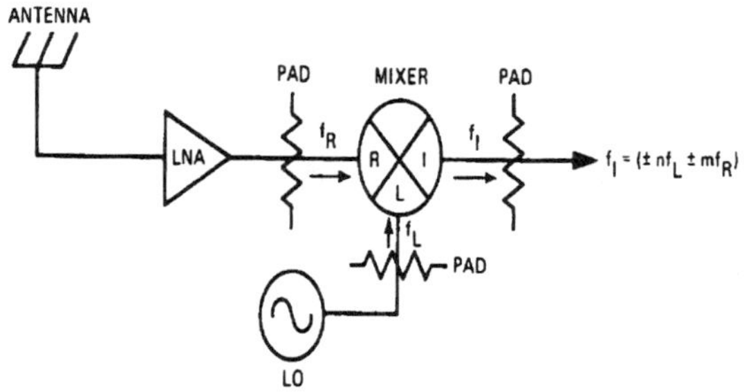

Figure 7.9 Schematic diagram showing mixer placement in a receiver front end.

term frequency converter usually implies a mixer/amplifier or mixer/oscillator combination. The term mixer more closely describes the mechanism through which frequency conversion occurs. Two inputs are mixed by means of nonlinearities and switching to produce a group of signals having frequencies equal to the sums and differences of the harmonies of the two input signals. The input signal to the mixer that has the desired information modulated onto it is called the received (input RF) signal, or f_R. The other input signal to the mixer, designated as f_L, is called the local oscillator (LO) signal, since it is generated by an oscillator physically located near the mixer in the system.

The LO signal is usually much stronger than the received signal; this causes the mixer to have better intermodulation suppression than would be possible if the LO and RF power levels were similar. The LO signal should be, in most cases, at least 20 dB higher in power than the RF input. The mixer output signal is the intermediate frequency and is designated f_I. The intermediate frequency is so termed because it falls between the RF and information frequencies.

This is simply stated as:

$$f_I = \pm m f_R \pm n f_L$$

where $m, n = 0, 1, 2, 3, ...$ The output products most generally desired are the sums and differences of the fundamentals of the received and LO signals. This is the case for which $m = n = 1$, giving

$$f_I = \pm f_L \pm f_R$$

While this formula implies that negative frequency products occur, these can be ignored in practical mixer applications in much the same way as incorrect roots can be ignored when calculating quadratic equations. For the case where $f_L > f_R$, which is called high-side LO, $f_I, = f_L \pm f_R$. When $f_L < f_R$, which is called low-side LO, $f_I = \pm f_L + f_R$. Hereafter, in this discussion, the definition of the intermediate frequency products will be restricted to include only the four intermodulation products for which $n = m = 1$. The higher-order products having $m, n = 1, 2, 3, 4...$ for which m and n are not simultaneously equal to 1, will be referred to as higher-order intermodulation products. Two other possible cases are $m = 1, 2, 3, ...,$ $n = 0$ and $m = 0$, $n = 1, 2, 3$. In these cases, the fundamental and harmonics of the received and local oscillator signals, respectively, leak through the mixer to appear at the IF output port. This is caused by finite interport isolation, and occurs to a varying extent in all mixers. A mixer is a three-port device, having two input ports and one output port. The port through which the received signal enters the mixer is called the R-port, and the port through which the local oscillator signal enters the mixer is called the L-port. The port through which all the output products exit the mixer is called the I-port. A mixer can also be a four-port device if it uses a dc bias for starved LO operation, which generally means LO input power is in the range of 0–6 dBm. Normal mixers using only the LO power to turn on the diodes require +6 to +20 dBm of LO power. Most mixers use Schottky barrier diodes, but GaAs diodes are sometimes utilized for operation in the millimeter-wave frequency range. Mixers also use bipolar transistors, J-FETs, and GaAs FETs, all of which require a fourth port for a dc voltage. There are many parameters to consider when choosing a mixer; an introduction to the most important of these follows.

Since a mixer converts power from one frequency to another, perhaps the most fundamental parameter is the measure of how efficiently frequency conversion occurs. This parameter is called conversion loss, and is defined as the difference in dB between the received signal power entering the R-port and the output IF power of the desired IF sideband exiting the I-port. Both the up and downconverted products, or sidebands, exit the L-port. Since normally only one of these products is desired, the other product is filtered out, causing half the downconverted power to be lost. Hence, there is an automatic 3-dB single sideband (SSB) conversion loss minimum. Further power losses during frequency conversion occur because some of the downconverted power is also lost in the form of unwanted higher-order mixing products, heat due to the series resistance of the diodes, and mismatches at the mixer. These all add to cause typical SSB conversion loss to range from 6 to 9 dB. Conversion loss is a strong function of LO power, which radically affects mismatch between the system and mixer.

VSWR is the measure of mismatch offered to the system by the mixer, and is usually specified over a given bandwidth as a function of LO power and temperature. It is calculated as follows.

$$VSWR = \frac{1 + |\rho|}{1 - |\rho|}$$

where $\rho = (Z_L - Z_O)/(Z_L + Z_O)$ and ρ is the reflection coefficient, Z_L is the input impedance of the mixer, Z_O is the characteristic impedance of the system. Since the VSWR does not include the phase of the reflection coefficient, the system designer does not know if the input impedance is above or below the normal 50-ohms characteristic impedance. For example, if the L-port VSWR is 2:1, measured in a 50-ohm system, the system designer does not know if the L-port input impedance is 25 ohms or 100 ohms. Actually, the input impedance of a broadband mixer swept over a frequency range of an octave or more, usually rotates through the low and high impedances, roughly producing a circle centered at 50 ohms, as viewed on a Smith chart. So a given mixer having L-VSWR of 2:1 over an octave bandwidth will have an input impedance varying from 25 ohms to 100 ohms, passing through an infinite number of complex impedance combinations as the LO frequency changes. R,

L, and I VSWRs are direct functions of LO power, which establishes the operating point of the diodes. Changing the LO power alters the diode operating point, resulting in a different impedance for all mixer ports, causing a corresponding change in VSWR. RF input power, which is at least 20 dB lower than LO input power, does not appreciably alter the diode bias point and, consequently, has little affect on VSWR. When the diode impedance changes, the input impedances of all three ports change. Hence, varying the LO power level will affect the VSWR of all three ports.

One mark of a good mixer design is that its VSWRs are optimized for the LO power that is in the middle of the normal operating power range of the mixer diodes used. This allows for good VSWRs over the maximum range of LO power levels. When designing a mixer, the L-VSWR is first optimized by adjusting the L-port circuit, allowing the LO power to properly bias the diodes and set the R- and I-port VSWRS. Then, the R- and I-port circuits are adjusted to properly match the diodes to the RF input and IF output loads.

Interport isolation is the measure of insertion loss between any two mixer ports. It is measured in dB and usually specified over a given bandwidth as a function of LO drive and temperature. Maximizing isolation between ports in mixers is necessary because unwanted signal feed-through wastes RF power and can obscure the desired IF output, as well as cause electromagnetic interference. Normally, only the isolation between L- and R-, and L- and I-ports is specified, because the LO input power, after leaking through the mixer, is comparable to the output IF power, whereas the RF input power usually is not. For instance, if the RF input power is −10 dBm and the R-to-I isolation is 20 dB (both are typical numbers), −30 dBm of RF power leaks out the I-port. If SSB conversion loss is 6 dB, the desired IF signal level is −16 dBm, which is 14 dB higher in power than the undesired RF feed through signal, also exiting the I-port. Such a relative power difference is usually sufficient. If LO power is +20 dBm and L-to-I isolation is 30 dB, −10 dBm of LO power leaks out the I-port, which is 6 dB higher in power than the −16 dBm IF product. If the LO frequency falls inside the IF band, this feed through can seriously obscure the desired IF output product. Hence, L-to-I isolation is more important to specify than R-to-I isolation.

R-to-I isolation is specified only when the relative power level of RF feedthrough and IF output power is critical, and, only for mixers having broadband IF outputs, thus allowing the frequency of the RF feed through power to fall in the IF band.

If the LO input power is +20 dBm and the L-to-R isolation is 30 dB, −10 dBm of LO power leaks out the R-port to become incident at the amplifier or antenna feeding the R-port. When the R-port has no buffer between it and the receiving antenna, the LO feedthrough power can radiate the receiving antenna. Hence, L-to-R and L-to-I isolations are most important, and normally the only ones specified.

SSB Noise Figure

Mixer SSB noise figure is measured in dB, and is the amount of noise added by the mixer to the converted signal plus the SSB conversion loss, Noise figure is the difference in dB between the input RF SNR ratio and the output IF SNR (IF power out includes either the up- or downconverted IF product):

$$\begin{aligned}\text{SSB NF (dB)} &= \text{RF Power In (dBm)} - \text{IF Power Out (dBm)}\\ &\quad + [\text{Noise Power Out (dBm)} - \text{Noise Power In (dBm)}]\\ &= [\text{SSB Conversion Loss(dB)}]\\ &\quad + [\text{Output-to-Input Noise Ratio (dB)}].\end{aligned}$$

Like SSB conversion loss, SSB noise figure is normally specified instead of double sideband (DSB) noise figure, because the mixing process produces both up- and downconverted IF products, and normally only one of these products is desired, so the other product is discarded. This causes half the input power to be lost, making the SSB noise figure 3 dB higher than the DSB noise figure. Additive noise has three main components: Johnson (thermal), shot, and flicker noise. Johnson noise is generated by Brownian motion of electrons in the series bulk resistance of the diode, causing random voltage fluctuations to appear across it. As diode temperature increases, the electrons move faster and over a longer distance, increasing the amplitude of the noise power generated. Another source of noise is the shot effect. This noise contribution is generated by random fluctuations in diode current. Both shot and thermal noises are generated randomly, and produce relatively constant noise power (white noise) over a given bandwidth. When calculating the amount of noise added by these two sources, it is important to specify the bandwidth over which the noise power is measured, since the noise power is proportional to bandwidth. Flicker noise is also generated in diodes. Its rms power is proportional to 1/frequency, so it becomes appreciable at lower frequencies. When diodes are operated at much below 400 kHz, flicker noise may become a problem. Thermal, shot, and flicker noises are always generated, and combine with SSB conversion loss to yield the overall mixer SSB noise figure. The SSB noise figure is usually about 0.5 dB higher than SSB conversion loss.

Phase-Noise Definitions

There are a number of ways to measure oscillator phase noise. The SSB phase noise is the most common measure of oscillator phase instability. It can be directly measured on a spectrum analyzer, providing that the oscillator has low amplitude noise modulation and the spectrum analyzer local oscillators are lower in phase noise than the unit under test. This latter condition is usually the limiting factor in the sensitivity of this measurement method. When this measurement is done using an analog spectrum analyzer, the result is usually 2–3 dB better than what the oscillator is actually achieving. The prime reasons for this error are averaging done in the spectrum analyzer after the log detector and the difference between the resolution bandwidth of the analyzer and its noise bandwidth. Other methods can be used to measure phase noise with greater sensitivity and accuracy at the cost of considerably more complexity in instrumentation.

$\mathcal{L}(f)$ in dBc/Hz

SSB phase noise: This is the phase instability of the oscillator measured in the frequency domain. It is the most commonly used measurement of phase noise. A spectrum analyzer can be used to measure it if the oscillator has no amplitude noise modulation and the phase noise of the spectrum analyzers oscillators is less than the measured oscillator. The units of dBc/Hz refer to dB below the carrier measured in a 1-Hz bandwidth.

$S_{\delta f}(f)$ in Hz^2/Hz

Spectral density of the frequency fluctuations: This is the power spectral density of a frequency discriminator's output. It can be directly measured by connecting an audio spectrum analyzer to the output of a frequency discriminator whose input is the oscillator under measurement.

$S_{\delta\phi}(f)$ in radians$^2/Hz$

Spectral density of the phase fluctuations: This is the power spectral density of a phase discriminator's output. It can be directly measured by connecting an audio spectrum analyzer to the output of a phase demodulator which has its input connected to the oscillator under test.

$S_y(f)$ in $1/Hz$

Spectral density of the fractional frequency fluctuations: This is $S_{\delta f}(f)$ divided by the oscillator frequency squared. The main advantage of this unit of measurement is that it is invariant under frequency multiplication and may therefore be used to judge the relative quality of oscillators at different frequencies.

$\sigma_y(\tau)$

Two-point Allen variance: This is a time domain measure of oscillator instability. It can be directly measured using a frequency counter to repetitively measure the oscillator frequency over a time period. The Allen variance is the expected value of the rms change in frequency with each sample normalized by the oscillator frequency.

β_f in Hz

Incidental frequency modulation: This is a measure of the rms frequency instability over a band of offset frequencies. It can be calculated by taking the square root of the spectral density of the frequency fluctuations integrated from a lower frequency limit to an upper frequency limit. It can be directly measured by passing the output of a frequency discriminator, whose input is the oscillator under test, through a bandpass filter and determining the rms frequency variation.

β_ϕ in radians

Incidental phase modulation: This is a measure of the total rms phase instability over a band of offset frequencies. It can be calculated by taking the square root of the spectral density of the phase fluctuations integrated from a lower frequency limit to an upper frequency limit. It can be directly measured by passing the output of a phase discriminator, whose input is the oscillator under test, through a bandpass filter and determining the rms phase variation.

f in Hz

Offset frequency: This is the frequency of the phase or frequency fluctuations. When the oscillator is directly viewed on a spectrum analyzer, this becomes the offset from the carrier frequency.

f_o in Hz

Frequency of carrier: This is the frequency of the oscillator which is being measured.

Incidental frequency modulation (IFM) is often used to specify overall oscillator instability. For IFM to be well defined, it should always be specified with a lower and upper frequency limit. For FM receivers, these limits are normally set to the lower and upper limits of the video pass-band. For other types of receivers the upper limit should be set equal to the IF bandwidth. If no upper limit is set in an IFM specification, then its magnitude tends to become very large. For phase-modulated signals, incidental phase modulation is preferred over IFM, since it provides a better measure of overall oscillator instability for that type of signal. All of these measures of phase noise can be related to each other as shown in Figure 7.10. Some of these formulas only apply under special conditions. SSB phase noise can only be converted from the various spectral density measures if the power in the phase fluctuations at frequencies greater than the offset frequency is much less than 1 radian. The offset frequency at which this condition becomes valid can vary from tens of hertz to tens of kHz, depending on the quality of the oscillator. The Allen variance can be directly computed from the fractional frequency fluctuations. However, the reverse is not true unless an assumption is made about the power law slope of the spectral density of the fractional frequency

FUNDAMENTAL RELATIONSHIPS

$$S_y(f) = \frac{1}{f_o^2} S_{\delta f}(f)$$

$$S_{\delta\phi}(f) = \frac{1}{f^2} S_{\delta f}(f)$$

$$\pounds(f) = \frac{1}{2} S_{\delta\phi}(f) \qquad \text{if} \int_f^{\infty} S_{\delta\phi}(f')\, df' \ll 1 \text{ rad}^2$$

$$\delta y^2(\tau) = 2\int_0^{\infty} S_y(f) \left[\frac{\sin^2(\pi f\tau)}{(\pi f\tau)^2} - \frac{\sin^2(2\pi f\tau)}{(2\pi f\tau)^2}\right] df$$

$$\beta_f = \sqrt{\int_{f_a}^{f_b} S_{\delta f}(f)\, df}$$

$$\beta_\phi = \sqrt{\int_{f_a}^{f_b} S_{\delta\phi}(f)\, df}$$

If $\int_{f_a}^{\infty} S_{\delta\phi}(f)\, df \ll 1 \text{ rad}^2$, then

$$\beta_f = \sqrt{2\int_{f_a}^{f_b} f^2 \pounds(f)\, df}$$

$$\beta_\phi = \sqrt{2\int_{f_a}^{f_b} \pounds(f)\, df}$$

FREQUENCY MULTIPLICATION RULE

$$\pounds_{Mf_o}(f) = \pounds_{f_o}(f) + 20 \log (M)$$

Figure 7.10 Phase-noise relationships.

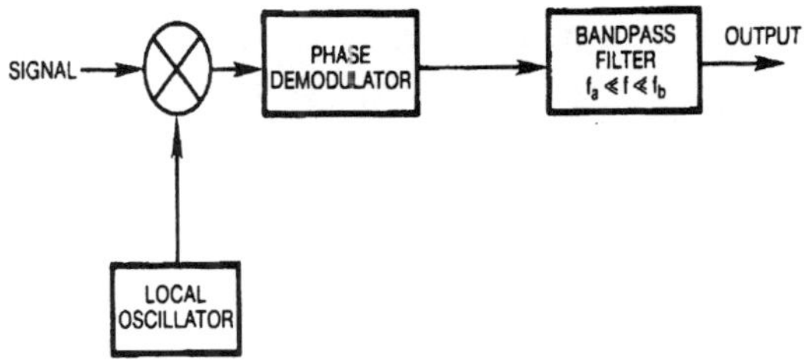

Figure 7.11 Simplified block diagram of a phase modulation receiver.

fluctuations. The frequency multiplication rule relates the increase in the SSB phase noise to multiplication integer, M. If an oscillator is multiplied in frequency by a factor of 10 in an ideal multiplier, the oscillator's SSB phase noise will increase by 20 dB. Similarly, if the oscillator's frequency is divided by 10 in an ideal frequency divider, its SSB phase noise will decrease by 20 dB.

Incidental Phase Modulation Effect

The local oscillator phase noise can limit the SNR of a phase-modulated signal to which the receiver is tuned. A simplified block diagram of a receiver is shown in Figure 7.11. In this case, the limiting SNR is determined by the power in the phase modulation divided by the incidental phase modulation squared,

$$\mathrm{SNR}_\theta = \frac{K_p^2 \int_{f_a}^{f_b} S_y(f)df}{\beta_\phi^2}$$

where K_p is the phase modulator gain constant.

For biphase shift keying, the phase shift is integer multiples of 180 degrees; for quadriphase shift keying, it is multiples of 90 degrees; and for eight-phase shift keying, it is multiples of 45 degrees. Other forms of digital transmission are used, but these usually involve both amplitude and phase-shift keying of the carrier. Local oscillator phase noise will effect the BER performance of a phase-shift keyed digital transmission system. A transmission error will occur any time the local oscillator phase, due to its noise, becomes sufficiently large that the digital phase detection makes an incorrect decision as to the transmission phase. For instance, a QPSK transmission system will make a transmission error if the instantaneous oscillator phase is offset by more than 45 degrees since the phase detector will determine that baud to be in the incorrect quadrant. Digital transmission systems with smaller phase multiples are more sensitive to degradation due to local oscillator phase noise.

The BER degradation due to local oscillator phase noise can only be determined if the probability distribution of the local oscillator phase is known. This cannot be determined uniquely from the measurement of the phase noise without using a detailed model of the oscillator. Furthermore, if the oscillator is within a phase-locked loop, the probability distribution of the phase will be modified by the parameters of the phase-locked loop. For these reasons, it is not practical to attempt to predict the exact degradation in the BER for a nonspecific case. However, a rule of thumb can be used to predict the system performance. The rule states that for BERs greater than 10^{-6}, the system performance can be maintained to within a few dB of theoretical BER for that modulation type if the incidental phase modulation of the local oscillator is less than one-tenth of the minimum phase step of the phase-shift keyed carrier. The incidental phase modulation should be computed from the natural frequency of the carrier recovery phase-lock loop to one-half of the IF bandwidth. For instance, in a QPSK system, the incidental phase modulation should be less than 9 degrees rms to meet this rule.

Example

A receiver has a local oscillator SSB phase noise given in the table below. What is the incidental phase modulation of the local oscillator integrated from 100 Hz to 1 MHz ?

- 100 Hz: −70 dBc/Hz;
- 1 kHz: −70 dBc/Hz;
- 10 kHz: −70 dBc/Hz;
- 100 kHz: −90 dBc/Hz;
- 1 MHz: −120 dBc/Hz.

The numerical evaluation of this integral yields

$$\beta_\phi = \sqrt{2\int_{100Hz}^{1MHz} \mathcal{L}(f)df} = 0.062 \text{ radians} = 3.6^\circ\text{rms}.$$

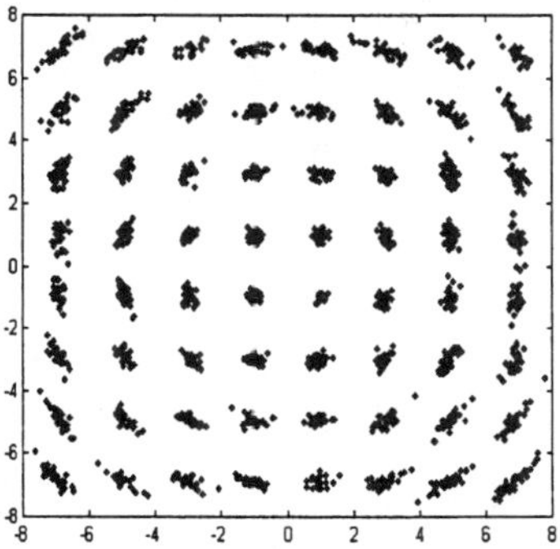

Figure 7.12 Phase-noise effects on a 64-QAM signal.

The minimum phase error that will cause a symbol error in QAM is 16.9, 7.7, 3.7, and 2.6 degrees for 16-QAM, 64-QAM, 256-QAM, 512-QAM, respectively.

The effect of the phase noise in the example on a 64-QAM signal is shown in Figure 7.12. When the signal has a multicarrier format the analysis is more complicated but approachable. There is a formula which derives γ_{PN} from the phase noise power density $|\Phi(f)|^2$ of the local oscillator (typically available from measurements) in a generic multicarrier system. For the lth subcarrier with a subcarrier spacing f_s, we have

$$\gamma_{\mathrm{PN}} = -\left[\sum_{k\neq l}\int_{-\infty}^{+\infty} \mathrm{sinc}^2(f/f_s)\left|\Phi\left(f-(k-l)f_s\right)\right|^2 df\right]_{\mathrm{dB}} \qquad (7.32)$$

where $\mathrm{sinc}(x)$ is a $\frac{\sin(x)}{x}$ weighting function. This formula is based on the fact that the MSE in the phase due to the phase noise is given by

$$\sigma_\theta^2 = \int_0^\infty 2\cdot\mathcal{L}(f)|W(f)|^2 df,$$

where $W(f)$ is the transfer function from the oscillator to the demodulator output and $\mathcal{L}(f)$ is the one-sided phase noise spectrum. Equation (7.32) is useful in the specification of oscillators phase noise for a multicarrier or single carrier system [of course, in (7.32) the summation $\sum_{k\neq l}$ disappears in the single-carrier case]. Basically from

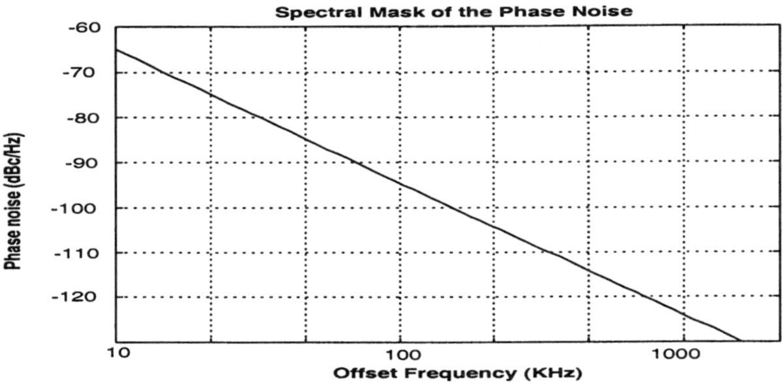

Figure 7.13 Typical spectral mask for oscillator phase noise.

a spectral mask for the phase noise, it is possible to compute (7.32) for a given frequency spacing (or bandwidth). An example of a typical spectral mask for phase noise is shown in Figure 7.13.

Reciprocal Mixing

Reciprocal mixing will cause the receiving system to lose sensitivity when there is a strong signal near the frequency to which the receiver is tuned. This effect is due to the phase noise of the local oscillator modulating the carrier of the strong signal. The carrier is spread in frequency by the phase noise modulation, which results in a power spectral density that is proportional to the local oscillator's SSB phase noise. When the receiver is tuned to a frequency near the strong carrier, the power density in the strong carrier's noise sidebands may exceed the noise floor of the receiver. If it does exceed the noise floor, then the receiver sensitivity is limited by reciprocal mixing. This effect is illustrated by Figure 7.14. The receiver is tuned into the frequency range of three carriers. The strongest carrier is in the center, with a weaker carrier on each side. The LO has an SSB phase noise that decreases with increasing offset frequency. The three carriers will appear at the IF output, and each carrier will have been modulated by the local oscillator phase noise. The phase noise on each carrier will dominate until the noise power is reduced to the noise floor of the receiver. Otherwise, the noise floor is flat across the IF passband. The central carrier is the strongest, and therefore exhibits the strongest phase-noise component. The weaker carrier on

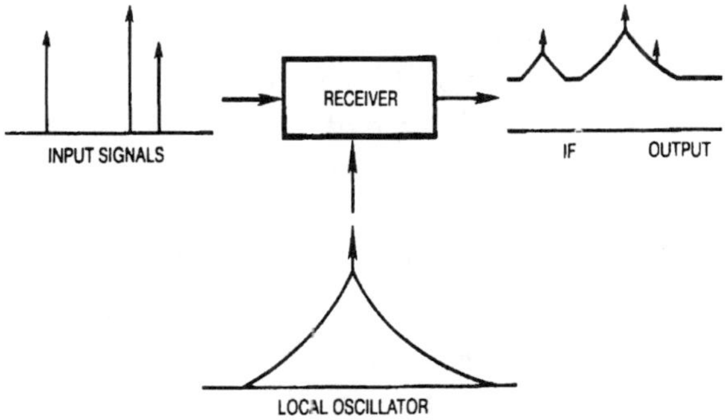

Figure 7.14 Reciprocal mixing model.

the left has a smaller range over which its phase noise dominates. The weaker carrier on the right is nearly masked by the phase noise from the strong carrier. If the receiver was tuned to this carrier, it would achieve a much worse SNR performance than would be predicted from the receiver's noise figure. This poorer performance is due exclusively to the local oscillator phase noise.

The increase in the noise floor of the receiver can be computed using the following methodology: The receiver noise floor in a 1-Hz bandwidth is the sum of the receiver's noise figure, NF, in dB and −174 dBm/Hz, $P_n = \mathrm{NF} - 174$ (in dBm/Hz). The noise generated in the receiver from a nearby carrier is the sum of the carrier power, P_c, in dBm and the SSB phase noise of the local oscillator at an offset frequency equal to the difference between the carrier frequency and the frequency to which the receiver is tuned.

$$P_o = P_c + \mathcal{L}(f).$$

The apparent noise floor of the receiver is the sum of these two powers. To compute this sum, the powers must be converted to the absolute power, summed, and then converted back to dBm. If it is necessary to compute the apparent noise floor at different frequencies, this process can be repeated at the desired offset frequencies. The net effect is that the receiver's apparent noise floor decreases as the receiver is tuned away from the carrier until it reaches the underlying noise floor generated by the receiver's front end.

Example

A receiver with a 15-dB noise figure is tuned to a carrier with a −20 dBm power level. What is the equivalent receiver noise figure 1 MHz from the carrier when $\mathcal{L}(1 \text{ MHz}) = -120$ dBc/Hz? The noise floor due to the receiver's front end is $P_n = NF - 174 = -159$ in dBm/Hz. At a 1-MHz offset frequency, the noise power due to the local oscillator phase noise is $P_o = P_o + \mathcal{L}(1 \text{ MHz}) = -140$ dBm/Hz.

The sum of these two powers at this offset from the carrier is an apparent noise floor of −140 dBm/Hz. The equivalent noise figure of the receiver is the difference between the apparent noise floor and −174 dBm/Hz, $\text{NF}_{\text{eq}} = 34$ dB. The receiver noise figure is increased 19 dB when it is tuned 1 MHz away from the −20 dBm carrier.

7.5.4 Self-Generated Interference

Of course, we have to remember that the signal received is generated by multiple antennas. As we saw in Chapter 4, we can analyze the effect of the multiantenna system accounting for the *residual* interference after processing of the multichannel baseband signal. A possible approach to assess the degradation would be to consider the expressions derived in Chapter 4. The CNR γ should be defined as

$$\gamma = \left[\frac{1}{\frac{(N+U-1)}{N\mathbf{v}^H\mathbf{P}\mathbf{v}} 10^{-0.1\gamma_{\text{NF}}} + 10^{-0.1\gamma_{\text{PN}}} + 10^{-0.1\gamma_{\text{IM}}}} \right]_{\text{dB}},$$

where $\mathbf{v}$ was defined in (4.31) and $\mathbf{P}$ in (4.34). Of course, this approach has limited applicability because it assumes that the channel is nonfrequency selective and slowly varying. It is, however, a valid assessment of the performance of the RF front ends.

7.5.5 Power Amplifier Nonlinearities

Nonlinear distortions in a high capacity microwave digital radio system are essentially due to the transmitter (high) power amplifier (HPA) which is typically driven close to saturation for power efficient transmission. The nonlinear effects on the transmitted signals are

1. Spectral spreading;
2. Intermodulation effects on the subcarriers (if any);
3. Warping of the signal constellation.

The input/output relation on the complex baseband signal $\tilde{s}(t)$ for a power amplifier is $s_u(t) = \tilde{s}(t)\mathcal{G}(|\tilde{s}(t)|)$ with $\mathcal{G}(|\tilde{s}(t)|) = \mathcal{A}[|\tilde{s}(t)|]\, e^{j\Psi[\tilde{s}(t)]}/|\tilde{s}(t)|$. The functions $\mathcal{A}[\cdot]$ and $\Psi[\cdot]$ characterize the AM/AM and AM/PM conversions in the HPA, respectively.

For the traveling-wave tube amplifier (TWTA), according to the Saleh model [10], these functions are

$$\mathcal{A}[r] = \frac{\nu r}{[1+\beta_a r^2]} \quad \Psi[r] = \frac{\alpha_\phi r^2}{[1+\beta_\phi r^2]}$$

where ν is the small-signal gain, and $1/\sqrt{\beta_a}$ is the amplifier input saturation voltage. The maximum output amplitude is $A_0 = \nu A_s/2$. α_ϕ/β_ϕ is the maximum phase-displacement that may be induced on the amplified signal. A common choice of the above parameters is the $\nu = 1$, $\beta_a = 0.25$, $\alpha_\phi = \pi/12$, $\beta_\phi = 0.25$. The model proposed in [11] for the solid state power amplifier (SSPA) is

$$\mathcal{A}[r] = \frac{\nu r}{[1+(\nu r/A_0)^{2p}]^{1/2p}}, \qquad \Psi[r] \simeq 0$$

where $A_0 = \nu A_s$ is the saturating amplitude, ν is the small-signal gain and p is an integer. A_s is typically equal to $\frac{1}{\sqrt{0.25}}$. Nonlinear distortions depend on the input backoff (IBO) or the output backoff (OBO)

$$\mathrm{IBO} = 10\log_{10}\frac{A_s^2}{\langle P_{\mathrm{IN}}\rangle}, \quad \mathrm{OBO} = 10\log_{10}\frac{A_s^2}{\langle P_{\mathrm{OUT}}\rangle}.$$

Following [12] we can characterize the distortions (seen at the receiver on the kth symbol, a_k) due to power amplification at the transmitter as

$$z_k = \alpha a_k + n_k + \eta_k$$

where

$$\alpha = R^*_{s_u}(0)/R_{\tilde{s}}(0)$$

with $R_{\tilde{s}}(\tau) = E\{\tilde{s}(t+\tau)\tilde{s}(t)^*\}$ and, as proved in [12],

$$\begin{aligned} R_{s_u}(\tau) &= E\{s_u(t+\tau)s_u(t)^*\} \\ &= \frac{\rho(\tau)}{2\sigma^2}\sum_{k=0}^{\infty}\frac{1}{k(k+1)}\left|\int_0^\infty dr r^2 \frac{r}{\sigma^2(1-\rho^2(\tau))}G(r^2)\right. \end{aligned}$$

$$\cdot\exp\left\{-0.5\frac{r^2}{\sigma^2(1-\rho^2(\tau))}\right\}\cdot\left[\frac{0.5\rho(\tau)r^2}{\sigma^2(1-\rho^2(\tau))}\right]^k\Bigg|^2. \tag{7.33}$$

The function $\rho(\tau)$ is

$$\rho(\tau)=\frac{R_{\tilde{s}}(\tau)}{R_{\tilde{s}}(0)}.$$

The multiplicative term α can be (optimistically speaking) removed at baseband. The additive term η_k is a Gaussian term whose in-phase and in-quadrature variances are given by

$$\sigma^2_{I,\eta_k}=\sigma^2_{Q,\eta_k}=\frac{\mathcal{F}[R_{s_u}(\tau)](0)+\mathcal{F}[R_{\tilde{s}}(\tau)](0)}{2|\alpha|^2},$$

where $\mathcal{F}[x(t)](f)$ is the Fourier transform of a generic signal $x(t)$. The fundamental assumption is that the complex envelope of the signal is well approximated by a Gaussian random process. This assumption is particularly true when $\tilde{s}(t)$ is a multicarrier signal with large number of carriers. Evidently an equivalent loss in CNR can be quantified from the numerical values of $\sigma^2_{I,\eta_{k,l}}+\sigma^2_{Q,\eta_{k,l}}$.

7.6 Description of the Hardware Architecture

A hardware prototype developed by WJ Communications enabled up to 6–12 transmit antennas and 12–24 receive antennas over a bandwidth up to 5 MHz, with an approximate power at each transmit antenna equal to 25 dBm and at 2.5 GHz. We point out that the prototype presented in [6] covers approximately 1% of the bandwidth of our prototype and it explores the indoor flat channel at PCS frequencies ($\simeq$ 1.9 GHz). Figures 7.15 and 7.16 show the architecture of the transceiver. The transmit and receive antennas (Gain = 7 dBi, 70 degrees azimuthal 3-dB beamwidth, 60 degrees vertical beamwidth) are connected to the site RF distribution (not shown for simplicity). The wideband RF modules (one per antenna, see Figure 7.15) are in charge of filtering, amplifying, downconverting, and digitizing the RF spectrum of interest. The wideband receiver is designed to be driven from a site RF distribution network and minimizes this interface requiring a single, low-gain RF feed per

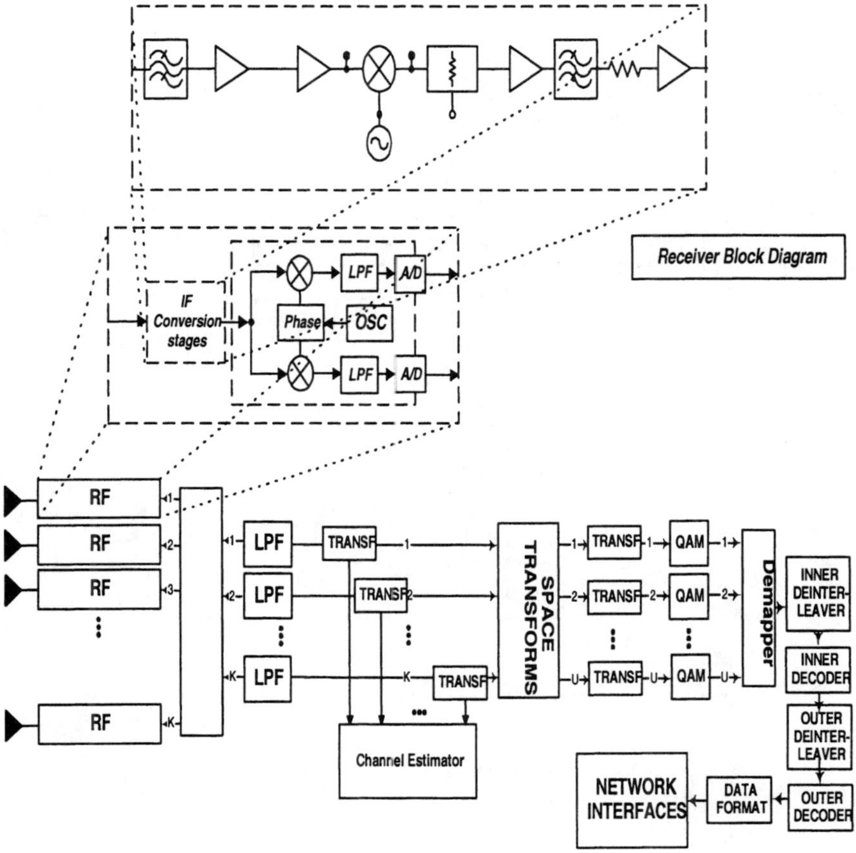

Figure 7.15 The hardware architecture of the transceiver (receiver section) developed by WJ Communications. DSP data flow is only shown at the functional level.

antenna element. High dynamic range amplifiers and mixers are utilized to obtain a large instantaneous dynamic range preserving signal fidelity. The signal as collected by the antenna is filtered and amplified. It is then mixed by a first local oscillator and filtered, amplified, and mixed down to low IF by an analog quadrature downconverter. The clock to the analog to digital converters is

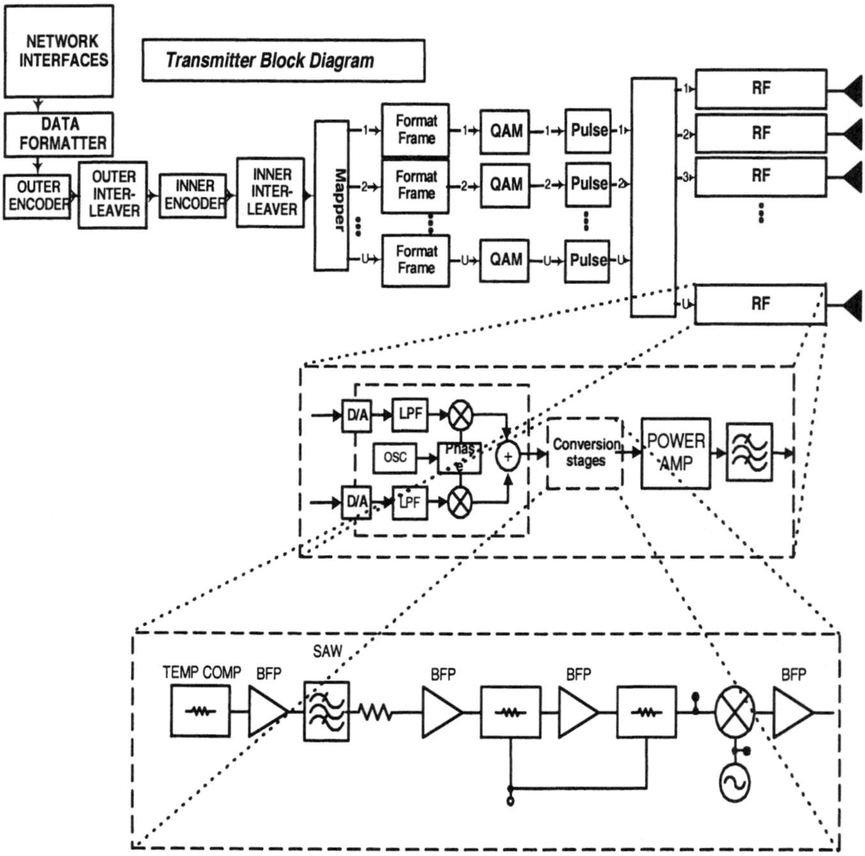

Figure 7.16 The hardware architecture of the transceiver (transmitter section) developed by WJ Communications. DSP data flow is only shown at the functional level.

generated by a frequency source locked to the site frequency reference. The wideband RF transmitter module is illustrated in

Figure 7.16. The samples from the DSP Bus are routed to the digital-to-analog converters in complex format and quadrature upconverted to an IF (300 MHz). The RF signal is further filtered, amplified, and upconverted. After proper amplification and filtering the signal goes to a power amplifier that is responsible for providing enough RF power to the signal to reach the remote receiver after antenna radiation. The characteristics of the radio are as follows:

- Noise figure 5 dB;
- 1-dB bandwidth +/−2.05 MHz;
- 3-dB bandwidth +/−2.5 MHz;
- Group delay variation: 60 ns p/p over 1-dB bandwidth;
- Phase noise:
    - −80 dBc at 1 kHz;
    - −95 dBc at 10 kHz;
    - −120 dBc at 100 kHz;
    - −138 dBc at 1 MHz;
- Receiver dynamic range: −98 dBm to −20 dBm;
- Reference stability: +/− 2.5 ppm;
- Transmitter 1-dB compression point +27 dBm.

The transceivers are locked to a common frequency reference. DSP is shown only at the functional level: In reality the outputs of the A/D and inputs to the D/A are connected to a high-speed memory module. Such module allows back-end processing relaxing some real-time constraint. For the experimental setup frames are post-processed off-line. The signal processing functions at the transmitter include a frame formatter, QAM modulators, and pulse-shaping filters and spatial signal conditioners. The signal processing functions at the receiver include spatial signal deconditioners, front-end filters, unitary time domain transforms spatial domain processing, unitary time-domain antitransforms QAM demodulators, and pulse-shaping filters. The channel coding strategy is inspired to the concatenated approach. Encoders and interleavers are specified in [13]. Data formatting refers to the organization of the information bitstream in a way that is compatible with the particular application of the communication system. A picture of the transmitter and receiver

Figure 7.17 The transceiver rack front end for six transmit and twelve receive antennas.

rack is reported in Figures 7.17 and 7.18, while the antennas are shown in Figure 7.19.

7.7 Experimental Results

The achieved spectral efficiencies are reported in Figure 7.20. The signal modulation employed at each antenna is 64-QAM and the baseband filters $P_k(\omega)$ are raised cosine filters with roll-off equal to 0.15. The peak-to-average ratio remains within 6–9 dB, which makes the technique significantly more advantageous than OFDM. SNR is calculated as described in [6, 7]. A map of the locations used to perform the transmission experiments in the former WJ Communications work site in Palo Alto, California, is shown in Figure 7.21. Such an environment is representative of a wireless local loop application with average range equal to 0.5 miles. The delay spread reaches a worst case condition of 2 μsec. Typical channels for such an environment are shown in Figure 7.22.

The antennas are arranged in a rectangular planar configuration

Figure 7.18 The transceiver racks.

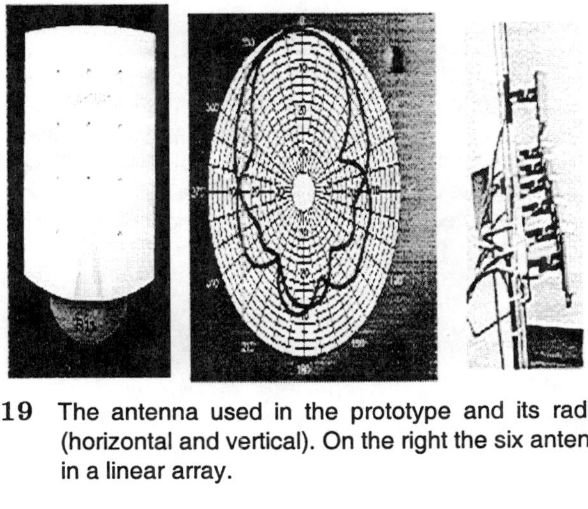

Figure 7.19 The antenna used in the prototype and its radiation pattern (horizontal and vertical). On the right the six antennas arranged in a linear array.

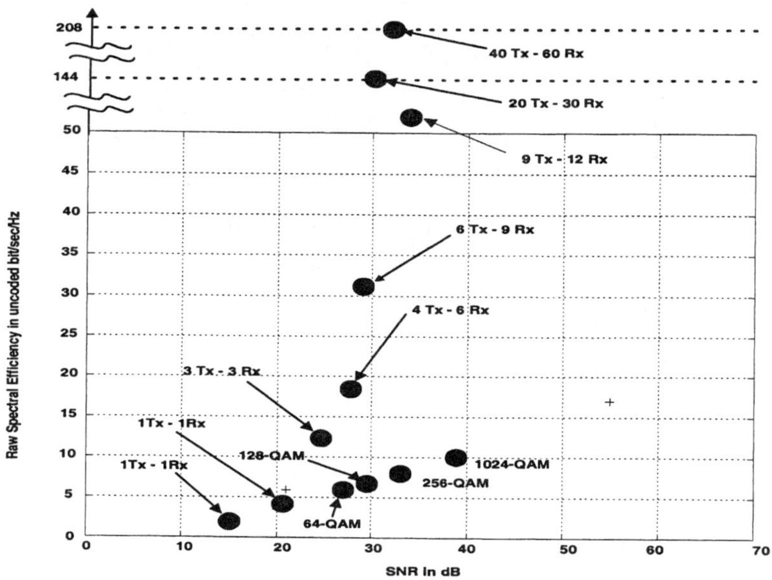

Figure 7.20 Progress of achieved raw spectral efficiencies in non-LOS wireless transmissions. For each point "NTx-MRx" indicates the use of N transmit antennas and M receive antennas. The indicated SNRs are relative to a BER $< 10^{-6}$ with an overhead equal to 28% (channel coding is a punctured convolutional code with rate 5/6 with soft-decision decoding).

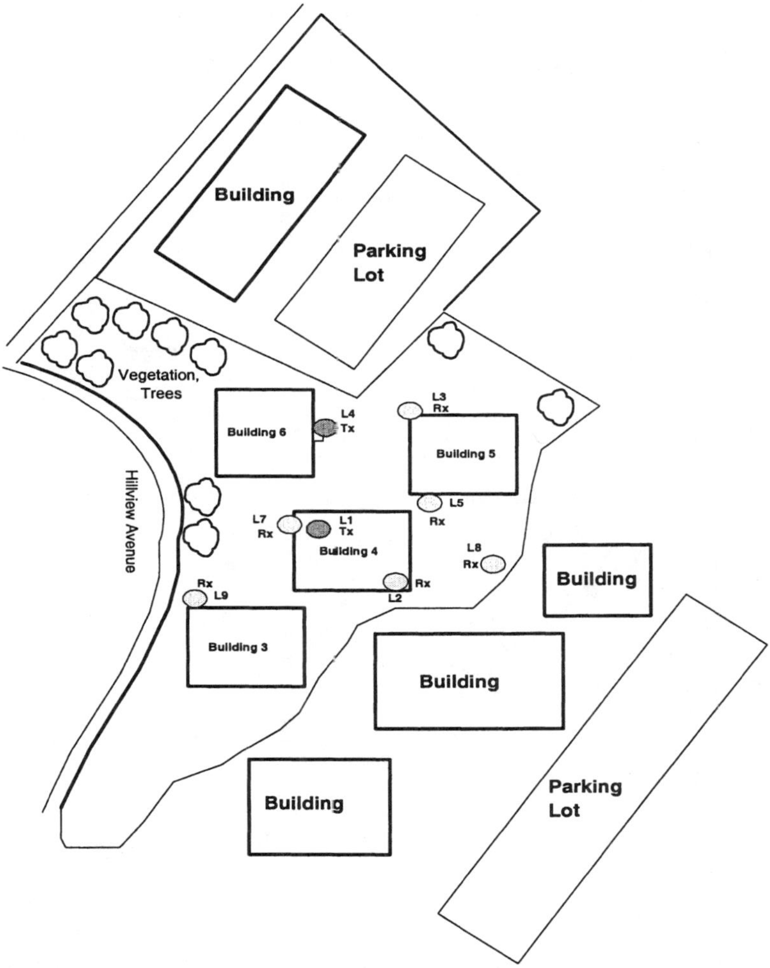

Figure 7.21 Map for the experimental outdoor measurements.

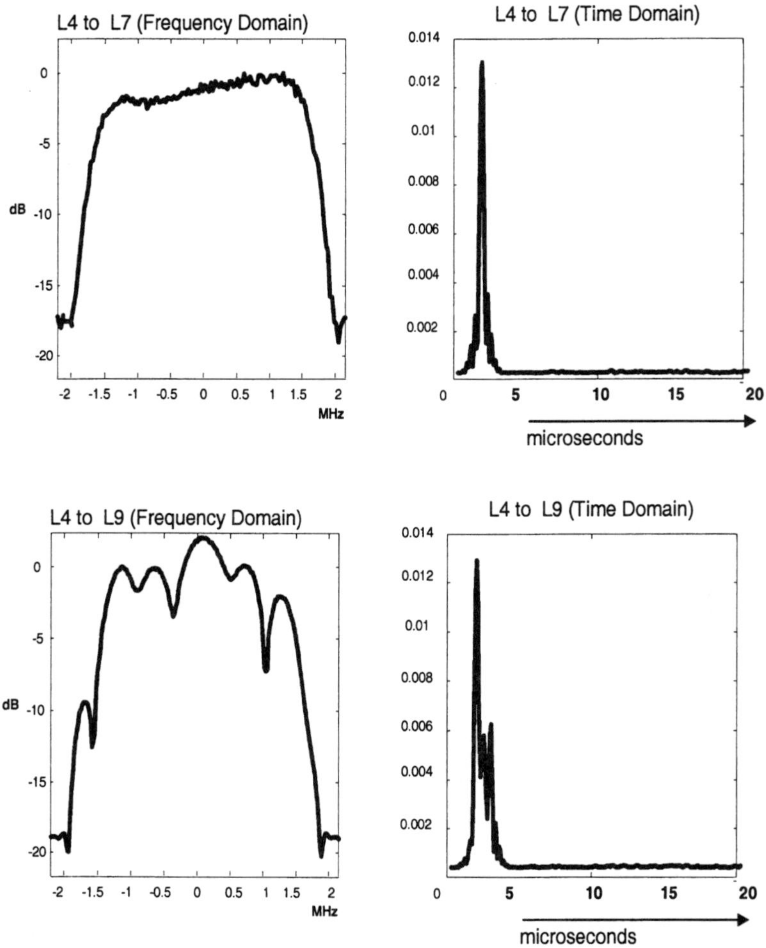

Figure 7.22 Typical responses show frequency selectivity of the channels for the experimental outdoor measurements. Refer to the map for the locations.

($\simeq$ 1 wavelength spacing), $6 \times M$, M variable. We also report results for a number of antennas larger than the ones available in hardware using the following methodology. Define M_T positions for the transmit N_{tx}-antenna array $\mathbf{r}_i$ $i = 1, 2, ..., N_{\text{tx}}$ and M_R positions for the receive N_{rx}-antenna array $\mathbf{r}'_k$ $k = 1, 2, ..., N_{\text{rx}}$. A channel for a $M_T N_{\text{tx}} \times M_R N_{\text{rx}}$ system (we mean $M_T N_{\text{tx}}$ transmitters and $M_R N_{\text{rx}}$ receivers) $\mathbf{H}_{M_T N_{\text{tx}} \times M_R N_{\text{rx}}}(\omega)$ can be thought as composed of channels $\mathbf{H}^{(i,k)}_{N_{\text{tx}} \times N_{\text{rx}}}(\omega)$ for $N_{\text{tx}} \times N_{\text{rx}}$ systems, where the transmit array is at some position $\mathbf{r}_i$ and the receive array is at some position $\mathbf{r}'_k$. So the assumption of a reasonably static channel[4] for the $\mathbf{H}_{M_T N_{\text{tx}} \times M_R N_{\text{rx}}}(\omega)$ allows us to perform the estimation of the channel with only N_{rx} elements at the receiver and N_{tx} elements at the transmitter by estimating (at different times) *smaller* channels $\mathbf{H}^{(i,k)}_{N_{\text{tx}} \times N_{\text{rx}}}(\omega)$ $i = 1, ..., N_{\text{tx}}$ and $k = 1, ..., N_{\text{rx}}$. After having performed the $M_T M_R$ channel estimations, we compose the *large* channel $\widehat{\mathbf{H}}_{M_T N_{\text{tx}} \times M_R N_{\text{rx}}}(\omega)$ using $\widehat{\mathbf{H}}^{(i,k)}_{N_{\text{tx}} \times N_{\text{rx}}}(\omega)$ $i = 1, ..., N_{\text{tx}}$ and $k = 1, ..., N_{\text{rx}}$. Subsequently we can process QAM data using DSP baseband processing on the channel $\widehat{\mathbf{H}}_{M_T N_{\text{tx}} \times M_R N_{\text{rx}}}(\omega)$. We have fundamentally exploited the channel time invariance and emulated signal collections in absence of moving objects (most of these collections were performed during night hours). A 1-Gbps experiment is performed using $M_T = 10$, $N_{\text{tx}} = 4$, $M_R = 10$ $N_{\text{rx}} = 6$, for a total of 40 transmit antennas and 60 receive antennas. We achieve a raw data rate of 960 Mbps in 4.6 MHz at SNR=30 dB (at this SNR a punctured convolutional coding scheme, rate 5/6 provides BER $< 10^{-6}$). In Figure 7.23 we show hardware results for a particular non-LOS situation. Looking at the map in Figure 7.21, the transmitter is located at location L4, while the receiver is located at location L9. The time spread is about 2.5 μsec. In Figure 7.24 we summarize the achieved data rates for BER less than 10^{-6} (with a 5/6 punctured convolutional encoder) in 4.6 MHz averaged over different locations. The data rates shown account for 28% overhead due to channel coding and frame overhead for training and time-frequency synchronization. Figures 7.25 and 7.26 show a comparison between multicarrier and single-carrier modulations.

4. The sequential channel measurements are performed coherently by locking oscillators at transmitter and receiver.

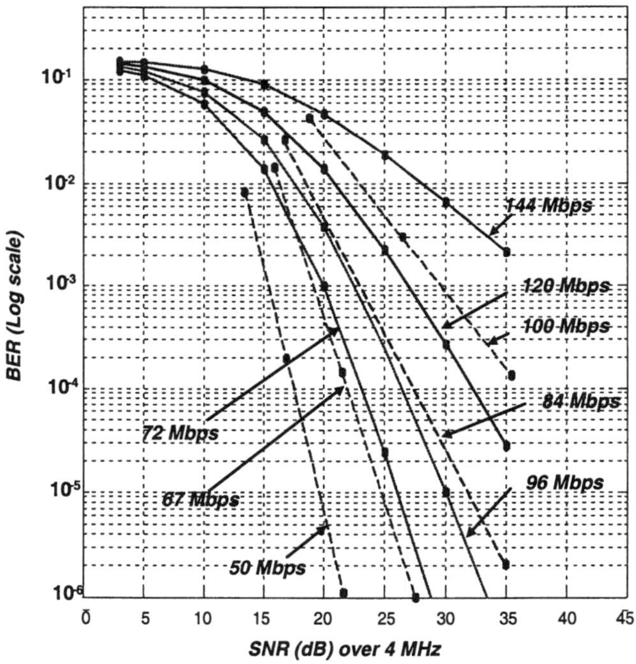

Figure 7.23 BER results for outdoor obstructed (non-LOS) in 4.6-MHz environment with three, four, and five transmit antennas and six receive antennas. Dashed curves are for punctured 3/4 convolutionally encoded 64-QAM. Solid curves are for uncoded 64-QAM.

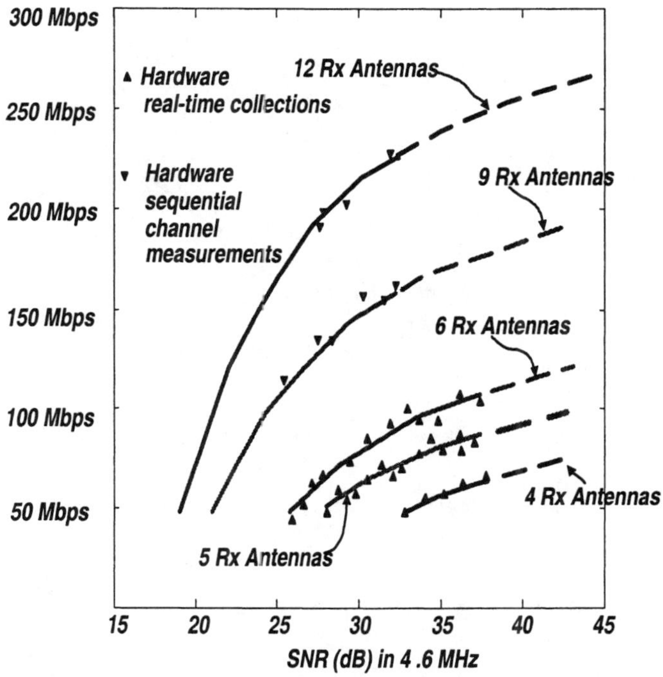

Figure 7.24 BER results averaged over different locations in 4.6 MHz. Curves are for convolutionally coded (rate 5/6) 64-QAM.

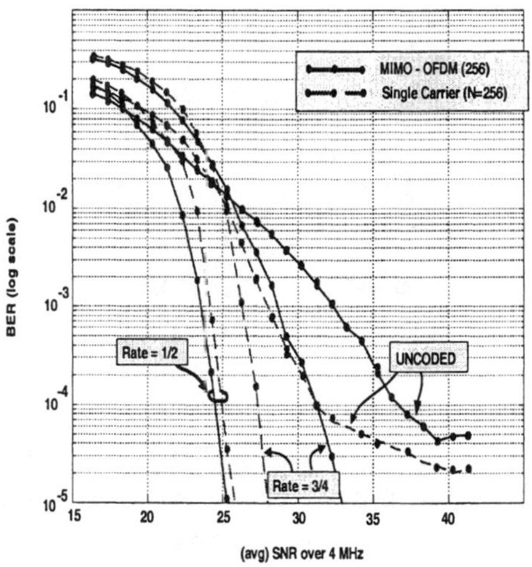

Figure 7.25 BER results for outdoor obstructed (non-LOS) in 4.6 MHz, multicarrier versus single-carrier.

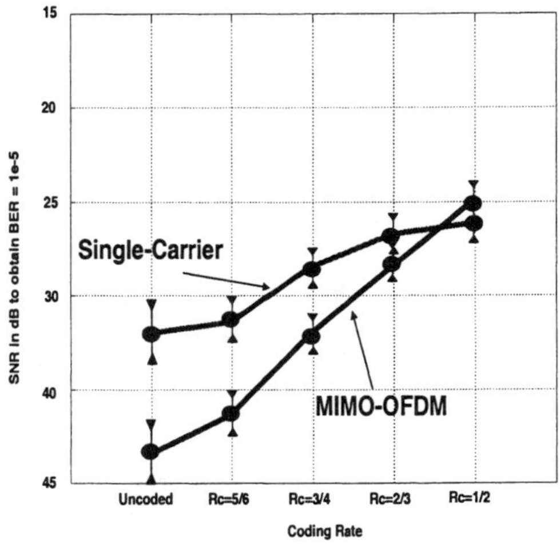

Figure 7.26 BER results for outdoor obstructed (non-LOS) in 4.6 MHz, multicarrier versus single-carrier.

7.8 Key Points of This Chapter

In this chapter we have discussed several design guidelines for the development of spatio-temporal digital radio systems. The most important RF design parameters were described and, in many instances, the steps neeeded to relate them to the end-to-end performance of the link were shown. As a case study we have considered the 12-transmit, 12-receive 2.5 GHz radio system developed by WJ Communications. We also described the results of some experimental outdoor field trials which have demonstrated the hardware feasibility of fixed broadband wireless radio links with spectral efficiencies in excess of 50 bps/Hz in severe multipath environments. Wireless data transmission in the range of 60–200 Mbps occupying a bandwidth of 2–4.6 MHz at 2.5 GHz was achieved. Using calibrated antenna collections and sequential spatio-temporal channel estimates in a static environment, we have evaluated transmission with antenna arrays with up to 40 transmit elements and up to 60 receive elements. This last experiment demonstrates the possibility of a 1-Gbps wireless transmission with a raw spectral efficiency of about 208 bps/Hz.

References

[1] P. M. Morse and H. Feshback, *Methods of Theoretical Physics, Part I*, New York: McGraw-Hill, 1953.

[2] S. L. Ariyavisitakul, J. H. Winters, and I. Lee, "Optimum Space-Time Processors with Dispersive Interference: Unified Analysis and Required Filter Span," *IEEE Trans. Comm.*, Vol. 47, pp. 1073–1083, July 1999.

[3] M. Martone, "An Adaptive Algorithm for Adaptive Antenna Array Low-Rank Processing in Cellular TDMA Base-Stations," *IEEE Trans. Commun.*, Vol. 46, pp. 627–643, May 1998.

[4] M. Martone, "On MMSE Real-Time Antenna Array Processing Using Fourth-Order Statistics in the US Cellular TDMA System," *IEEE Journal Sel. Areas in Comm.*, Vol. 16, pp. 1396–1410, Oct. 1998.

[5] M. Martone, "Cumulant-Based Adaptive Multichannel Filtering for Wireless Communication Systems with Multipath RF Propagation Using Antenna Arrays," *IEEE Trans. Veh. Tech.*, Vol. 47, No. 2, pp. 377–391, May 1998.

[6] G. D. Golden, et al., "Detection Algorithms and Initial Laboratory Results Using V-BLAST Space-Time Communication Architecture," *IEE Electronics Letters*, Vol. 35, Jan. 1999.

[7] G. J. Foschini, "Layered Space-Time Architecture for Wireless Communication in a Fading Environment When Using Multi-Element Antennas," *Bell Labs Tech. J.*, Vol. 1, No. 2, pp. 41–59, 1996.

[8] L. L. Scharf, *Statistical Signal Processing*, Reading, MA: Addison Wesley, 1991.

[9] P. P. Vaidyanathan, *Multirate Systems and Filter Banks*, Englewood Cliffs, NJ: Prentice Hall, 1993.

[10] A. A. Saleh, "Frequency Independent and Frequency Dependent Nonlinear Models of TWT Amplifiers," *IEEE Trans. Commun.*, Vol. COM-29, pp. 1715–1720, Nov. 1981.

[11] C. Rapp, "Effects of HPA-Nonlinearity on 4-DPSK-OFDM-Signal for a Digital Sound Broadcasting System," in *Proc. 2nd European Conference on Satellite Communications*, Liege, Belgium, Oct. 1991, ESA-SP-332.

[12] E. Costa, M. Midrio, and S. Pupolin, "Impact of Amplifier Nonlinearities on OFDM Transmission System Performance," *IEEE Communications Letters*, Vol. 3, Feb. 1999.

[13] "Digital Video Broadcasting (DVB): Framing Structure, Channel Coding and Modulation for Digital Terrestrial Television (DVB-T)," *European Telecommunications Standard*, ETS-300-744, March 1997.

8

Solutions for High Mobility

8.1 Introduction

In this chapter we introduce two different strategies for detection in high mobility environments. The first one is based on maximum likelihood detection (described in Chapter 5 in its traditional implementation) and the use of new wavelet-based subspaces that approximate the original (time-varying) signal subspace. The nested sequence of linear vector spaces, defined by a wavelet-based multiresolution decomposition of the fading channel time variations, provides a set of subspaces that, at increasingly high level of detail, are "efficient" representations of the original signal subspace. For each of these representations sequence detection at different levels of *resolution* can be performed. The second strategy is described within the framework of multicarrier modulation. Typically, digital Fourier transform schemes make the implementation of the multicarrier system feasible and attractive with respect to single-carrier systems. However, when the channel is doubly selective (that is, time-frequency selective) as it usually happens in the rapidly fading wireless channel, this traditional methodology fails. Since the channel frequency response is rapidly time-varying, the optimal transmission/reception methodology should be able to process nonstationary signals. In other words, the subchannel carrier frequencies should be time-varying. They should ideally decompose the frequency distortion of the channel perfectly at any instant in time. However, this ideally optimal approach presents significant challenges both in terms of conceptual and computational complexity. The idea disclosed in this work is that a nonstationary approach can be approximated using signal bases that are especially suited for the analysis/synthesis of nonstationary signals. We propose in fact the use of a multicarrier system that employs orthogonal signal bases of the chirp type that in practice correspond to the fractional Fourier transform signal basis. The significance of the methodology relies on the important practical consideration that analysis/synthesis methods of the fractional Fourier type can be implemented with a complexity that is equivalent to the traditional fast Fourier transform (FFT).

Both approaches are described in a single-antenna transmission system to simplify notation and avoid unnecessary complexity in the derivations. It is, however, important to point out that the two techniques can be applied with marginal modifications to the multi-antenna case.

8.2 Generalized MLSD

As already explained in Chapter 2, maximum likelohood sequence detection (MLSD) is the optimum strategy for sequence detection in the presence of a perfectly known channel and (white) Gaussian noise [1, 2]. When the channel is a priori unknown, a symbol decision-directed method is typically used to perform estimation [3]. This scheme implicitly assumes reliable symbol decisions and requires that the estimate is performed with a certain delay (the decision delay) with respect to the current state of the trellis search. This is a viable solution in case of a slowly fading channel. An alternative solution is to accomplish blind independent branch coefficients computation with no delay for each survivor path (PSP) [4] within the trellis search. This approach basically implements the generalized likelihood detector (GLD) [5] and is based on the assumption that the channel is static (that is, time invariant) over a certain number of symbols [4, 6, 7]. Slow variations of the channel are compensated using adaptive algorithms like the LMS or RLS [8]. If the channel coefficients variations in time are fast with respect to the convergence time of the adaptive algorithm, significant degradation may result. Other approaches model the time-varying components of the impulse response of the multipath channel as low-pass Gaussian processes with their correlation representing ensemble average characteristics. These average characteristics are in turn used to fit parametric models (typically autoregressive moving average systems) so that a bank of Kalman filters [9] can be used for tracking [10, 11]; similar in spirit is also the innovations-based approach, even if Kalman filters are not employed [12]. These ideas rely on a *global* characterization of the time-varying nature of a multipath channel: the well-known WSSUS model [13].

The WSSUS model is an approximation of the radio frequency propagation mechanism in a dense multipath environment. Indeed, the channel at any delay has smooth (low-frequency) variations with occasional rapid changes. A channel model that is uniquely based on ensemble average characteristics will inevitably lose *local* characteristics that are caused, for example, by rapid transitions and sudden changes of the impulse response dynamics. To connect this consideration with a simple and intuitive example, consider a signal made of two pure distinct oscillations occurring at non-overlapping distinct time intervals. A system of the type used in [9] can correctly model the two tones, but with *no localization in time*, which forces the wrong representation for the signal. In other words, representations based on linear models of the ARMA type lack parsimony. The time-varying reflections of electromagnetic waves caused by a moving transmitter (or receiver) are in nature very similar to a signal where multiple propagation modes are present at different times. To represent the frequency behavior of a signal locally in time, the signal should be analyzed by functions that are localized in both time and frequency (i.e., signals that are compactly supported in the time and frequency domains). In reality no function can be compactly supported simultaneously in time and frequency so the scientific community has directed intense research efforts in the development of functions with "good" time-frequency localization: Wavelets are the best known tool for linear time-frequency analysis. In this work we propose new methods that can characterize parsimoniously the time-varying nature of the detection problem, and we focus on a multiresolution representation [14, 15] of the fading process in each component of the channel response, elaborating some ideas of [16, 17] and exactly as partially presented in [18]. We will use the WSSUS model *only* to tune our channel representation and to eliminate gross redundancy: In other words, second-order statistics will provide general guidelines on how to select the low-resolution components of the channel. However, and this is the fundamental feature of the new method, we retain some high-resolution components of the channel dynamics. This provides a more parsimonious representation of real-life multipath channels and in turn a more reliable detector. It is important to mention that [19] is based on a polynomial expansion that is, in principle, very similar to the idea we want to describe. However, polynomial signal representations are typically not parsimonious.

The GLD metric, derived for a time-varying channel, involves finding the orthogonal projection of the received signal onto the subspace of the signal distorted by multipath[1]: the transmitted signal (corresponding to one of the transmitted sequences) is in fact known to lie in a subspace, but its exact location is not known, because unknown are the channel parameters. Detectors in this class are also called in signal processing *matched subspace detectors* [20] because their statistic is "matched" to the a priori known signal subspace $\mathcal{S}$. Unfortunately, the computation of the perfectly matched orthogonal projection of the received signal onto the signal subspace is, in the general time-varying case, impossible. The basic idea introduced in this work is in the use of wavelet-based subspaces $\mathcal{V}$ that approximate the original signal subspace $\mathcal{S}$ in some sense. These reduced-size signal subspaces $\mathcal{V}$ are obtained by means of a decomposition of the fading channel time variations using optimal unconditional bases such as orthonormal wavelet bases [21] in the same spirit of [14] (and in the more general framework presented in [22]).

Consider $\tilde{f}(t, \tau)$, the impulse response of a generic multipath channel at observation time t.

1. For simplicity we will denote as *signal subspace* the subspace of the signal distorted by multipath throughout the chapter.

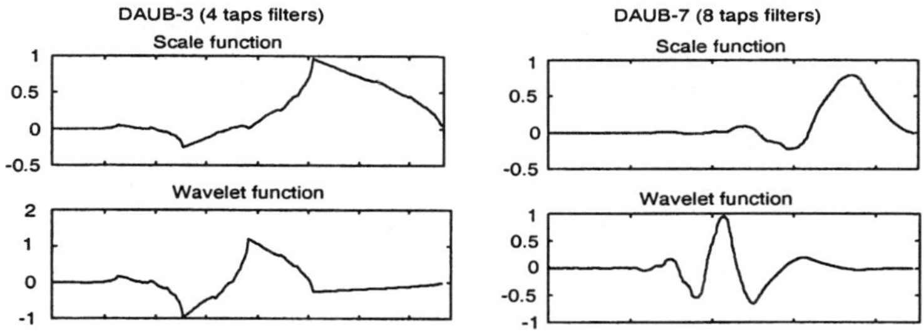

Figure 8.1 Examples of wavelet and scaling functions of the Daubechies type.

A multiresolution decomposition can be applied to $\tilde{f}(t,\tau)$ at any τ with respect to t as

$$\tilde{f}(t,\tau) = \sum_{m} \zeta(m,\tau)2^{-P/2}\phi(2^{-P}t - m) + \sum_{l=-\infty}^{P} \sum_{m} \xi(l,m,\tau)2^{-l/2}\psi(2^{-l}t - m), \quad (8.1)$$

where $\xi(l,m,\tau)$ are wavelet coefficients, $\zeta(m,\tau)$ are scaling coefficients, $\psi(t)$ is a prototype bandpass wavelet function and $\phi(t)$ is a lowpass scaling function [23, 24]. Examples of mother wavelets of the Daubechies type are reported in Figure 8.1. Observe that l indexes the *scale*, or *resolution* (the smaller the l, the higher the resolution), while m indexes the *spatial* location of analysis. If the mother wavelet is centered at time 0 and frequency f_c, $\xi(l,m,\tau)$ measures the content of $\tilde{f}(t,\tau)$ around time $2^l m$ and frequency $2^{-l}f_c$, while $\zeta(m,\tau)$ represents the local mean around time $2^P m$. The celebrated merits of this decomposition are motivated by an outstanding localization capability in the time-frequency plane (see [21]). The set of linear vector spaces defined by such decomposition provides also a nested sequence of subspaces that, ordered at increasingly high level of detail $l = P, P-1, ..., 1, 0$, are efficient representations of the original signal subspace $\mathcal{S}$. We show a method to focus these subspace representations to the original signal subspace using the concept of Kolmogorov $n$width for signal classes, and we modify this basis selection technique for the case where the multipath channel is characterized statistically according to the WSSUS assumption.

According to the model (4.42), the maximum likelihood (ML) principle the optimal metric corresponding to the hypothesis that the sequence $\boldsymbol{\alpha} = [\alpha_0, \alpha_1, ..., \alpha_{NR-1}]^T$ was transmitted[2], is

$$L_N = -\sum_{n=0}^{NR-1} |y_n - \boldsymbol{\alpha}_n^T \mathbf{h}_n|^2 = -\sum_{n=0}^{NR-1} |y_n - s_n|^2 = -||\mathbf{y} - \mathbf{s}||^2 \quad (8.2)$$

where $\mathbf{y} = [y_0, y_1, ..., y_{NR-1}]^T$, $\mathbf{s} = [s_0, s_1, ..., s_{NR-1}]^T$,

$$\boldsymbol{\alpha}_n = [\alpha_n, ...\alpha_{n-D}]^T,$$

$$\mathbf{h} = \left[\mathbf{h}_0^T, \mathbf{h}_1^T, ..., \mathbf{h}_{NR-1}^T\right]^T.$$

Since the channel is assumed unknown, the optimal metric cannot be exactly computed and the application of the generalized likelihood (GL) argument[3] results in the statistic

$$\mathcal{L}_N = ||\mathbf{P}_{\mathcal{S}}[y]||^2, \quad (8.3)$$

where $\mathbf{P}_{\mathcal{S}}[y]$ is the orthogonal projection of $\mathbf{y}$ onto the signal subspace, denoted $\mathcal{S}$ and formally expressed[4] as

$$\mathcal{S} = \left\{s_n = \boldsymbol{\alpha}_n^T \mathbf{h}_n,\ \mathbf{h}_n \in \mathcal{C}^{D+1},\ n = 0, 1, ..., NR-1\right\}. \quad (8.4)$$

In (8.3)

$$\mathbf{P}_{\mathcal{S}}[y] = \widehat{\mathbf{s}} = \mathbf{s}(\boldsymbol{\alpha}, \mathbf{h}_{[ML]}^{\boldsymbol{\alpha}}) \quad (8.5)$$

where $\mathbf{h}_{[ML]}^{\boldsymbol{\alpha}}$ is a maximum likelihood estimate of the channel coefficients $\mathbf{h}$ conditioned on the sequence $\boldsymbol{\alpha}$, the orthogonal projection $\mathbf{P}_{\mathcal{S}}[y]$ is an estimate of $\mathbf{s}$, say, $\widehat{\mathbf{s}}$, which evidently depends conditionally on the sequence $\boldsymbol{\alpha}$ and on the estimate of the channel $\mathbf{h}_{[ML]}^{\boldsymbol{\alpha}}$ given $\boldsymbol{\alpha}$. The GL statistic has been typically used under the assumption that the channel is static, $\mathbf{h}_n = \mathbf{h}_{\text{static}}$ for

$$n = 0, 1, ..., NR-1.$$

---

2. The metric is optimal only if the noise is white and Gaussian, and if the channel is perfectly known.
3. The generalized likelihood method, which implies a maximum likelihood estimate of the channel, is not the only possible approach in the joint channel-data estimation problem.
4. $\mathcal{C}^U$ is the space spanned by all complex U-vectors.

In this case the statistic (8.3) collapses into

$$\mathcal{L}_N = \mathbf{y}^H \mathbf{P}_A \mathbf{y} \tag{8.6}$$

where

$$\mathbf{P}_A = \mathbf{A}(\mathbf{A}^H \mathbf{A})^{-1} \mathbf{A}^H$$

is the projection operator with

$$\mathbf{A} = [\boldsymbol{\alpha}_C, \boldsymbol{\alpha}_1, \ldots \boldsymbol{\alpha}_{NR-1}].$$

The statistic (8.6) is obtained writing the generalized log-likelihood function as

$$L_N = -||\mathbf{y} - \mathbf{A}\hat{\mathbf{h}}_{static}||^2 = -\mathbf{y}^H \mathbf{Q}_A \mathbf{y}, \tag{8.7}$$

where $\mathbf{Q}_A = \mathbf{I}_{NR} - \mathbf{P}_A$. Evidently the maximization of $-\mathbf{y}^H \mathbf{Q}_A \mathbf{y}$ is equivalent to the maximization of $\mathbf{y}^H \mathbf{P}_A \mathbf{y}$. Equation (8.6) has motivated many algorithms for joint channel-sequence estimation [4, 7, 8]. In most real-life situations the time-varying coefficients $\mathbf{h}_n$ cannot be assumed static, so the typical approach is to compensate for slow variations of the channel using adaptive algorithms that force the ML estimate of the channel to be constantly in search of a convergence point. If the channel coefficients variations in time are fast with respect to the convergence time of the adaptive algorithm, significant degradation may result.

On the other hand, modeling the channel as time-variant, if no specific structure about $\mathbf{h}_n$ $n = 0, 1, \ldots, NR-1$ is assumed, it may be impossible to find $\mathbf{h}^{\boldsymbol{\alpha}}_{[ML]}$. In fact there may be infinitely many orthogonal projectors onto the nonuniquely defined signal subspace[5]

5. Assuming the structure for $\mathbf{h} = \left[\mathbf{h}_0^T, \mathbf{h}_1^T, \ldots, \mathbf{h}_{NR-1}^T\right]^T$ means, for example, to assume a particular autocorrelation matrix $\mathbf{R}_h = E\left\{\mathbf{h}\mathbf{h}^H\right\}$. Then one could use a truncated Karhunen Loeve transform (KLT) expansion of the form $\mathbf{h} = \mathbf{T}^H \mathbf{h}_{T-KLT}$ where $\mathbf{T}^H$ is found from the eigenvectors of $\mathbf{R}_h$, and the truncation of the expansion is such that:
1) $\mathbf{h}_{T-KLT}$ is of size $K \leq NR$,
2) The eigenvectors in $\mathbf{T}^H$ contain any *dominant* eigenvector of $\mathbf{R}_h$.
The typical spectral characteristics of the random process $\mathbf{h}_n$ (under the well-known stationary fading assumptions [25]) are such that these two conditions are easily satisfied in any case of practical interest. The signal subspace is now unique and the orthogonal projector of the received signal is computable.

for each transmitted sequence (this problem is a rather fundamental problem in estimation theory whose continuous time counterpart is pointed out in [5]).

In any case the GL statistic involves finding the orthogonal projection of the received samples y_n onto the signal subspace $\mathcal{S}$: The signal is known to lie in the subspace $\mathcal{S}$ but its exact location is unknown because $\mathbf{h}$ is unknown. The detectors described by (8.3) are *matched subspace detectors* [20], because they are matched to the signal subspace $\mathcal{S}$. Moreover, since they form their statistic according to the energy of the orthogonal projection of the received signal onto the signal subspace, they can also be interpreted as *generalized energy detectors*. This last interpretation opens many interesting ramifications [26] of this conceptual framework and leads us to the following question: Can we find an alternative subspace $\mathcal{V}$ whose orthogonal projection is computed more easily than $\mathbf{P}_{\mathcal{S}}[y]$ and is at the same time a *good approximation* of the original signal subspace $\mathcal{S}$?

In other words, can we find generalized energy detectors that are not perfectly matched to $\mathcal{S}$ but still perform satisfactorily in the sense that

$$||\mathbf{P}_{\mathcal{S}}[y]||^2 \simeq ||\mathbf{P}_{\mathcal{V}}[y]||^2 \quad ?$$

And if yes, how do we tune or focus $\mathcal{V}$ to $\mathcal{S}$? Observe that any subspace $\mathcal{V}$ that is a good approximation of $\mathcal{S}$ and whose orthogonal projection is computed more easily than the original $\mathbf{P}_{\mathcal{S}}[y]$ is a good candidate for our detector. Multiresolution analysis of the channel coefficients $h(n,k)$ provides us with a valuable tool to answer these questions.

8.2.1 Multiresolution Decompositions for the Channel

An unknown deterministic or random signal can be decomposed according to its resolution (components at different levels). We report in Appendix 8A a self-contained brief introduction to wavelet theory. The channel $h(n,k)$ at any fixed k can be considered a signal at resolution 0 with an associated signal subspace $\mathcal{V}_0 = \mathcal{S}$. It is fundamental to observe that $h(n,k)$ can be represented at lower resolution applying a half band lowpass filter having impulse response $c_0(n)$, followed by a down-sampling-by-two operation. An added detail can be obtained by using a highpass filter with impulse response $c_1(n)$, then down-sampling by two. If the filters $c_i(n)$ $i = 0, 1$ satisfy some regularity constraints [23, 24] and form an orthonormal set, then $c_i(n)$ [in the z-domain $C_i(z)$] $i = 0, 1$ is a dyadic perfect reconstruction filter bank [27]. Using recursively this method (refer to Appendix 8A for details), one obtains at a generic resolution depth P (with associated signal subspace $\mathcal{V}_P$) the following representation for the channel

$$h(n,k) = \sum_{m=0}^{NR/2^P-1} \zeta_{P,m}(k) c_0^{(P)}(2^P m - n) + \sum_{l=1}^{P} \sum_{m=0}^{NR/2^l-1} \xi_{l,m}(k) c_1^{(l)}(2^l m - n), \tag{8.8}$$

where $\zeta_{P,m}(k)$ and $\xi_{l,m}(k)$ are coefficients of the wavelet-based decomposition (WBD), and the filters $c_0^{(P)}(n)$ and $c_1^{(l)}(n)$ (with real coefficients) are given in the z-domain by $C_0^{(P)}(z) = \prod_{i=1}^{P-1} C_0(z^{2^i})$ and $C_1^{(l)}(z) = C_1(z^{2^l-1}) \prod_{i=1}^{l-2} C_0(z^{2^i})$. Using vector notation it is possible to express (8A.11) simply as

$$\begin{aligned} h(n,k) &= \boldsymbol{\zeta}_P(k)^T \mathbf{c}_0^{(P)}(n) + \sum_{l=1}^{P} \boldsymbol{\xi}_l(k)^T \mathbf{c}_1^{(l)}(n) \\ &= \left[\mathbf{c}_0^{(P)}(n)^T, \mathbf{c}_1^{(P)}(n)^T, ..., \mathbf{c}_1^{(1)}(n)^T\right]^T \begin{bmatrix} \boldsymbol{\zeta}_P(k) \\ \boldsymbol{\xi}_P(k) \\ \vdots \\ \boldsymbol{\xi}_1(k) \end{bmatrix} \\ &= \mathbf{c}(P,n)^T \mathbf{w}(P,k), \end{aligned} \tag{8.9}$$

where the organization of the wavelet coefficients in $\mathbf{w}(P,k)$ is $\mathbf{w}(P,k) = \left[\boldsymbol{\zeta}_P(k)^T \boldsymbol{\xi}_P(k)^T \cdots \boldsymbol{\xi}_1(k)^T\right]^T$, with $[\boldsymbol{\zeta}_P(k)]_m = \zeta_{P,m}(k)$, $[\boldsymbol{\xi}_l(k)]_m = \xi_{l,m}(k)$. Evidently

$$\mathbf{c}(P,n) = \left[\mathbf{c}_0^{(P)}(n)^T, \mathbf{c}_1^{(P)}(n)^T, ..., \mathbf{c}_1^{(1)}(n)^T\right]^T$$

where

$$\left[\mathbf{c}_i^{(l)}(n)\right]_m = c_i^{(l)}(2^l m - n)$$

for $i = 0, 1$. Moreover, we can write

$$\mathbf{h}_n = \overbrace{\left(\mathbf{I}_{D+1} \otimes \mathbf{c}(P,n)^T\right)}^{\mathbf{C}(P,n)} \overbrace{\left[\mathbf{w}(P,0)^T, ..., \mathbf{w}(P,D)^T\right]^T}^{\mathbf{w}(P)}, \qquad (8.10)$$

where $\otimes$ is the Kronecker product. Then it should be clear that the idea is to represent the signal subspace $\mathcal{S}$ by means of the subspace

$$\mathcal{V}_P = \left\{\tilde{s}_n = \boldsymbol{\alpha}_n^T \mathbf{C}(P,n)\mathbf{w}(P),\ \mathbf{w}(P) \in \mathcal{C}^{(D+1)NR},\ n = 0, 1, ..., NR-1\right\}.$$

8.2.2 Wavelet Coefficients Shrinkage

Observe that (8A.11) is of limited practical interest, because it projects $\mathbf{h}$ onto $\mathbf{w}(P)$, in a subspace which is of the same size as the original. However, the excellent compacting properties (see [21]), or in other terms, the localization capabilities in the time-frequency plane of the wavelet transform, are able to concentrate most of the energy of the time variations of the channel in the low resolution coefficients of the fading process. It is possible to determine which components of the expansion (8A.11) can be neglected (i.e., shrunken) without compromising the parsimony of the wavelet-based representation. We will show, in fact, that it is possible to represent the channel in a reduced order dimensional space by matching the ensemble of signal subspaces (according to the scattering function of the multipath channel) to its multiresolution decomposition. This validation approach will imply that only a few coefficients of (8A.11) are necessary to represent the channel at any delay.

Define $\mathbf{w}(P,k)_M$ as the vector obtained zeroing the last

$$\sum_{m=1}^{M} NR/2^m$$

elements of $\mathbf{w}(P,k)$, $\mathbf{c}(P,n)_M$ as the vector obtained zeroing the last

$$\sum_{m=1}^{M} NR/2^m$$

elements of $\mathbf{c}(P,n)$. We can then try to approximate the signal subspace $\mathcal{S}$ by means of the reduced size approximation subspace

$$\tilde{\mathcal{V}}_P(M) \;=\; \left\{\tilde{s}_n = \boldsymbol{\alpha}_n^T \left(\mathbf{I}_{D+1} \otimes \mathbf{c}(P,n)_M^T\right)\right.$$

$$\left[\mathbf{w}(P,0)_M^T, ..., \mathbf{w}(P,D)_M^T\right]^T,$$
$$\mathbf{w}(P,k)_M \in \mathcal{C}^{NR\left(1-\sum_{m=1}^{M} 2^{-m}\right)},$$
$$k = 0, 1..., D, n = 0, 1, ..., NR-1\}. \quad (8.11)$$

8.2.3 Interpretations

The WBD of the NR-long vector

$$\mathbf{h}(k) = [h(0,k), h(1,k), ..., h(NR-1,k)]^T,$$

representing the dynamics of the channel at lag k, can be expressed as

$$\mathbf{h}(k) = \mathbf{W} \; \mathbf{w}(P,k), \quad (8.12)$$

where $\mathbf{W}$ is an $NR \times NR$ orthonormal linear transformation expressing the operation of the wavelet decomposition.[6] Consider $\overline{\mathbf{W}} = [\mathbf{W}_1, ..., \mathbf{W}_{\mathcal{N}}]^T$, the $NR \times \mathcal{N}$ matrix formed by the first $\mathcal{N} = NR\left(1 - \sum_{m=1}^{M} 2^{-m}\right)$ columns of $\mathbf{W}$ corresponding to the *shrinkage* (that is, the zeroing) of $\mathbf{w}(P,k)$. Applying the transformation

$$\boldsymbol{\theta}(k) = \overline{\mathbf{W}}^H \mathbf{h}(k) \quad (8.13)$$

we essentially define a subspace parameter $\boldsymbol{\theta}(k)$. The $\mathcal{N}$ orthonormal columns of $\overline{\mathbf{W}}$ span an $\mathcal{N}$-dimensional subspace defined $\mathcal{W}_{\mathcal{N}}$ such that $\mathcal{W}_{\mathcal{N}} \subseteq \mathcal{C}^{NR}$ (the space of any complex vector of length NR). Observe that when the transformation $\overline{\mathbf{W}}$ is selected with $\mathcal{N} < NR$, any estimate or representation of $\boldsymbol{\theta}(k)$ is insensitive to disturbance components of $\mathbf{h}(k)$ for $m = \mathcal{N}+1, \mathcal{N}+2, ..., NR$ and the resulting estimator (or representation) has a smaller variance than the full space estimator. The proposed channel representation method when incorporated in practical detectors will achieve a decreased variance

6. The matrix $\mathbf{W}$ is easily expressed from (8.9) as

$$\mathbf{W} = \begin{bmatrix} \mathbf{c}(P,0)^T \\ \vdots \\ \mathbf{c}(P,NR-1)^T \end{bmatrix}$$

of the channel estimate error. It is, however, important to emphasize that this advantage is obtained at the expense of estimation bias because in practice

$$\mathbf{h}(k) = \overline{\mathbf{W}}\boldsymbol{\theta}(k) = \mathbf{P}_W\mathbf{h}(k) \tag{8.14}$$

with $\mathbf{P}_W = \overline{\mathbf{W}} \;\; \overline{\mathbf{W}}^H$ holds only approximately true and while the variance increases with the dimensions of the subspace (that is, $\mathcal{N}$), the bias decreases [20]. Evidently the goal of selecting a transformation $\overline{\mathbf{W}}$ is to optimize the trade-off between bias, variance, and computational load.

8.2.4 Multiresolution Sequence Detection

Having defined appropriate representations for the signal subspace $\mathcal{S}$, we can use as detection statistic $\mathcal{L}_{N,P} = ||\mathbf{P}_{\mathcal{V}_P}[y]||^2$, or much more computationally efficient metrics

$$\tilde{\mathcal{L}}_{N,P}(M) = ||\mathbf{P}_{\tilde{\mathcal{V}}_P(M)}[y]||^2, \tag{8.15}$$

for any particular pair (P, M). Expression (8.15) defines an interesting concept: the concept of *multiresolution sequence detection.* We can in fact possibly detect the sequence at different resolution levels for different resolution depths P, and, although not treated in this chapter, potentially using different bases, for example wavelet packets. In the following section we show how to select $\tilde{\mathcal{V}}_P(M)$ in a way that it parsimoniously approximates $\mathcal{S}$.

8.2.5 Focusing the Generalized Energy Detector

We have seen how a multiresolution decomposition of the channel dynamics obtains a nested subspace representation for the signal

$$\mathcal{S} = \mathcal{V}_0 \subset \mathcal{V}_1 \subset ...\mathcal{V}_P \subset ... \tag{8.16}$$

at increasingly high level of detail. Of course we are interested in a decomposition that is the least computationally expensive, that is, we need the smallest representation subspace, for a reasonable approximation accuracy. We have to distinguish between two important cases:

- Assuming that the channel is unknown but characterized by deterministic variations, we can design a representation subspace that is close in some sense to $\mathcal{S}$ using the concept of Kolmogorov $n$width of signal classes (subspaces) [28].

- If the unknown channel variations are characterized statistically, $\mathcal{S}$ is not anymore a linear subspace but an *ensemble of linear subspaces* and we can design a representation subspace that is close on average to $\mathcal{S}$.

8.2.6 Deterministic Modeling of the Multipath Channel

We can use as a measure of the distance between two closed subspaces $\mathcal{S}$ and $\mathcal{V}$, the deviation [28] which is based on the concept of (constrained) Kolmogorov $n$width.

Definition[28] Let $\mathcal{X}$ be a normed linear space and $\mathcal{Y}$ a subset of $\mathcal{X}$. The constrained Kolmogorov $n$width is

$$d_N^I(\mathcal{Y};\mathcal{X}) = \inf_{\mathcal{X}_N \in I} \sup_{\mathbf{x}\in\mathcal{Y}} \inf_{\mathbf{y}\in\mathcal{X}_N} ||\mathbf{x}-\mathbf{y}||$$

where I is the index set for the available subspace set and $\mathcal{X}_N$ is a subspace of dimension N.

Since the signal class defined by $\mathcal{S}$ is well modeled by the bounded cone-class model [28], we define $S_\mathcal{S} = \{\mathbf{u} \in \mathcal{S}, ||\mathbf{u}||^2 = E_s\}$ (imposing a constraint on the energy of the signal) and the *deviation* of $\mathcal{S}$ from $\mathcal{V}$ as

$$\delta(\mathcal{S},\mathcal{V}) = \sup_{\mathbf{u}\in S_\mathcal{S}} \left\{ \inf_{\mathbf{v}\in\mathcal{V}} ||\mathbf{u}-\mathbf{v}|| \right\}. \tag{8.17}$$

We then compute $\delta(\mathcal{S},\mathcal{V})$ for the set of subspaces $\mathcal{V} = \tilde{\mathcal{V}}_P(M)$. We have

$$\left(\inf_{\tilde{s}_n\in\mathcal{V}} ||s_n - \tilde{s}_n||^2\right)^2 = ||s_n||^2 - ||\mathbf{P}_\mathcal{V}[s]||^2 \tag{8.18}$$

where

$$s_n = \boldsymbol{\alpha}_n^T \mathbf{h}_n \in \mathcal{S}$$

and

$$\tilde{s}_n = \tilde{\boldsymbol{\alpha}}_n^T \mathbf{w}(P,M) \in \mathcal{V}$$

with

$$\tilde{\boldsymbol{\alpha}}_n^T = \boldsymbol{\alpha}_n^T \mathbf{C}(P,M,n),$$

$$\mathbf{C}(P,M,n) = \left(\mathbf{I}_{D+1} \otimes \mathbf{c}(P,n)_M^T\right),$$

$$\mathbf{w}(P,M) = \left[\mathbf{w}(P,0)_M^T, ..., \mathbf{w}(P,D)_M^T\right]^T.$$

It is straightforward to obtain

$$\mathbf{P}_{\mathcal{V}}[s] = \widetilde{\mathbf{A}}(\widetilde{\mathbf{A}}^H\widetilde{\mathbf{A}})^{-1}\widetilde{\mathbf{A}}^H\mathbf{s} \tag{8.19}$$

with $\widetilde{\mathbf{A}} = [\widetilde{\boldsymbol{\alpha}}_0, \widetilde{\boldsymbol{\alpha}}_1, ...\widetilde{\boldsymbol{\alpha}}_{NR-1}]$, $\mathbf{s} = [s_0, s_1, ..., s_{NR-1}]^T$, so that $||\mathbf{P}_{\mathcal{V}}[s]||^2 = \mathbf{s}^H\widetilde{\mathbf{A}}(\widetilde{\mathbf{A}}^H\widetilde{\mathbf{A}})^{-1}\widetilde{\mathbf{A}}^H\mathbf{s}$. Also, $||s_n||^2 = \mathbf{h}^H\,\overline{\mathbf{A}}^H\,\overline{\mathbf{A}}\,\mathbf{h}$ where

$$\overline{\mathbf{A}} = \begin{bmatrix} \boldsymbol{\alpha}_0^T & \mathbf{0}^T & \cdots & \mathbf{0}^T \\ \mathbf{0}^T & \boldsymbol{\alpha}_1^T & \ddots & \mathbf{0}^T \\ \vdots & \ddots & \ddots & \vdots \\ \mathbf{0}^T & \cdots & \cdots & \boldsymbol{\alpha}_{NR-1}^T \end{bmatrix}$$

and $\mathbf{h} = \left[\mathbf{h}_0^T, \mathbf{h}_1^T, ..., \mathbf{h}_{NR-1}^T\right]^T$. Then, assuming the compactness of the set $\mathcal{S}$,

$$\begin{aligned} \delta(\mathcal{S},\mathcal{V})^2 &= \max_{||s_n||^2=E_s,\mathbf{s}\in\mathcal{S}} \left\{||s_n||^2 - \mathbf{s}^H\widetilde{\mathbf{A}}(\widetilde{\mathbf{A}}^H\widetilde{\mathbf{A}})^{-1}\widetilde{\mathbf{A}}^H\mathbf{s}\right\} \\ &= E_s - \min_{||s_n||^2=E_s,\mathbf{s}\in\mathcal{S}} \left\{\mathbf{s}^H\widetilde{\mathbf{A}}(\widetilde{\mathbf{A}}^H\widetilde{\mathbf{A}})^{-1}\widetilde{\mathbf{A}}^H\mathbf{s}\right\}. \end{aligned} \tag{8.20}$$

Observe that the deviation is a one-sided measure. That means that it addresses only the question of how well $\mathcal{S}$ is included in $\mathcal{V}$ but not vice versa, which is also an important issue. In fact, $\mathcal{V}$ may be too large to be of any practical interest. What we really need is the smallest subspace that minimizes the deviation. A small subspace implies that the number of coefficients needed to represent the orthogonal projection is small: This results in less degrees of freedom, reduced sensitivity to noise, and simplicity of computation. Since we are looking for the subspace representation for $\mathcal{S}$ over a set of predetermined basis vectors indexed by the integer P (and for each P approximated by M), we can do the following:

1. Find the deviation for each subspace with respect to $\mathcal{S}$ in the class of subspaces represented by (8.16).

2. Select the one (that is we can select the resolution depth P) that gives a tolerably small deviation.
3. Shrink the subspace retaining only part of the coefficients and select M in a way that the deviation is still close to the one selected in Step 2.

The problem of this approach is in the computation of the deviation $\delta(\mathcal{S},\mathcal{V})^2$ as in (8.20) which is remarkably complicated by the search of the minimum over the signals $\mathbf{s}$ that satisfy $||s_n||^2 = E_s, \mathbf{s} \in \mathcal{S}$.

8.2.7 Statistical Modeling of the Multipath Channel

If the unknown channel variations are characterized statistically, $\mathcal{S}$ is not anymore a linear subspace but an *ensemble of linear subspaces.* Statistical characteristics of the fading channel are typically available from experimental measurements in the form of the scattering function [13] or the autocorrelation function

$$\frac{1}{2}E\left\{f^*(t_1,\tau_1)f(t_2,\tau_2)\right\} = P_f(\tau_1, t_2 - t_1)\delta(\tau_2 - \tau_1)$$

where the expectation is over the channel ensemble. The scattering function of the multipath channel is related to $P_f(\tau_1, t_2 - t_1)$ by the Fourier transform, with respect to τ_1 and with respect to $t_2 - t_1$. Assuming knowledge of $P_f(\tau_1, t_2 - t_1)$, the discrete time autocorrelation function of the channel $\frac{1}{2}E\left\{h(n,k_1)^*h(n+l,k_2)\right\} = R_h(l,k_1,k_2)$ is also known; in fact, it can be computed as

$$\begin{aligned} R_h(l,k_1,k_2) &= \int_{-\infty}^{+\infty} p^*((n-k_1)T/R - \beta) \\ &\quad p((n+l-k_2)T/R - \beta)P_f(\beta, lT/R)d\beta. \end{aligned} \tag{8.21}$$

A simple approach to modify the definition of deviation, to incorporate statistical knowledge of the channel, is

$$d(\mathcal{S},\mathcal{V}) = \sqrt{\sup_{\mathbf{u}\in\mathcal{S}_S} E\left\{\left(\inf_{\mathbf{v}\in\mathcal{V}} ||\mathbf{u}-\mathbf{v}||\right)^2\right\}}, \tag{8.22}$$

where the expectation is also over the channel ensemble. We can now write (8.20) as

$$d(\mathcal{S},\mathcal{V})^2 = E_s - \min_{||s_n||^2=E_s,\mathbf{s}\in\mathcal{S}} E\left\{\mathbf{s}^H\widetilde{\mathbf{A}}(\widetilde{\mathbf{A}}^H\widetilde{\mathbf{A}})^{-1}\widetilde{\mathbf{A}}^H\mathbf{s}\right\}. \quad (8.23)$$

Using some properties of quadratic forms of Gaussian random variables, it is possible to show that

$$\begin{aligned} E\left\{\mathbf{s}^H\widetilde{\mathbf{A}}(\widetilde{\mathbf{A}}^H\widetilde{\mathbf{A}})^{-1}\widetilde{\mathbf{A}}^H\mathbf{s}\right\} &= E\left\{\mathbf{h}^H\,\overline{\mathbf{A}}^H\,\widetilde{\mathbf{P}}_A\,\overline{\mathbf{A}}\,\mathbf{h}\right\} \\ &= Tr\left[\overline{\mathbf{A}}^H\,\widetilde{\mathbf{P}}_A\,\overline{\mathbf{A}}\,2\mathbf{R}_h\right] \end{aligned} \quad (8.24)$$

where $\mathbf{R}_h = \frac{1}{2}E\left\{\mathbf{h}\,\mathbf{h}^H\right\}$, $\widetilde{\mathbf{P}}_A = \widetilde{\mathbf{A}}(\widetilde{\mathbf{A}}^H\widetilde{\mathbf{A}})^{-1}\widetilde{\mathbf{A}}^H$, and $Tr[...]$ is the trace operator. Evidently

$$d(\mathcal{S},\mathcal{V})^2 = E_s - Tr\left[\overline{\mathbf{A}}^H\,\widetilde{\mathbf{P}}_A\,\overline{\mathbf{A}}\,2\mathbf{R}_{\overline{h}}\right]$$

subject to $||s_n||^2 = \overline{\mathbf{h}}^H\;\overline{\mathbf{A}}^H\;\overline{\mathbf{A}}\;\overline{\mathbf{h}} = E_s$. The elements of the autocorrelation matrix $\mathbf{R}_{\overline{h}}$ can be obtained from (8.21). The strategy can be detailed as follows:

1. Find the deviation for each subspace with respect to $\mathcal{S}$ in the class of subspaces represented by (8.16) over the range $1 \le p \le P_{\max}$.
2. Select the smallest p such that $d(\mathcal{S},\mathcal{V}_p)^2 - d(\mathcal{S},\mathcal{V}_{p-1})^2 \le \Delta$.
3. Shrink the subspace retaining $(D+1)NR\left(1-\sum_{m=1}^{M}2^{-m}\right)$ coefficients for $M_{\min} \le M \le p$.

The steps of the approach can be summarized as in the following.

- $d(\mathcal{S},\mathcal{V}_0)^2 = E_s$
- `for` $i = P_{\max} : -1 : 1$
  — `compute` $d(\mathcal{S},\mathcal{V}_i)^2$
  — `if` $d(\mathcal{S},\mathcal{V}_i)^2 - d(\mathcal{S},\mathcal{V}_{i-1})^2 \ge \Delta \Rightarrow P = i$, `exit the for loop`
  — `if` $i = P_{\max} \Rightarrow P = P_{\max}$

- `end`
- $d(\mathcal{S}, \tilde{\mathcal{V}}_P(P))^2 = 0$
- `for` $m = P : -1 : M_{\min}$
  - — `compute` $d(\mathcal{S}, \tilde{\mathcal{V}}_P(m))^2$
  - — `if` $d(\mathcal{S}, \mathcal{V}_P(m))^2 - d(\mathcal{S}, \mathcal{V}_P(m-1))^2 \geq \Delta \Rightarrow M = m$, `exit the for loop`
  - — `if` $m = M_{\min} \Rightarrow M = M_{\min}$
- `end`

Observe that Δ is a preselected small number that determines the approximation accuracy of the representation subspace.

8.2.8 Real-Time Implementation

The metric in (8.15) can be expressed as

$$\tilde{\mathcal{L}}_N = \mathbf{y}^H \widetilde{\mathbf{P}}_A \mathbf{y}, \tag{8.25}$$

where $\widetilde{\mathbf{P}}_A$ is the projection matrix for the signal space $\widetilde{\mathcal{A}}$ spanned by the columns of $\widetilde{\mathbf{A}}$. The maximization of (8.25) is equivalent to the minimization of

$$\tilde{L}_N = ||\mathbf{y} - \widetilde{\mathbf{A}}\hat{\mathbf{w}}(P, M)||^2 = \mathbf{y}^H \widetilde{\mathbf{Q}}_A \mathbf{y} \tag{8.26}$$

where $\widetilde{\mathbf{Q}}_A = \mathbf{I}_{NR} - \widetilde{\mathbf{P}}_A$. Define the symbol sequence

$$\overline{\mathbf{a}} = [\overline{a}_0, \overline{a}_1, ..., \overline{a}_{N-1}]^T$$

as the sequence of symbols corresponding to the hypothesis that the sequence $\boldsymbol{\alpha} = [\alpha_0, \alpha_1, ..., \alpha_{NR-1}]^T$ has been transmitted. In other words: $\overline{\mathbf{a}}$ upsampled by R corresponds to $\boldsymbol{\alpha}$. Assume that $(D+1)/R = D_s + 1$ is an integer and that the modulation order for the symbols is $\mathcal{A}$ (size of the alphabet). The trellis defined by this channel has $\mathcal{A}^{D_s}$ nodes. It is convenient to define the state variables

$$\mu_n = [\overline{a}_{n-1}, ..., \overline{a}_{n-D_s}]^T$$

and

$$V_{n+1}(\mu_n, \mu_{n+1}) = \left| y_n - \widetilde{\boldsymbol{\alpha}}_n^T \hat{\mathbf{w}}(\mu_n) \right|^2,$$

where

$$\tilde{\boldsymbol{\alpha}}_n = \mathbf{C}(P, M, n)^T \boldsymbol{\alpha}_n,$$

$\mathbf{C}(P, M, n) = \left(\mathbf{I}_{D+1} \otimes \mathbf{c}(P, n)_M^T\right)$ and $\mathbf{c}(P, n)_M$ is the vector obtained zeroing the last $\sum_{m=1}^{M} NR/2^m$ elements of $\mathbf{c}(P, n)$. The vector $\mathbf{c}(P, n)$ is defined in (8.9). The sequence $\boldsymbol{\alpha}_n$ is obtained upsampling by R the sequence $[\overline{a}_n \overline{a}_{n-1}, ... \overline{a}_{n-D_s}]^T$ and it is defined by the particular state transition $\mu_n \rightarrow \mu_{n+1}$.

Define $\hat{\mathbf{w}}(\mu_n)$ as the estimated WBD vector $\hat{\mathbf{w}}(P, M)$ at time step n (or epoch n).[7] Possible approaches to a sample-by-sample estimation of $\hat{\mathbf{w}}(\mu_n)$ are gradient-based methods (like the LMS) and RLS methods. It is important to observe that the true WBD vector $\mathbf{w}(P, M)$ is time-invariant, so the task of the adaptive algorithm in the proposed approach is to converge to the channel parameters as opposed to tracking them. This is why the simple LMS algorithm is selected in the following. $\hat{\mathbf{w}}(\mu_n)$ is recursively computed from each survivor, as indicated in the following steps.

- *Step 0:* Initialization.
  - — $\mathbf{c}(P, n)$ is defined in (8.9);
  - — $\mathbf{c}(P, n)_M$ is the vector obtained zeroing the last $\sum_{m=1}^{M} NR/2^m$ elements of $\mathbf{c}(P, n)$;
  - — $\mathbf{C}(P, M, n) = \left(\mathbf{I}_{D+1} \otimes \mathbf{c}(P, n)_M^T\right)$;
  - — β is the adaptation step of the LMS algorithm.
- *Step 1:* Extend survivor paths. Let $n = n + 1$ and compute $V_{n+1}(\mu_n, \mu_{n+1})$ for each pair of states μ_n, μ_{n+1} such that the transition from μ_n to μ_{n+1} is allowed by the trellis structure:
  - — $(\mu_n \rightarrow \mu_{n+1}) \Rightarrow [\overline{a}_n, \overline{a}_{n-1}, ... \overline{a}_{n-D_s}]^T$;
  - — $\boldsymbol{\alpha}_n^T = [\alpha_n, ... \alpha_{n-D}]^T = \left[\overline{a}_n, \overbrace{0, ..., 0}^{R-1}, \overline{a}_{n-D_s}, \overbrace{0, ..., 0}^{R-1}\right]^T$;
  - — $\tilde{\boldsymbol{\alpha}}_n = \mathbf{C}(P, M, n)^T \boldsymbol{\alpha}_n$;
  - — $V_{n+1}(\mu_n, \mu_{n+1}) = \left| y_n - \tilde{\boldsymbol{\alpha}}_n^T \hat{\mathbf{w}}(\mu_n) \right|^2$.

7. Observe that even if $\tilde{\alpha}_n$ is a $q = NR2^{-M}$ vector, it still depends on $D_s + 1$ symbols only.

- *Step 2:* Drop paths. For each state μ_{n+1} compute the accumulated metric

$$\Gamma_{n+1}(\mu_{n+1}) = \min_{\mu_n} \{\Gamma_n(\mu_n) + V_{n+1}(\mu_n, \mu_{n+1})\}, \quad (8.27)$$

  and extend the survivors by incorporating only those transitions that comply with the rule (8.27).

- *Step 3:* Update WBD coefficients. For each pair of states μ_n, μ_{n+1} such that the transition from μ_n to μ_{n+1} complies with (8.27), update the WBD coefficients:

  — $\hat{\mathbf{w}}(\mu_{n+1}) = \hat{\mathbf{w}}(\mu_n) + \beta \left[y_n - \tilde{\boldsymbol{\alpha}}_n^T \hat{\mathbf{w}}(\mu_n) \right] \tilde{\boldsymbol{\alpha}}_n^*.$

- *Step 4:* Iterate. Go to Step 1 if not end of slot.

8.2.9 Performance Analysis Results

The slots are generated of dyadic length with $R = 2$, $(NR = 256)$ the symbol period is 41.2 μsec and the modulation format is $\pi/4 - DQPSK$. $C_0(z)$ $C_1(z)$ are Daubechies filters of length 3 (DAUB-3 filters with four taps). We assume [29] that

$$R_h(l, k_1, k_2) = \begin{cases} \sigma_k^2 J_0(2\pi l f_D T/R) & for\ k_1 = k_2 = k \\ 0 & \text{elsewhere} \end{cases}$$

where $J_0(x)$ is the Bessel function of order zero, f_D is Doppler spread which depends on the velocity of the mobile transmitter, and σ_k^2 is the variance of the fading process. The dispersive channel is created using a hardware fading emulator that models a two-ray Rayleigh fading channel [30]. The delay interval is the difference in time of arrival between the two rays at the antenna. It is possible to verify, using the described focusing technique with $\Delta = 0.001$, that for speeds of the mobile up to 300 km/hr and $P = 3$ or $P = 4$, it is possible to impose $\xi_{l,m}(k) = 0$ in (8A.11) for any l, m, k. This means that we can have only $NR/2^M = NR/2^P = 32$ or $NR/2^M = NR/2^P = 16$ WBD coefficients. Figure 8.2 shows an experiment where we perform multiresolution analysis $(P = 5)$ of a fading coefficient $h(n, 0)$, for a time varying multipath event (512 samples symbol-spaced, $R = 1$).

Considering (8A.11) for $P = 5$ and $NR = 512$

$$h(n,k)|_{k=0} = \sum_{m=0}^{512/2^5 - 1} \zeta_{5,m}(0) c_0^{(5)}(2^5 m - n)$$

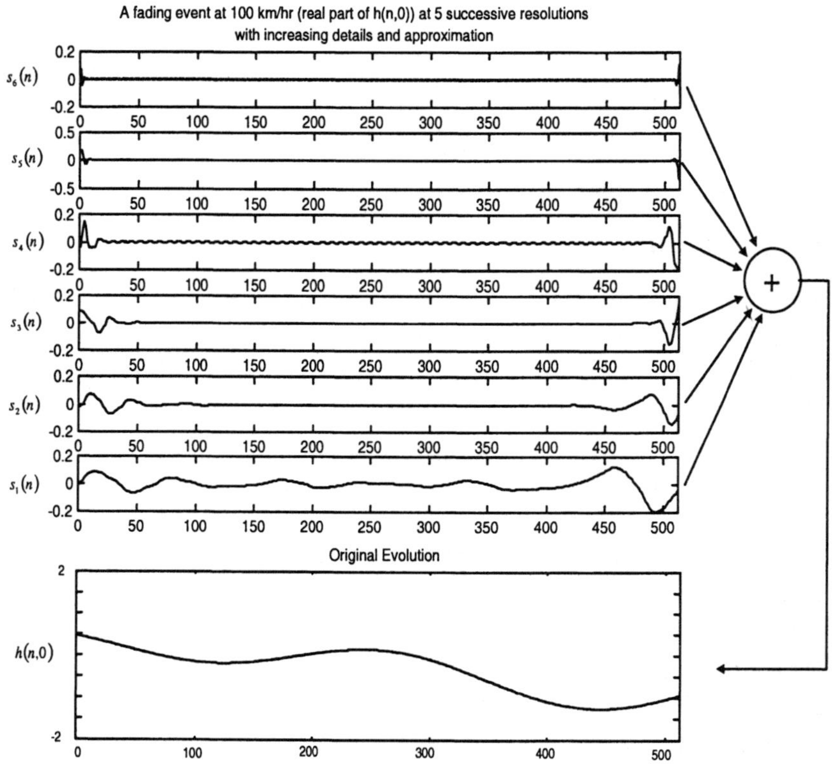

Figure 8.2 Multiresolution wavelet-based representation of a fading event (real part of the coefficient for a 512 symbol spaced samples snapshot) for a time-varying multipath channel representing a mobile transmitter traveling at 100 km/hr. The maximum resolution depth is P=5 and the mother wavelet used is generated from a Daubechies filter of length 8 (DAUB-7).

$$
\begin{aligned}
&+\sum_{m=0}^{512/2^5-1} \xi_{5,m}(0)c_1^{(5)}(2^5m-n) \\
&+\sum_{m=0}^{512/2^4-1} \xi_{4,m}(0)c_1^{(4)}(2^4m-n) \\
&+\sum_{m=0}^{512/2^3-1} \xi_{3,m}(0)c_1^{(3)}(2^3m-n) \\
&+\sum_{m=0}^{512/2^2-1} \xi_{2,m}(0)c_1^{(2)}(2^2m-n) \\
&+\sum_{m=0}^{512/2^1-1} \xi_{1,m}(0)c_1^{(1)}(2^1m-n) \\
=\ & s_1(n)+s_2(n)+s_3(n) \\
&+s_4(n)+s_5(n)+s_6(n). \qquad (8.28)
\end{aligned}
$$

Figure 8.2 shows $s_1(n)$, $s_2(n)$, $s_3(n)$, $s_4(n)$, $s_5(n)$, and $s_6(n)$: the multiresolution components of $h(n,0)$.

In Figures 8.3 and 8.4 we emphasize the concept of approximated decomposition by shrinkage of wavelet coefficients. From top to bottom of Figures 8.3 and 8.4: WBD for the real part of the tap variations, dynamics of the real part and reconstructed dynamics using 16 and 32 wavelet coefficients, respectively, in Figures 8.3 and 8.4. The excellent compacting properties (see [21]) of the wavelet transform make the approach very attractive in practice.

Since an analytical approach to the BER performance of the outlined detector is beyond the scope and the length of this chapter, we present the results of an experimental analysis. There are two antennas (which is a typical antenna configuration for base stations) and the metric has to be modified for the two independent fading branches as

$$L_N = \sum_{n=0}^{NR-1} \sum_{i=1}^{2} |y_n^{(i)} - \tilde{\boldsymbol{\alpha}}_n^T \hat{\mathbf{w}}^{(i)}(\mu_n)|^2,$$

where $y_n^{(i)}$, $\hat{\mathbf{w}}^{(i)}(\mu_n)$ are signal samples and WBD channel vector from the ith antenna $i = 1, 2$, respectively, at epoch n. More details on the hardware setup of the experiments and the hardware architecture of

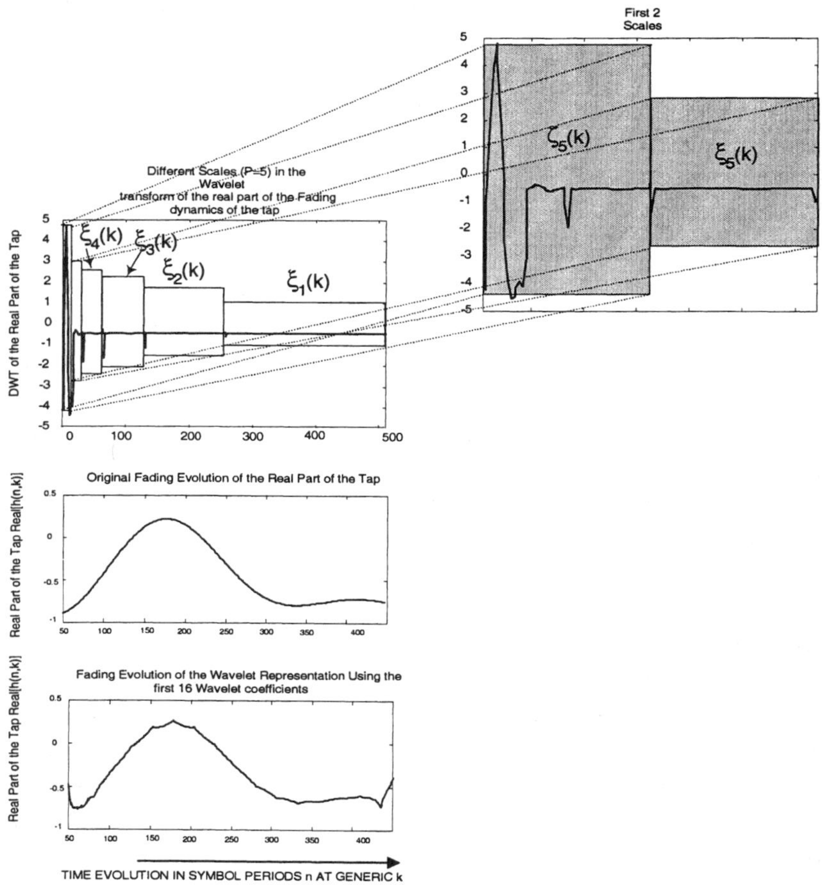

Figure 8.3 An example of time-varying multipath WBD decomposition for a 500 samples (symbol spaced) snapshot. From top to bottom: WBD (P=5) for the real part of the tap variations, dynamics of the real part and reconstructed dynamics using only 16 wavelet coefficients. The mother wavelet used is generated from a Daubechies filter of order 3 (DAUB-3).

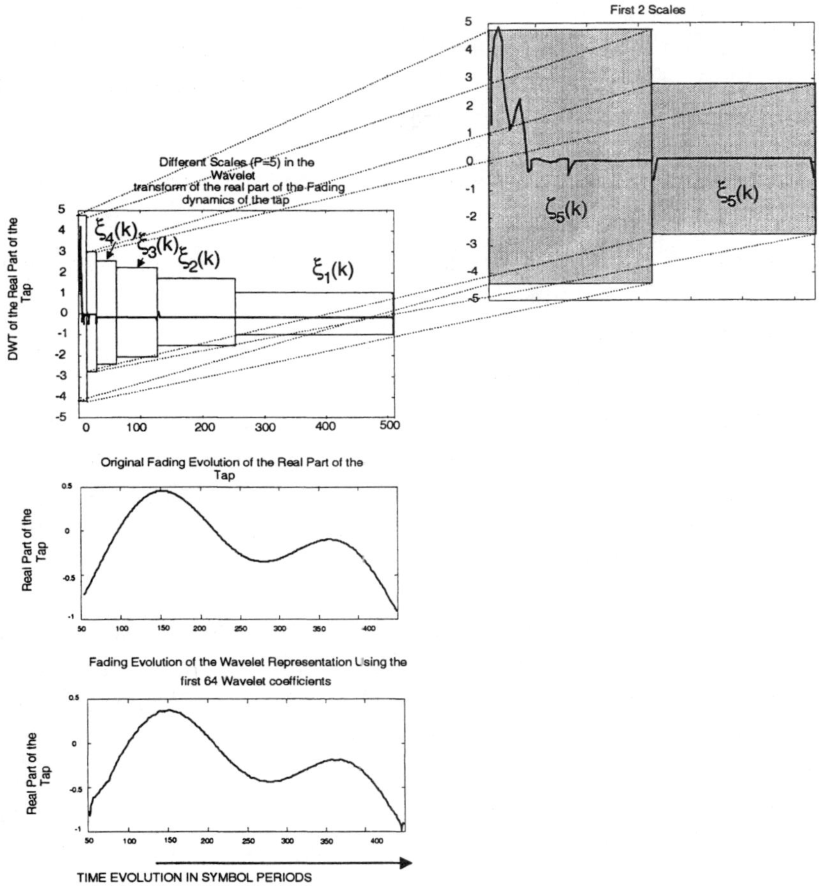

Figure 8.4 A second example of time-varying multipath WBD decomposition for a 500 samples (symbol spaced) snapshot where the reconstruction of fading is obtained with more detail. From top to bottom: WBD (P=5) for the real part of the tap variations, dynamics of the real part and reconstructed dynamics using 64 wavelet coefficients. The mother wavelet used is generated from a Daubechies filter of order 3 (DAUB-3).

the receiver can be found in [31]. The pulse $p(t)$ is a raised cosine filter with roll-off 0.35. The trellis has four states because the time span of the channel is 41.2 μsec (one symbol period). We compare the following approaches to the technique presented in this work: PSP with LMS, QR decomposition-based RLS, PSP with Kalman filters [9]. More specifically, the results obtained using the different strategies are labeled in the figures as:

- `PSPLMS`: One LMS channel estimator per survivor using the PSP technique [4];
- `PSPKF`: One Kalman channel estimator[8] per survivor using the PSP technique [9];
- `PSPQRRLS`: One QR-RLS channel estimator per survivor using the PSP technique;
- `WBDPSPLMS(P)`: One WBD-channel LMS estimator per survivor using the PSP technique, resolution depth equal to P, and representing the channel in the low-resolution $NR/2^P$-dimensional space, which means retaining $NR/2^P$ WBD coeffcients.

The complexity of each algorithm is reported in Table 8.1 in MIPS for a two-antenna receiver. The complexity is relevant to the processing of one single reverse digital traffic channel (RDTC). The above algorithms have been designed for Analog Devices ADSP-2181 processors (40-MHz clock rate, 16 bits fixed point arithmetic) and executed on baseband samples collected from the hardware platform. Figure 8.5 shows results at 100 km/hr versus delay interval with an SNR per bit equal to 25 dB. Observe that this experiment is also testing the robustness of the various algorithms. In fact, the channel model assumed by the MLSD algorithm is still a two-ray channel where the two rays are spaced one symbol period apart. Figure 8.6 is for delay interval equal to 41.2 μsec, varying the mobile speed and SNR per bit equal to 22 dB.

Traditional approaches are significantly more sensitive than the wavelet-based approach to rapidly fading channels. One important

8. The channel evolution is modeled as $\mathbf{h}_n = \mathbf{F}\mathbf{h}_{n-1} + \mathbf{G}\mathbf{w}_n$, where $\mathbf{F}$ and $\mathbf{G}$ are related to second-order statistics of the channel as described in [9], while $\mathbf{w}_n$ is a zero-mean white Gaussian vector process.

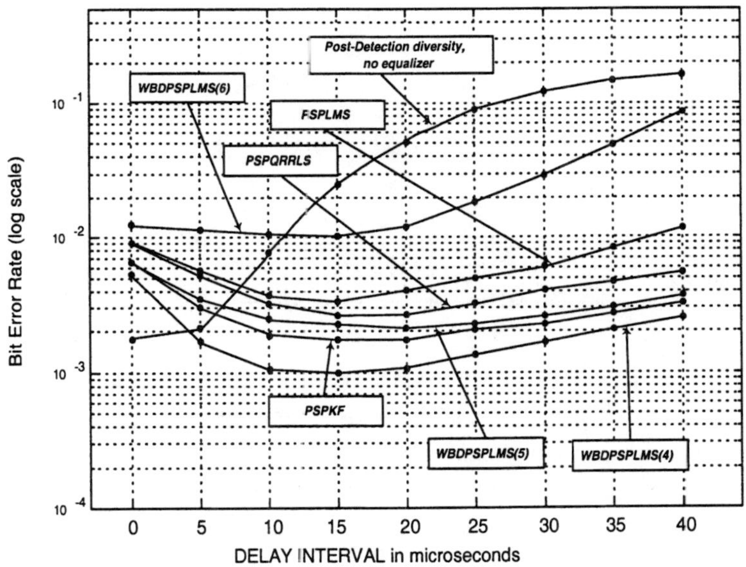

Figure 8.5 BER at SNR per bit equal to 25 dB and 100 km/hr, versus delay interval (of the two-ray Rayleigh fading hardware channel). Training is used for the first 14 symbols (IS-136 standard) and slots have an RDTC-like format with 128 $\pi/4$-DQPSK modulated symbols. The mother wavelet used is generated from a Daubechies filter of order 3 (DAUB-3).

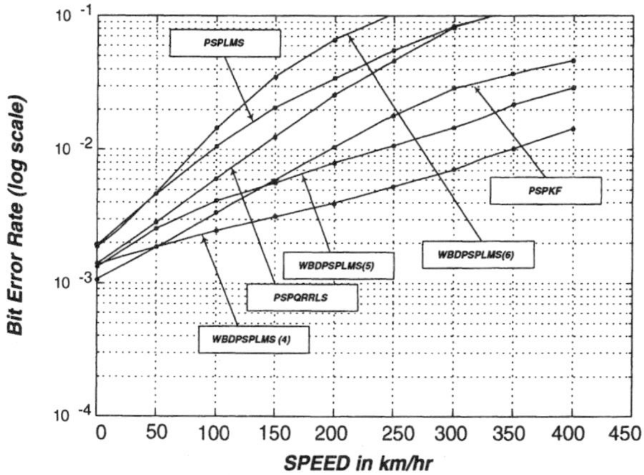

Figure 8.6 BER at SNR per bit equal to 25 dB at 41.2 μsec (1 symbol period), versus speed of the mobile (two-ray Rayleigh fading hardware channel). Training is used for the first 14 symbols (IS-136 standard) and slots have an RDTC-like format with 128 $\pi/4$-DQPSK modulated symbols. The mother wavelet used is generated from a Daubechies filter of order 3 (DAUB-3).

point that deserves more investigation is the issue of the sensitivity of the algorithm to imperfect knowledge of channel statistics. At this point we can only conjecture that an accurate knowledge of the statistics of the channel is not very critical to the selection of P and M, if Doppler spread and time-dispersion are exactly known.

The algorithms have been designed for Analog Devices ADSP-2181 processors (40-MHz clock rate, 16 bits fixed-point arithmetic) and executed on baseband samples collected from the hardware platform. Figure 8.5 shows results at 100 km/hr versus delay interval with an SNR per bit equal to 25 dB. Observe that this experiment also is testing the robustness of the various algorithms. In fact, the channel model assumed by the MLSD algorithm is still a two-ray channel where the two rays are spaced one symbol period apart. Figure 8.6 is for delay interval equal to 41.2 μsec, varying the mobile speed and SNR per bit equal to 22 dB. Traditional approaches are significantly more sensitive than the wavelet-based approach to rapidly fading channels (see also Figure 8.7). One important point that deserves more investigation is the issue of the sensitivity of the algorithm to imperfect knowledge of channel statistics. At this point

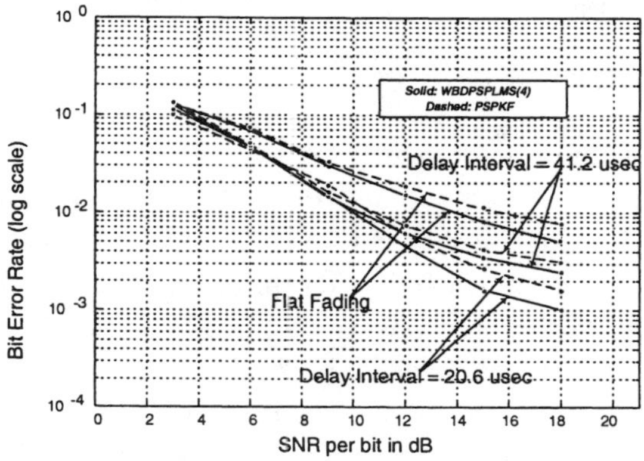

Figure 8.7 BER at 0-20.6-41.2 μsec delay interval for the WBDPSP(4) and PSPKF techniques (at 80 km/hr). Training is used for the first 14 symbols (IS-136 standard) and slots have an RDTC-like format with 128 symbols. The mother wavelet used is generated from a Daubechies filter of order 3 (DAUB-3 in Fig. 1).

Table 8.1
Computational Complexity with Two Antennas, Four States Viterbi Algorithm
$R = 2, D = 3, NR = 256.$

Algorithm	MIPS
DWTPSPLMS(3)	54
DWTPSPLMS(4)	49
DWTPSPLMS(5)	45
DWTPSPLMS(6)	39
PSPLMS	37
PSPKF	52
PSPQRRLS	48

we can only conjecture that an accurate knowledge of the statistics of the channel is not very critical to the selection of P and M, if Doppler spread and time dispersion are exactly known.

8.3 Multicarrier Modems

Multicarrier techniques transmit data by dividing the stream into several parallel bit streams. Each of the subchannels has a much lower bit rate and is modulated onto a different carrier. OFDM is a special case of multicarrier modulation with equally spaced subcarriers and overlapping spectra [32]. The OFDM time-domain waveforms are chosen such that mutual orthogonality is ensured in the frequency domain. Time dispersion is easily handled by such systems because the substreams are essentially free of intersymbol interference (ISI). This last aspect of multicarrier schemes has contributed significantly to increase the popularity of OFDM in many wireless and wire-line applications [33, 34] and has opened a competition with more traditional single carrier time-domain schemes [31, 35–37]. When the channel is frequency-dispersive (that is, it is afflicted by non-negligible Doppler spread as in the wireless channel), inter-channel interference may degrade an OFDM system performance to untolerable levels. Consider the baseband equivalent signal generated by a generic N-channel multicarrier system [35] expressed as

$$s(t) = \sum_{k=-\infty}^{+\infty} \sum_{l=0}^{N-1} a_{k,l} g_{k,l}(t) = \sum_{k=-\infty}^{+\infty} \sum_{l=0}^{N-1} a_{k,l} \phi_l(t - kT), \qquad (8.29)$$

where T is the symbol period and $a_{k,l}$ is the generally complex valued information-bearing symbol, $g_{k,l}(t) = \phi_l(t - kT)$, $l = 0, 1, ..., N - 1$ are the fundamental basis waveforms.

In wireless links the transmitted signal $s(t)$ is linearly distorted by the multipath fading channel operator[9] $\mathbf{H}$ as

$$y(t) = (\mathbf{H}s)(t) = \sum_{k=-\infty}^{+\infty} \sum_{l=0}^{N-1} a_{k,l} f_{k,l}(t) = \sum_{k=-\infty}^{+\infty} \sum_{l=0}^{N-1} a_{k,l} f_l(t - kT)$$

where $f_l(t) = (\mathbf{H}\phi_l)(t)$. The fundamental problem is selecting the transmission basis $\phi_l(t)$ in such a way that the projection of the signal onto an identically structured signal set $f_{k,l}(t)$ gives the transmitted symbols as

$$\int_{-\infty}^{+\infty} y(t) f_{k,l}^*(t)dt = \int_{-\infty}^{+\infty} y(t) f_l^*(t - kT)dt = a_{k,l}.$$

This condition implies relative simplicity of the receiver and robustness to AWGN.

The selection of a channel-dependent signal basis in the case of a static channel $(\mathbf{H}s)(t) = \int h(\tau)s(t-\tau)d\tau$ is well-understood and corresponds to functions $\phi_l(t)$ equal the eigenmodes of the channel autocorrelation. The eigenmodes of a frequency selective static channel can be well approximated by Fourier bases and transmitter and receiver easily implemented by means of Fourier transform methods. The popularity of OFDM schemes stems exactly from this fundamental property. It is important to realize that if the channel is doubly dispersive, as in the rapidly fading wireless channel, the entire conceptual framework of the basic Fourier-domain channel partitioning scheme loses its optimality. A similar point of view was expressed in [38], where orthogonality was somehow sacrificed for time-frequency localization of the transmitted signal set: an increased robustness to doubly dispersive channel distortions was the main goal. However, the problem was not attacked at the root, because exponential Fourier-like signal sets were still used, both at the transmitter and at the receiver. In this work we introduce a

9. The operator $\mathbf{H}$ represent the effects of multipath RF wave propagation. One standard way of representing the channel is

$$(\mathbf{H}s)(t) = \int_{-\infty}^{+\infty} \int_{-\infty}^{+\infty} S(\tau, \nu)s(t-\tau)e^{j2\pi\nu t} d\tau d\nu$$

where $S(\tau, \nu)$ is the spreading function of the wireless channel [13].

new methodology that employs a signal set specifically designed for the synthesis/analysis of time-varying (nonstationary) signals.

We can think of the doubly dispersive channel as a channel with time-varying frequency response. The optimal transmission/reception methodology should be able to diagonalize nonstationary signals, in a sense employing Karhunen-Loeve eigenfunctions as bases. In general, the subchannel carrier frequencies should be time-varying and ideally decompose the frequency distortion of the channel perfectly at any instant in time. This optimal approach presents significant challenges both in terms of conceptual and computational complexity.

The idea disclosed in this work is that the optimality of a nonstationary approach can be somehow *approximated*, without any increase of computational complexity with respect to the classical OFDM scheme. We propose the use of a multicarrier system that employs orthogonal signal bases of the chirp type. Such bases are related to the fractional Fourier transform whose time-frequency properties [39] are well-known in the signal processing community. Exactly as the Fourier harmonic analysis employs sinusoidal function to decompose periodic signals, fractional Fourier techniques employ chirp harmonics for the decomposition of signals with time-varying periodicity. The significance of the methodology we introduce relies on the important practical consideration that analysis/synthesis methods of the fractional Fourier type are implemented with a complexity that is equivalent to traditional FFT computational procedures.

8.3.1 The Fractional Fourier Transform

The fractional Fourier transform (FRFT) was introduced in [40, 41] as a generalization of the Fourier transform. The transform immediately appeared useful in many signal processing applications, but remained essentially unknown until recently, when several new results, interpretations, and applications were found [39, 42]. The interpretation of the FRFT as a "rotation" operator in the time-frequency plane [39] has generated significant interest in the signal processing community. The tranformation kernel of the continuous FRFT is defined as follows

$$K_\alpha(t,u) = A_\alpha e^{j\pi(t^2+u^2)\cot\alpha - j2\pi tu\csc\alpha} \tag{8.30}$$

where α is the rotation angle of the transformed signal and

$$A_\alpha = \frac{e^{\{-j\pi \mathrm{sign}[\sin\alpha]/4 + j\alpha/2\}}}{\sqrt{|\sin\alpha|}}.$$

The forward and inverse FRFT are defined as

$$\mathcal{F}_\alpha \{x(t)\}(u) = \mathcal{X}_\alpha(u) = \int_{-\infty}^{+\infty} x(t) K_\alpha(t,u) dt, \quad (8.31)$$

$$x(t) = \int_{-\infty}^{+\infty} \mathcal{X}_\alpha(u) K_{-\alpha}(t,u) du. \quad (8.32)$$

The domains of the signal for $0 < |\alpha| < \pi$ are defined *fractional Fourier domains* and $\alpha = \frac{\pi}{2}$ in (8.31)–(8.32) gives the traditional Fourier transform (FT). The FT can be considered a projection of a given signal in the time-frequency plane on the frequency axis, or in other words a rotation of $\pi/2$ with respect to the time axis. The FRFT can be considered a projection of the signal on an axis that forms an angle α with the time axis: a rotation in the time-frequency plane. The reader interested in more interpretations and explanations regarding the FRFT transformation is directed to [39].

The problem of interest of most recent literature dealing with the FRFT is the research of efficient digital methods for the computation of the transforms in (8.31)–(8.32). In [42] a method with NlogN (N is the number of samples) complexity was recently discovered that also provides significant accuracy of representation. This interesting result opens the possibility of using the FRFT at no extra cost anywhere that traditional FFT schemes reveal their inefficiency in dealing with nonstationary signals. Shortly thereafter, the same authors [43] presented a method of filtering in the fractional Fourier domain equivalent to the traditional Fourier-domain filter. While Fourier-domain filters have the dual time-domain interpretation as convolutions, fractional Fourier domain filters correspond to complicated operations of time and frequency. Nevertheless it was shown in [43] that in many cases this type of filtering achieves an MSE which is lower than traditional Fourier-domain filters. This is particularly true when recovery from nonstationary distortions is attempted.

Traditional techniques for multicarrier modulation attempt a frequency decomposition of the transmission available bandwidth. The effect of the time invariant channel distortions can be compensated on a subchannel-by-subchannel basis by single-tap frequency domain equalizers. As a consequence the overall traditional multicarrier system can be seen as an optimal Fourier-domain filter. However, if the channel is time-varying, the traditional multicarrier system loses optimality because the optimal recovery operator is, in general, time-variant. This means that it can not be implemented in the conventional Fourier domain and is exactly the reason that motivates the use of a FRFT-based technique.

8.3.2 System Model

Consider the baseband representation of a multicarrier system as in (8.29) with

$$g_{k,l}(t) \quad = \quad g(t-kT)e^{j2\pi lFt}, \tag{8.33}$$

where F is the carrier frequency spacing. The use of pulses as in (8.33) results in a rectangular *tiling* of the time-frequency plane. The product $TF \geq 1$ defines the time-frequency product of each independent function in the signal set. In the OFDM case the pulse $g(t)$ in (8.33) is a rectangular window of duration T and $F = 1/T$. A *coarsification* of the time-frequency grid is typically employed using a guard-time between temporal adjacent symbols for mitigation of the time-dispersive characteristic of a frequency selective channel. However, properly shaping the basic symbols in each subchannel by using a pulse different from the rectangular one mitigates frequency dispersion effects on the channel caused by Doppler spreads. This approach is, however, fundamentally suboptimum and likely to be ineffective in presence of large Doppler spreads. Suppose that we decided to use a signal basis that does not operate a rectangular tiling of the time-frequency plane, for example [44],

$$g_{k,l}(t) = \phi_l(t-kT) = g(t-kT)f_{-\alpha,l}(t-kT), \tag{8.34}$$

where $f_{\alpha,n}(t)$ is given by

$$f_{\alpha,n}(t) = \sqrt{\frac{\sin\alpha + j\cos\alpha}{T}}\, e^{-j\frac{t^2+(\sin(\alpha)n2\pi/T)^2}{2}\cot\alpha + jn(2\pi/T)t},$$
$$-T/2 \leq t \leq T/2.$$

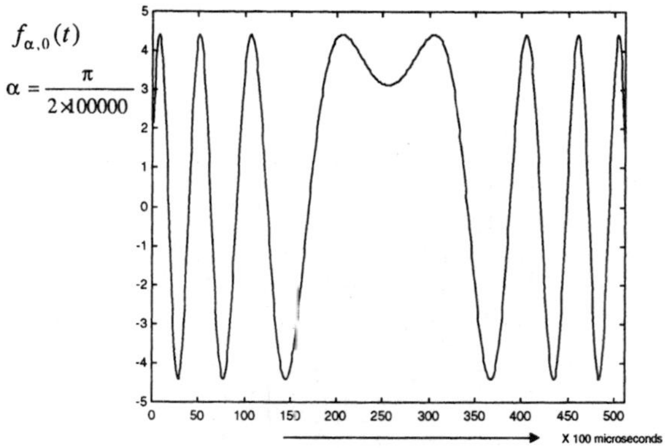

Figure 8.8 The basis waveform $f_{\alpha,n}(t)$ with $N = 256$ for $n = 0$.

The selection of the function $f_{\alpha,n}(t)$ is motivated by the fact that exactly as the exponential (sinusoidal) basis of the Fourier transform corresponds in the Fourier domain to an impulse, the basis $f_{\alpha,n}(t)$ corresponds (up to a constant factor) in the fractional Fourier domain to an impulse.

In other words [44], $f_{\alpha,n}(t)$ has been derived in such a way that

$$FRFT_{-\alpha}\left[\delta\left(u - n(\sin\alpha/T)\right)\right] = \text{const} \times f_{\alpha,n}(t),$$

for $n = -\infty, ..., -1, 0, 1, ... + \infty$. The derivation of this interesting fact is reported in [44]. It is also important to observe that the set of functions $f_{\alpha,l}(t)$ is an orthonormal set. In fact we have [44]

$$\int_{-\infty}^{+\infty} f_{\alpha,m}(t) f^*_{\alpha,n}(t)dt = \begin{cases} 1 & \text{for } m = n \\ 0 & \text{for } m \neq n \end{cases}. \tag{8.35}$$

In Figures 8.8 and 8.9 we show two members of the set $f_{\alpha,n}(t)$ for $\alpha = \frac{\pi}{2\,100000}$ and $N = 256$ sampled at 10 kHz. The frequency domain representation is shown in Figure 8.10. If the pulse $g(t)$ is a rectangular window of duration T, the transmission system (8.29) with (8.34) is equivalent to a new method of modulation where the generic basis signal is a chirp signal with chirp rate $-\cot\alpha$ and the frequency spacing between adjacent basis signals is $\frac{1}{T}$. The frequency of the nth basis signal is (linearly) dependent on time and it is equal to

$$\omega_{\alpha,n} = n\frac{2\pi}{T} - t\cot\alpha.$$

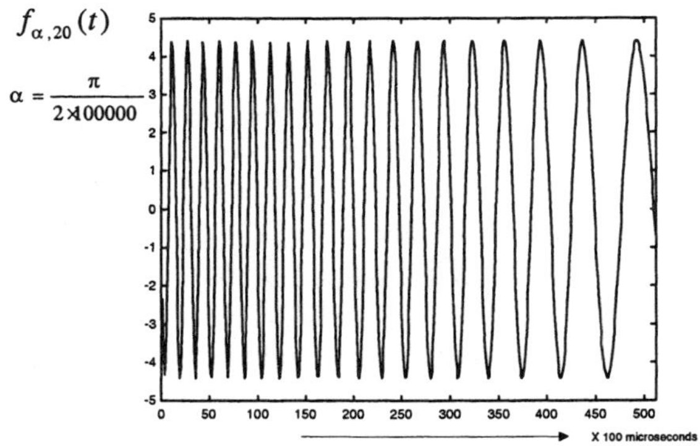

Figure 8.9 The basis waveform $f_{\alpha,n}(t)$ with $N = 256$ for $n = 20$.

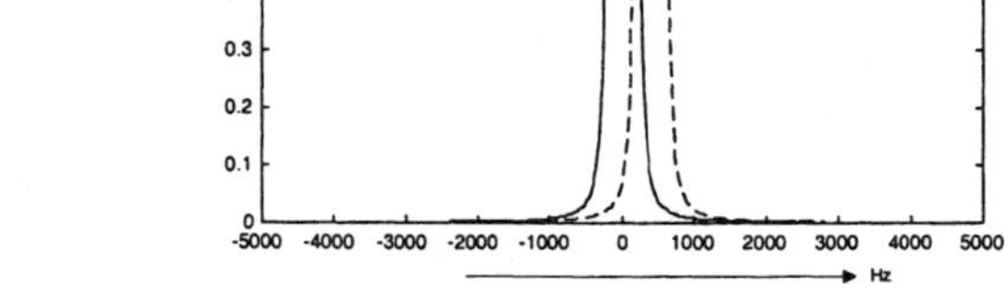

Figure 8.10 Spectral energy distribution of the two basis waveforms shown in the Figures 8.8 and 8.9.

Traditional OFDM systems are just special cases of (8.29) with (8.34) and $\alpha = \frac{\pi}{2}$.

The condition (8.35) shows that the transmitted symbols (in the absence of channel distortions) can be recovered by performing the following matched filtering operation

$$a_{k,l} = \int_{-\infty}^{+\infty} s(t)g_{k,l}^{*}(t)dt = \int_{-\infty}^{+\infty} s(t)g^{*}(t-kT)f_{\alpha,l}^{*}(t-kT)dt.$$

8.3.3 The Transceiver as a Fractional Fourier Domain Filter

Assume that a single-shot transmission for the kth block of symbols has to be achieved and that the number of subchannels (carriers) is infinite.[10] Define $\mathcal{F}_{\alpha}\left\{x(t)\right\}(u)$ as the continuous FRFT of $x(t)$ with angle equal to α and $s_k(t)$ as the modulated signal $s(t)$ relative to the symbols in the kth block $a_{k,-\infty},\ldots,a_{k,+\infty}$,

$$s_k(t) = \sum_{n=-\infty}^{+\infty} a_{k,n}f_{-\alpha,n}(t) = \mathcal{D}_{-\alpha}\left\{\mathbf{a}_k\right\}(t), \tag{8.36}$$

where the symbols in the kth block are organized in a vector $\mathbf{a}_k$. It was shown in [44] that

$$a_{k,n} = \mathcal{K} \times \mathcal{F}_{\alpha}\left\{s_k(t)\right\}\left(\frac{n}{T}\sin\alpha\right) \tag{8.37}$$

in (8.36) with $\mathcal{K}$ as a real constant. We can then interpret the transmission system (8.29) with (8.34) as a block inverse fractional Fourier transform and the receiver as a block fractional Fourier transform. This means that at the output of the receiver a multiplicative filter of the type decribed in [43] can be used in presence of arbitrary time-varying distortion for the channel $\mathbf{H}$. The signal $s_k(t)$ undergoes linear distortion and additive noise $n_k(t)$ as

$$r_k(t) = \left(\mathbf{H}s_k\right)(t) + n_k(t).$$

Since in general the channel distortion is time-varying (because the kernel of the degradation model represented by the spreading

10. This is an ideal abstraction that helps the derivation of the method and that was also motivated in [38].

function is nonstatic), the optimal recovery operator for the signal $s_k(t)$ is also a general linear time-varying estimate of the form

$$\widehat{s}_k(t) = (\mathbf{W}r_k)(t) = \int_{-\infty}^{+\infty} w(t,\tau) r_k(\tau) d\tau. \qquad (8.38)$$

The MSE for nonstationary signals that have finite energy is defined as

$$\mathrm{MSE}(\mathbf{W}) = E\left\{||s_k(t) - \widehat{s}_k(t)||^2\right\},$$

where $||x(t)||^2 = \int_{-\infty}^{+\infty} x(t)x^*(t)dt$. Minimization of MSE $(\mathbf{W})$ can be obtained by the operator kernel $w(t,\tau)$ that satisfies

$$E\{s_k(t)r_k^*(\tau)\} = \int_{-\infty}^{+\infty} w(t,u) E\{r_k(u) r_k^*(\tau)\} du. \qquad (8.39)$$

The solution of (8.39) is computationally intensive and not practical. The idea proposed in [43] is the use of a multiplicative filter that multiplies the signal $\mathcal{F}_\alpha\{r_k(t)\}(u)$ in the αth fractional Fourier-domain, so that

$$\widehat{s}_k(t) = (\mathbf{W}r_k)(t) = \mathcal{F}_{-\alpha}\{w_\alpha(u)\mathcal{F}_\alpha\{r_k(t)\}(u)\}(t). \qquad (8.40)$$

It is important to recognize that the class of the αth fractional Fourier filters in (8.40) is a subclass of the filters in (8.38). The fractional Fourier-domain filters are well-suited for the recovery of time-varying degradations and are much more effective [43] than traditional time-invariant techniques.

In practice, due to (8.37), there is no need to perform the inverse transformation, because the transmitted symbols can be considered samples of the continuous FRFT. We can then write

$$\begin{aligned}\hat{a}_{n,k} &= \mathcal{K}^{-1}\left[w_\alpha(u)\mathcal{F}_\alpha\{r_k(t)\}(u)\right]_{u=\frac{n}{T}\sin\alpha} \\ &= \mathcal{K}^{-1}\left[w_\alpha(u)\mathcal{F}_\alpha\{(\mathbf{H}\mathcal{D}_{-\alpha}\{\mathbf{a}_k\})(t) + n_k(t)\}(u)\right]_{u=\frac{n}{T}\sin\alpha}.\end{aligned} \qquad (8.41)$$

The block diagram of the system that implements the operations in (8.41) is shown in Figure 8.8.

The solution to the filter estimation problem can be obtained using the calculus of variations method described in [43]. Following this method, we obtain that

$$w_\alpha(u) = \frac{E\{s_k(u)r_k^*(u)\}}{E\{r_k(u)r_k^*(u)\}}.$$

It is interesting that the optimal value of α (which defines in what fractional Fourier domain transmission will occur) can be found from the off-line minimization of the following cost function [43]

$$\begin{aligned} J &= \int_{-\infty}^{+\infty} E\{s_k(u)s_k^*(u)\} \\ &\quad -2Re\,\{w_\alpha^*(u)E\{s_k(u)r_k^*(u)\}\} \\ &\quad +w_\alpha(u)w_\alpha^*(u)E\{r_k(u)r_k^*(u)\}du. \end{aligned} \tag{8.42}$$

In practice selecting the angle α that minimizes (8.42) is equivalent to selecting the fractional Fourier domain that best matches the time-varying characteristics of the channel. So this formulation also reduces to a classical best basis selection.

8.3.4 Discrete-Time Implementation

There are many possible ways to discretize the operations in (8.41). We select the methodology proposed in [42], because it is attractive from the computational point of view. We will start our research of a discrete-time operator for the operations in (8.41) from the derivation of the discrete-time model for an FFT-based system.

Notation: Observe that in the following derivations the notation $[\mathbf{v}]_k$ is used for the kth element of vector $\mathbf{v}$ and the notation $[\mathbf{M}]_{k,n}$ is used for the k, n element of the matrix $\mathbf{M}$.

Assume that $T = RT_s$ for R positive integer and write (8.29) as

$$s(t) = \sum_{k=-\infty}^{+\infty} \sum_{l=0}^{N-1} a_{k,l}\phi_l(t - kRT_s).$$

Sampling at $T_e \leq T_s$ we obtain

$$s_k(n) = \sum_{m=-\infty}^{+\infty} \sum_{l=0}^{N-1} a_{m,l}\phi_l\left((nN - mR)T_s - kT_e\right)$$

with $s_k(n) = s\left((nR - k)T_e\right)$ $k = 0, 1, ...N - 1$ where R is such that $RT_e = NT_s$. The particular case $R = N$ and $T_e = T_s$,

$\phi_l(t) = g(t)e^{2\pi j f_l t}$, $g(t) = \begin{cases} \frac{1}{\sqrt{NT_s}} & 0 \leq t \leq NT_s \\ 0 & \text{elsewhere} \end{cases}$ and $f_l = \frac{l}{NT_s}$
collapses to the traditional FFT-based multicarrier system

$$\begin{aligned} s_k(n) &= \sum_{m=-\infty}^{+\infty} \sum_{l=0}^{N-1} a_{m,l} g\left((n-m)NT_s - kT_s\right) \exp\left\{\frac{j2\pi lk}{N}\right\} \\ &= \frac{1}{\sqrt{NT_s}} \sum_{l=0}^{N-1} a_{n-1,l} \exp\left\{\frac{j2\pi lk}{N}\right\}. \end{aligned} \tag{8.43}$$

If we define

$$\mathbf{a}(n) = \left[a_{n-1,0}, a_{n-1,1}, ..., a_{n-1,N-1}\right]^T$$

and

$$\mathbf{s}(n) = \left[s_0(n), s_1(n), ..., s_{N-1}(n)\right]^T,$$

(8.43) can be rewritten in vector form as

$$\mathbf{s}(n) = \mathbf{F}^{[-1]}\mathbf{a}(n). \tag{8.44}$$

where $\mathbf{F}^{[-1]}$ represents the (orthonormal) mapping in the traditional Fourier domain and we have assumed without loss of generality a unitary sample period T_s.

The instantaneous frequency of the nth waveform in (8.34) is $\frac{-t\cot\alpha}{2\pi} + \frac{n}{T}$. So it is evident that since $|t| \leq T/2$ and since $n \leq N-1$ (because we want to use N of these functions), the maximum frequency extent of the signal

$$s_k(t) = \sum_{n=0}^{N-1} a_{k,n} f_{-\alpha,n}(t) \tag{8.45}$$

is $\frac{T\cot\alpha}{4\pi} + \frac{N-1}{T}$ (if the symbols $a_{k,n}$ are assumed identically and independently distributed, i.i.d.). Limiting the value of α in the range $0.5\pi/2 \leq |\alpha| \leq 1.5\pi/2$, we also have that the maximum frequency for $s_k(t)$ is

$$\frac{T|\cot\alpha|}{4\pi} + \frac{N-1}{T} \leq \frac{F\text{max}}{2} = \frac{N}{2T},$$

for any value of N, T. The relationship between T and N is enforced by the fact that the time-frequency product of $s_k(t)$ must be equal to N and that the sample period is chosen in such a way that $T = \sqrt{N}T_s$. If this is the case $F_{\max} = \frac{\sqrt{N}}{T_s}$ and the signal set occupies an "equal" portion of time and frequency.

Now we try to obtain a formula for the discrete-time computation of (8.41). Since we need $\mathcal{F}_\alpha\{r_k(t)\}(u)$ we study the problem of obtaining $\mathcal{F}_\alpha\{s_k(t)\}(u)$. From the definition

$$\mathcal{F}_\alpha\{s_k(t)\}(u) = A_\alpha e^{j\pi\phi_\alpha u^2}\int_{-\infty}^{+\infty} e^{-j2\pi\beta_\alpha tu}e^{j\pi\phi_\alpha t^2}s_k(t)dt, \quad (8.46)$$

for $\phi_\alpha = \cot\alpha$ and $\beta_\alpha = \csc\alpha$. The frequency "shearing" for the signal $s_k(t)$ when multiplied by the chirp $e^{j\pi\phi_\alpha t^2}$ for $0.5\pi/2 \leq |\alpha| \leq 1.5\pi/2$ results in a bandwidth which is at most $2F_{\max}$ (contained in $[+F_{\max}, -F_{\max}]$). So we can represent the signal $e^{j\pi\phi_\alpha t^2}s_k(t)$ using the Shannon interpolation formula

$$\begin{aligned} e^{j\pi\phi_\alpha t^2}s_k(t) &= \sum_{n=-N}^{N} e^{j\pi\phi_\alpha\left(\frac{n}{2F_{\max}}\right)^2} s_k\left(\frac{n}{2F_{\max}}\right) \\ &\cdot\text{sinc}\left(2F_{\max}\left(t-\frac{n}{2F_{\max}}\right)\right) \end{aligned} \quad (8.47)$$

where $N = \frac{F_{\max}^2}{T_s^2}$, and the summation is between N and $-N$ because $s_k(t)$ is band-limited to $[+F_{\max}/2, -F_{\max}/2]$. Substitution of (8.47) in (8.46) and performing the integration[11] as in [42] gives

$$\begin{aligned} \mathcal{F}_\alpha\{s_k(t)\}(u) = {} & \frac{A_\alpha}{2F_{\max}}e^{j\pi\phi_\alpha u^2} \\ & \cdot\sum_{n=-N}^{N} e^{-j2\pi\beta_\alpha(un)/(2F_{\max})}e^{j\pi\phi_\alpha\left(\frac{n}{2F_{\max}}\right)^2} s_k\left(\frac{n}{2F_{\max}}\right) \end{aligned}$$

which can be also discretized as

11. The interested reader is referred to [42] for more explanations on this derivation.

$$\mathcal{F}_\alpha \{s_k(t)\} \left(\frac{m}{2T}\right) = \frac{A_\alpha}{2F_{\max}} \times \sum_{n=-N}^{N} e^{j\pi\phi_\alpha \left(\frac{m}{2T}\right)^2 - j\beta_\alpha \frac{2\pi m}{2T}\frac{n}{2F_{\max}}} \cdot e^{j\pi\phi_\alpha \left(\frac{n}{2F_{\max}}\right)^2} s_k\left(\frac{n}{2F_{\max}}\right).$$

Using the relations among T, $F_{\max}$ and N with again a unitary sample period T_s we obtain the operator for the fractional Fourier domain operator

$$[\mathbf{K}_\alpha]_{m,n} = K_\alpha(m-N-1, n-N-1) \quad m,n = 1,2,...,2N+1,$$

with

$$K_\alpha(m,n) = \frac{A_\alpha}{2\sqrt{N}}\exp\left\{j\pi\phi_\alpha\frac{m^2}{4N} - j2\pi\beta_\alpha\frac{mn}{4N} + j\pi\phi_\alpha\frac{n^2}{4N}\right\}.$$

A multicarrier system in the (discrete) fractional Fourier domain can be proposed that uses the operator

$$\mathbf{F}^{[a]} = \mathbf{D}\mathbf{K}_{a\frac{\pi}{2}}\mathbf{J} \tag{8.48}$$

at the receiver. The matrix $\mathbf{D}$ is a decimation by two matrix[12] and $\mathbf{J}$ is an interpolation by two matrix[13] [27]. At the transmitter a natural choice is to replace $\mathbf{F}^{[-1]}$ with $\mathbf{F}^{[-a]}$ in (8.44). We then obtain the desired block operation

$$\mathbf{s}(n) = \mathbf{F}^{[-a]}\mathbf{a}(n).$$

12. A decimation by P matrix $\mathbf{D}$ acts on an NP-vector to give the decimated N-vector. $\mathbf{D}$ is obtained from the identity $NP \times NP$ matrix removing $P-1$ rows from consecutive P rows. A decimation by 2 ($P=2$) matrix with $N=3$ is

$$\mathbf{D} = \begin{bmatrix} 1 & 0 & 0 & 0 & 0 & 0 \\ 0 & 0 & 1 & 0 & 0 & 0 \\ 0 & 0 & 0 & 0 & 1 & 0 \end{bmatrix}$$

13. An interpolation by P matrix acts on an N vector to give the upsampled and interpolated NP-vector. An interpolation by 2 ($P=2$) matrix with

Observe that this mapping is valid[14] for $0.5 \leq |a| \leq 1.5$. The operator is such that

$$\mathbf{F}^{[a]}\mathbf{F}^{[-a]} \simeq \mathbf{I},$$

and the Fourier domain that results in the traditional FFT-based scheme is obtained for $a = 1$.

The operator $\mathbf{F}^{[a]}$ allows an algorithm implementation with complexity $\mathcal{O}\left[N \log N\right]$, which is a remarkable property [42]. This is easily seen rewriting the block operation performed by the operator $\mathbf{F}^{[-a]}$ as

$$\mathbf{s}(n) = \mathbf{D}\mathbf{K}_{-a\frac{\pi}{2}}\mathbf{J}\mathbf{a}(n) = \mathbf{D}\tilde{\mathbf{a}}_{-a}(n)$$

where $\tilde{\mathbf{a}}_{-a}(n)$ can be obtained as

$$\begin{aligned}
\left[\tilde{\mathbf{a}}_{-a}(l)\right]_{m+N+1} &= \sum_{n=-N}^{N} K_{-\alpha}(m,n)\left[\mathbf{J}\mathbf{a}(l)\right]_{n+N+1} \\
&= \frac{A_{-\alpha}}{2\sqrt{N}} \sum_{n=-N}^{N} \exp\left\{j\pi\phi_{-\alpha}\frac{m^2}{4N} - j2\pi\beta_{-\alpha}\frac{mn}{4N}\right\} \\
&\quad \cdot\exp\left\{j\pi\phi_{-\alpha}\frac{n^2}{4N}\right\}\left[\mathbf{J}\mathbf{a}(l)\right]_{n+N+1} \\
&= \frac{A_{-\alpha}}{2\sqrt{N}}\exp\left\{j\pi(\phi_{-\alpha} - \beta_{-\alpha})\frac{m^2}{4N}\right\} \sum_{n=-N}^{N} x_n y_{m-n}
\end{aligned} \tag{8.49}$$

where

$$x_n = \exp\left\{j\pi(\phi_{-\alpha} - \beta_{-\alpha})\frac{n^2}{4N}\right\}\left[\mathbf{J}\mathbf{a}(l)\right]_{n+N+1},$$

$N = 3$ and h_n as the interpolation filter is

$$\mathbf{J} = \begin{bmatrix} h_0 & h_1 & 0 & 0 & 0 & 0 \\ 0 & h_0 & h_1 & 0 & 0 & 0 \\ 0 & 0 & h_0 & h_1 & 0 & 0 \\ 0 & 0 & 0 & h_0 & h_1 & 0 \\ 0 & 0 & 0 & 0 & h_0 & h_1 \\ 0 & 0 & 0 & 0 & 0 & h_0 \end{bmatrix} \begin{bmatrix} 1 & 0 & 0 \\ 0 & 0 & 0 \\ 0 & 1 & 0 \\ 0 & 0 & 0 \\ 0 & 0 & 1 \\ 0 & 0 & 0 \end{bmatrix}$$

14. Extensions of this interval can be easily determined as derived in [42].

and

$$y_m = \exp\left\{j2\pi\beta_{-\alpha}\frac{m^2}{4N}\right\}.$$

Equation (8.49) shows that the summation is a simple convolution of the elements of the vector $\mathbf{Ja}(l)$ premultiplied by a chirp sequence. The result of the convolution is again chirp-modulated. Evidently the majority of the computational effort is required by the convolution $\sum_{n=-N}^{N} x_n y_{m-n}$. A well-known efficient method to perform convolutions is the FFT which results in a complexity $\mathcal{O}[N \log N]$. Since the chirp modulations only contribute linearly to the computational load, the overall computational complexity is $\mathcal{O}[N \log N]$, approximately the same complexity required to implement the traditional N-carrier OFDM system based on the use of the FFT.

At the receiver the signal block $\mathbf{r}(n)$, which is the sampled baseband equivalent of the channel-distorted signal $(\mathbf{H}s)(t) + n(t)$, is transformed back to the fractional Fourier domain and a multiplicative filter $\mathbf{w}_a$ is applied

$$\hat{\mathbf{s}}(n) = \mathbf{\Lambda}_{w,a}\mathbf{F}^{[a]}\mathbf{r}(n).$$

The matrix $\mathbf{\Lambda}_{w,a}$ is a diagonal $N \times N$ matrix whose diagonal elements consist of the elements of the complex weight vector $\mathbf{w}_a$. The solution that minimizes the MSE defined as

$$J = E\left\{||\mathbf{s}(n) - \hat{\mathbf{s}}(n)||^2\right\},$$

with $||\mathbf{x}(n)||^2 = \sum_i |[\mathbf{x}(n)]_i|^2$ is obtained following the same methodology developed in [43] which gives

$$[\mathbf{w}_\alpha]_i = \frac{E\{[\mathbf{s}(n)]_i\,[\mathbf{r}^*(n)]_i\}}{E\{[\mathbf{s}(n)]_i\,,[\mathbf{r}^*(n)]_i\}}. \tag{8.50}$$

The block diagram of the discrete-time baseband FRFT-based system is shown in Figure 8.12, where we also show the application of the circular prefix [35]. $\mathbf{F}^{[a]}$ (and $\mathbf{F}^{[-a]}$) are defined in (8.48). $\mathbf{\Lambda}_{w,a}$ is a diagonal matrix with complex diagonal elements. The traditional OFDM system is simply obtained replacing the transformation $\mathbf{F}^{[-a]}$ with $\mathbf{F}^{[-1]}$ defined in (8.43) and (8.44) at the transmitter and $\mathbf{F}^{[a]}$ with $\mathbf{F}^{[-1]^H}$ at the receiver. The diagonal matrix $\mathbf{\Lambda}_{w,a}|_{a=1}$ then becomes a bank of single-tap frequency domain equalizers.

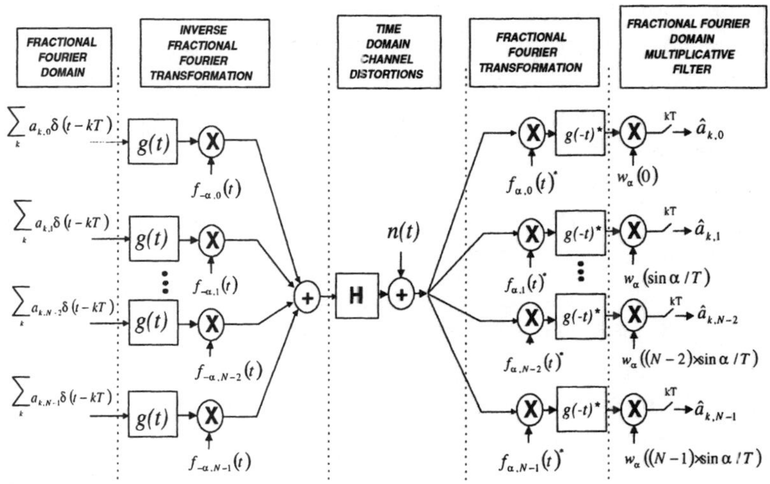

Figure 8.11 Block diagram of a FRFT general multicarrier system. Observe that for $\alpha = \frac{\pi}{2}$ we obtain the traditional multicarrier system.

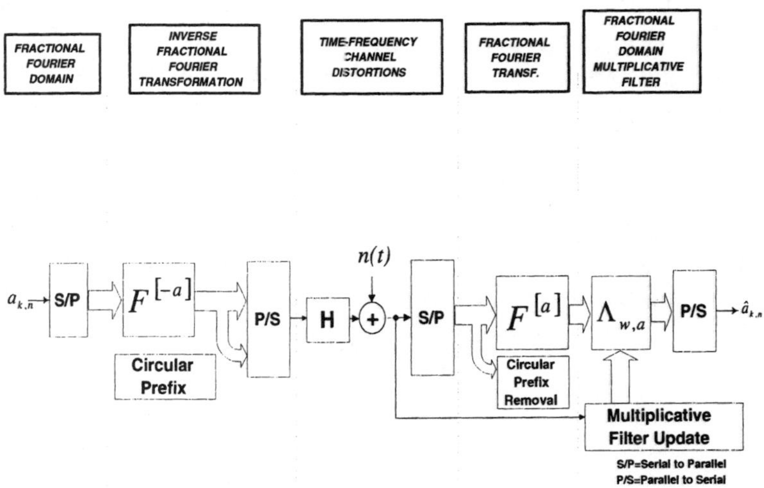

Figure 8.12 Block diagram of the proposed FRFT-based multicarrier system. Observe that for $\alpha = \frac{\pi}{2}$ we obtain the traditional multicarrier system.

8.3.5 Performance Assessment

In this section we study the performance of the scheme in a wireless high mobility environment. The symbol period is 160 μsec, the delay spread is up to 40 μsec, the modulation scheme is QPSK coherently detected, the DMT symbol duration (N) is 128, the carrier frequency is 5.2 GHz, the Doppler spread f_D is 200–1000 Hz, and the propagation environment is a non-LOS (Rayleigh) RF link. We include 40 μsec of guard time and assume a frame structure with 15 reference symbols dedicated to training of the multiplicative filter. The measure of performance is raw (uncoded) BER averaged over 10,000 multicarrier blocks. The baseline performance is an OFDM system that employs the same parameters with traditional IFFT/FFT processing and the ideal diagonalizing operator, which assumes perfect knowledge of the channel (both at the transmitter and at the receiver). In discrete time a generic time-varying linear operator is represented by $\mathbf{H}$, and $\mathbf{H}$ is, in fact, a matrix with no "structure" (it becomes a very structured convolution matrix in the static case, easily diagonalized by the FFT with cyclic prefix). Define the singular value decomposition of $\mathbf{H}$ as $\mathbf{H} = \mathbf{U}\boldsymbol{\Sigma}\mathbf{V}^H$, where $\boldsymbol{\Sigma}$ is diagonal. To diagonalize $\mathbf{H}$ one can use the left singular vectors $\mathbf{V}$ [instead of $\mathbf{F}^{[-a]}$] at the transmitter and the right singular vectors $\mathbf{U}^H$ [instead of $\mathbf{F}^{[a]}$] at the receiver (refer to Figure 8.12). The optimum one-tap equalizers are given by the diagonal elements of $\boldsymbol{\Sigma}^{-1}$.

The channel model is based on the wide sense stationary uncorrelated scattering (WSSUS) assumption [25]. The complex weights are generated as filtered Gaussian processes fully specified by the scattering function $S(t,\tau)$. The α angle (or equivalently the parameter a) for the fractional Fourier domain is selected as to minimize the following cost function

$$\begin{aligned}\tilde{J}(\alpha) &= \sum_i E\{[\mathbf{s}(n)]_i\,[\mathbf{s}^*(n)]_i\} \\ &\quad -2Re\,\{[\mathbf{w}_\alpha]_i^*\,E\{[\mathbf{s}(n)]_i\,[\mathbf{r}^*(n)]_i\}\} \\ &\quad +[\mathbf{w}_\alpha]_n\,[\mathbf{w}_\alpha^*]_n\,E\{[\mathbf{r}(n)]_i\,[\mathbf{r}^*(n)]_i\}. \qquad (8.51)\end{aligned}$$

Observe that the computational cost of computing $\tilde{J}(\alpha)$ is essentially equivalent to the cost associated with the calculation of $E\{[\mathbf{s}(n)]_i\,[\mathbf{r}^*(n)]_i\}$ and $E\{[\mathbf{r}(n)]_i\,[\mathbf{r}^*(n)]_i\}$ and $E\{[\mathbf{s}(n)]_i\,[\mathbf{s}^*(n)]_i\}$ for each value of α. For a given Doppler spread α can be optimized

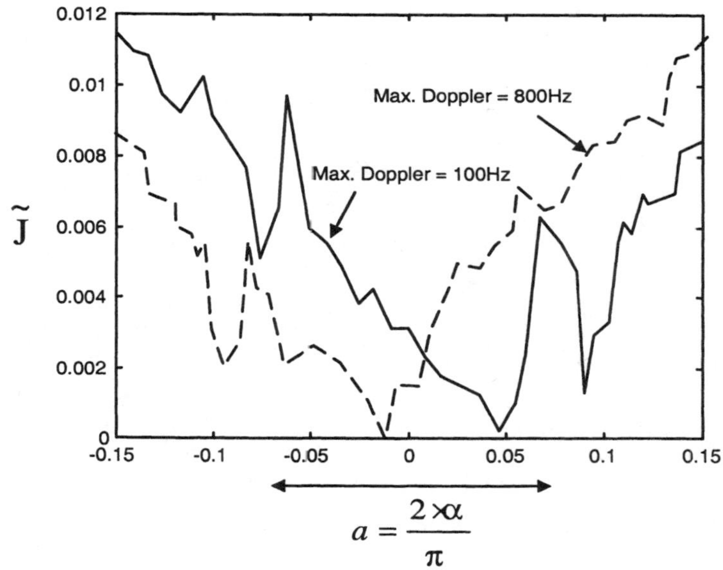

Figure 8.13 The cost function $\tilde{J}$ versus $a = \frac{2\,\alpha}{\pi}$ shown in normalized form for two different channel environments.

minimizing (8.51). The procedure we employ in the simulation analysis is an off-line calibration method. Figure 8.13 shows $\tilde{J}$ with respect to $a = \frac{2\,\alpha}{\pi}$, for a Rayleigh faded channel with $f_D = 100$ Hz and $f_D = 800$ Hz.

There is an optimum value of α in each case. In the examples below we optimize the value of the angle α for the fractional Fourier processing for any given Doppler spread scenario.

Figure 8.14 reports performance for $f_D = 0 - 1000$ Hz. Observe that as the rate of change in the channel decreases, the performance improvement with respect to the FFT-based system also decreases. For a static channel the classical scheme is more effective than the FRFT scheme. As we increase Doppler spread, the suboptimality of the FFT-based system is evidently outbalanced by the performance of the FRFT method. Figure 8.15 reports performance for $f_D = 0 - 900$ Hz of the FRFT scheme, the FFT scheme and the single carrier QR-RLS DFE equalized scheme [37]. The SNR per bit is 10 dB and the delay spread is 20 μsec.

As already pointed out, we optimize off-line the α angle.

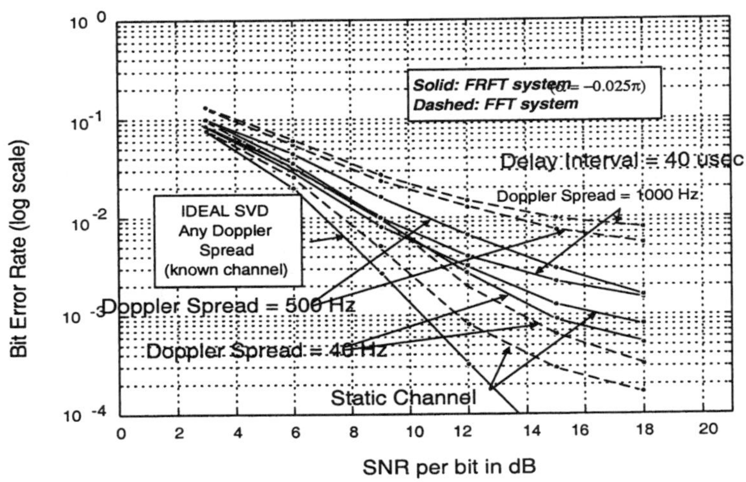

Figure 8.14 BER performance of the fractional Fourier scheme as it compares to the classical traditional scheme based on IFFT/FFT processing at different Doppler spreads and to the ideal SVD-based system with perfectly known channel.

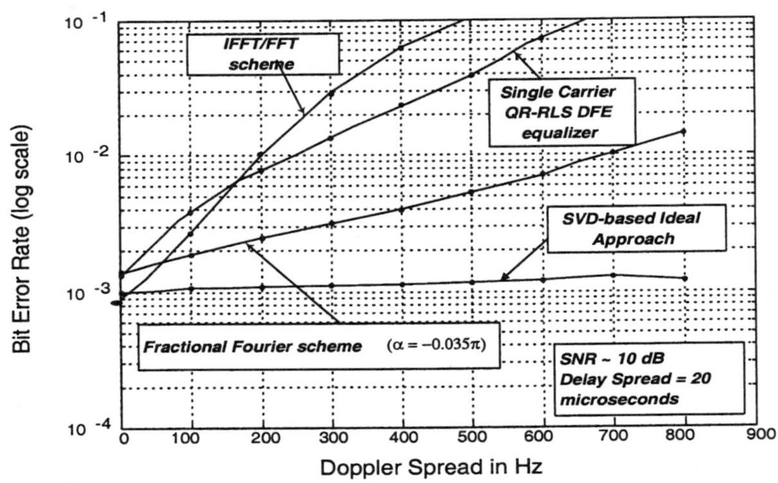

Figure 8.15 BER performance versus Doppler spread for the fractional Fourier scheme as it compares to the classical traditional schemes based on IFFT/FFT processing, to the single carrier equalized scheme and to the ideal SVD-based system with perfectly known channel.

However, an improvement can be obtained if we incorporate an on-line estimation procedure (not studied in this chapter). We can envision a frame format organized in FRFT blocks, where in each frame a few blocks are dedicated to training of the multiplicative filter. Training the multiplicative filter requires the following steps:

- For each value of α quantized on a grid with $\mathcal{G}$ steps, transmission of the inverse FRFT of $\mathcal{N}$ training signal blocks corresponding to each particular α.
- Using $\mathcal{N}$ training received signal blocks, estimate

  $$E\{[\mathbf{s}(n)]_m, [\mathbf{r}(n)]_m^*\},$$

  $$E\{[\mathbf{r}(n)]_m, [\mathbf{r}(n)]_m^*\}$$

  [$E\{[\mathbf{s}(n)]_m, [\mathbf{s}(n)]_m^*\}$ is precomputed and stored in memory] for each value of α in the grid.
- Compute (8.51) for each value of α in the grid and select α_{opt} as the value which minimizes $\tilde{J}$.
- Compute $\mathbf{w}_{\alpha_{opt}}$ using (8.50).

8.4 Key Points of This Chapter

We have explained in this chapter two solutions to the high mobility wireless channel. The first method applies to the problem of detecting a sequence in rapidly fading multipath, making use of a model that parsimoniously characterizes the time-variant nature of the detection problem. The metric used by the generalized likelihood detector involves finding the orthogonal projection of the received signal onto the subspace of the signal afflicted by multipath fading: The faded signal corresponding to one of the transmitted sequences is in fact known to lie in a subspace but its exact location is not known, because the channel parameters are unknown. This *matched subspace detector* has a significant drawback: The computation of the orthogonal projection of the received signal onto the signal subspace may be impossible if the channel is arbitrarily time-varying. We have presented in this chapter the idea of using new subspaces that approximate the original signal subspace, and whose orthogonal projection is easily computed in practice. By applying a wavelet-based multiresolution decomposition to the fading channel time variations, one obtains a nested sequence of linear vector spaces, that, at increasingly high level of detail, are efficient representations of the original signal subspace. We have shown a method to focus these representations to the original signal subspace using the concept of deviation between two closed subspaces and the knowledge of the scattering function of the multipath channel. Observe that the method can also be interpreted as a way to incorporate a priori knowledge on the structure of the signal subspace and as a method to design low-rank generalized energy detectors [20] where one evidently sacrifices bias for variance in the channel estimates. The results of hardware experiments proved that the method is effectively a viable candidate to sequence detection in rapidly fading channels.

References

[1] G. D. Forney, "Maximum Likelihood Sequence Estimation of Digital Sequences in the Presence of Intersymbol Interference," *IEEE Trans. Inform. Theory*, Vol. IT-18, pp. 363–378, May 1972.

[2] G. Ungerboeck, "Adaptive Maximum-Likelihood Receiver for Carrier-Modulated Data Transmission Systems," *IEEE Trans. Comm.*, Vol. 22, May 1974.

[3] M.-C. Chiu and C.-C. Chao, "Analysis of LMS-Adaptive MLSE Equalization on Multipath Fading Channels," *IEEE Trans. Comm.*, Vol. 44, pp. 1684–1692, Dec. 1996.

[4] R. Raheli, A. Polydoros, and C.-K. Tzou, "Per-Survivor Processing: A General Approach to MLSE in Uncertain Environments," *IEEE Trans. Comm.*, Vol. 43, pp. 354–364, Feb./March/April 1995.

[5] L. Van Trees, *Detection, Estimation, and Modulation Theory*, Part I, New York: Wiley, 1968.

[6] K. M. Chugg and A. Polydoros, "MLSE for an Unknown Channel-Part I: Optimality Considerations," *IEEE Trans. Comm.*, Vol. 44, pp. 836–846, July 1996.

[7] M. Ghosh and C. L. Weber, "Maximum Likelihood Blind Equalization," *Opt. Eng.*, Vol. 31, No. 6, pp. 1224–1228, June 1992.

[8] N. Seshadri, "Joint Data and Channel Estimation Using Trellis Search Techniques," *IEEE Trans. Comm.*, Vol. 42, pp. 80–89, Feb./March/April 1994.

[9] Q. Dai and E Shwedyk, "Detection of Bandlimited Signals over Frequency Selective Rayeigh Fading Channels," *IEEE Trans. Comm.* Vol. 42, pp. 941–950, March 1994.

[10] R. A. Iltis, "A Bayesian Maximum-Likelihood Sequence Estimation Algorithm for A Priori Unknown Channels and Symbol Timing," *IEEE Journal Sel. Areas in Comm.*, Vol. 10, No. 3, pp. 579–588, April 1992.

[11] M. E. Rollins and S. J. Simmons, "Simplified Per-Survivor Kalman

Processing in Fast Frequency-Selective Fading Channels," *IEEE Trans. Comm.*, Vol. 45, pp. 544–553, May 1997.

[12] X. Yu and S. Pasupathy, "Innovations-Based MLSE for Rayleigh Fading Channels," *IEEE Trans. Comm.*, Vol. 43, No. 2/3/4, pp. 1534–1544, Feb./March/April 1995.

[13] J. G. Proakis, *Digital Communications*, New York: McGraw-Hill, 1989.

[14] M. K. Tsatsanis and G. B. Giannakis, "Time-Varying System Identification and Model Validation Using Wavelets," *IEEE Trans. Sig. Proc.*, Vol. 41, No.12, pp. 3512–3523, Dec. 1993.

[15] J. Zhang and G. Walter, "A Wavelet-Based KL-Like Expansion for Wide-Sense Stationary Random Processes," *IEEE Trans. Sig. Proc.*, Vol. 42, No. 7, pp. 1737–1745, July 1994.

[16] R. W. Dijkerman and R.R. Mazumdar, "Wavelet Representations of Stochastic Processes and Multiresolution Stochastic Models," *IEEE Trans. Sig. Proc.*, Vol. 42, No. 7, pp. 1640–1652, July 1994.

[17] M. I. Doroslovacki and H. Fan, "Wavelet-Based Linear System Modeling and Adaptive Filtering," *IEEE Trans. Sig. Proc.*, Vol. 44, No. 5, pp. 1156–1167, May 1996.

[18] M. Martone, "Wavelet-Based Separating Kernels for Sequence Estimation with Unknown Rapidly Time-Varying Channels," *IEEE Comm. Lett.*, March 1999.

[19] D. K. Borah and B. D. Hart, "Receiver Structures for Time-Varying Frequency-Selective Fading Channels," *IEEE Journal Sel. Areas in Comm.*, Vol. 17, No. 11, pp. 1863–1875, Nov. 1999.

[20] L. L. Scharf, *Statistical Signal Processing*, Reading, MA: Addison Wesley, 1991.

[21] D. L. Donoho, "Unconditional Bases Are Optimal Bases for Data Compression and for Statistical Estimation," *App. Comp. Harmonic Anal.*, pp. 100–115, Dec. 1993.

[22] G. B. Giannakis and C. Tepedelenlioglu, "Basis-Expansion Models and Diversity Techniques for Blind Identification and Equalization of Time-Varying Channels," *Proc. of the IEEE*, Vol. 86, No. 10, pp. 1969–1986, Oct. 1998.

[23] I. Daubechies, *Ten Lectures in Wavelets*, New York, SIAM, 1992.

[24] I. Daubechies, "Orthogonal Bases of Compactly Supported Wavelets," *Commun. Pure Appl. Math.*, Vol. 41, pp. 909–996, 1988.

[25] P. A. Bello, "Characterization of Randomly Time Variant Linear Channels," *IEEE Trans. Comm. Syst. Tech.*, Vol. 11, pp. 360–393, Dec. 1963.

[26] A. C. Bovik, P. Maragos, and T. F. Quatieri, "AM-FM Energy Detection and Separation in Noise Using Multiband Energy Operators," *IEEE Trans. Sig. Proc.*, Vol. 41, pp. 3245–3265, Dec. 1993.

[27] P. P. Vaidyanathan, *Multirate Systems and Filter Banks*, Upper Saddle River, NJ: Prentice Hall, 1993.

[28] J. Liang and T. W. Parks, "Kolmogorov N-Widths and Wavelet Representation for Signal Classes," *IEEE Trans. Sig. Proc.*, Vol. 44, pp. 1693–1703, July 1996.

[29] W. C. Jakes, Jr., *Microwave Mobile Communications,* New York: Wiley, 1974.

[30] TIA/EIA/IS-136.1-A, "TDMA Cellular/PCS–Radio Interface–Mobile Station–Base Station Compatibility–Digital Control Channel," Oct. 1996.

[31] M. Martone, "An Adaptive Algorithm for Adaptive Antenna Array Low-Rank Processing in Cellular TDMA Base-Stations," *IEEE Trans. Commun.*, Vol. 46, pp. 627–643, May 1998.

[32] R. W. Chang, "Orthogonal Frequency Division Multiplexing," *U.S. Patent* 3,488,445, filed 1966, issued Jan. 6, 1970.

[33] L. J. Cimini, "Analysis and Simulation of a Digital Mobile Channel using Orthogonal Frequency Division Multiplexing," *IEEE Trans. Comm.*, Vol. 33, pp. 665–675, July 1985.

[34] B. Hirosaki, "An Orthogonally Multiplexed QAM System Using the Discrete Fourier Transform," *IEEE Trans. Comm.*, Vol. 29, pp. 982–989, July 1981.

[35] J. S. Chow, J. C. Tu, and J. M. Cioffi, " A Discrete Multitone Transceiver System for HDSL Application," *IEEE J. Sel. Areas Comm.*, Vol. 9, No. 6, pp. 895–908, Aug. 1991.

[36] M. Martone, "On MMSE Real-Time Antenna Array Processing Using Fourth-Order Statistics in the US Cellular TDMA System," *IEEE Journal Sel. Areas in Comm.*, Vol. 16, No. 8, pp. 1396–1410, Oct. 1998.

[37] M. Martone, "Complex Scaled Tangent Rotations (CSTAR) for Fast Space-Time Adaptive Equalization of Wireless TDMA," *IEEE Trans. Comm.*, Vol. 46, No. 12, Dec. 1998.

[38] W. Kozek and A. Molisch, "Nonorthogonal Pulseshapes for Multicarrier Communications in Doubly Dispersive Channels," *IEEE Journal Selected Areas Comm.*, Vol. 16, No. 8, pp. 1579–1589, Oct. 1998.

[39] L. B. Almeida, "The Fractional Fourier Transform and Time-Frequency Representations," *IEEE Trans. Sig. Proc.*, Vol. 42, No. 11, pp. 3084–3091, Nov. 1994.

[40] V. Namias, "The Fractional Fourier Transform and Its Application in Quantum Mechanics," *J. Inst. Math Applications*, Vol. 25, pp. 241–265, 1980.

[41] A. C. Mc Bride and F. H. Kerr, "On Namia's Fractional Fourier Transform," *IMA J. Appl. Math.*, Vol. 39, pp. 159–175, 1987.

[42] H. M. Ozatakas, et al., "Digital Computation of the Fractional Fourier Transform," *IEEE Trans. Sig. Proc.*, Vol. 44, pp. 2141–2150, Sept. 1996.

[43] M. A. Kutay, et al., "Optimal Filtering in Fractional Fourier Domains," *IEEE Trans. Sig. Proc.*, Vol. 45, pp. 1129–1143, May 1997.

[44] S. C. Pei, M. H. Yeh, and T.-L. Luo, "Fractional Fourier Series Expansion for Finite Signals and Dual Extension to Discrete-Time Fractional Fourier Transform," *IEEE Trans. Sig. Proc.*, Vol. 47, pp. 2883–2891, Oct. 1999.

[45] S. Mallat, "A Theory for Multiresolution Signal Decomposition the Wavelet Representation," *IEEE Trans. Pattern Anal. and Machine Intell.*, Vol. 11, pp. 674–693, July 1989.

[46] O. Rioul, "A Discrete-Time Multiresolution Theory," *IEEE Trans. Sig. Proc.*, Vol. 41, No. 8, pp. 2591–2606, Aug. 1993.

Appendix 8A: Essential Wavelet Theory

8A.1 Discrete Wavelet Transformations

A generic signal $f(t)$ can be represented in terms of translates and dilates of a single prototype (typically bandpass) wavelet $\psi(t)$

$$f(t) = \sum_{l,m=-\infty}^{+\infty} \xi(l,m) 2^{-l/2} \psi(2^{-l}t - m), \tag{8A.1}$$

or equivalently for some $P > 0$

$$\begin{aligned} f(t) &= \sum_{m} \zeta(m) 2^{-P/2} \phi(2^{-P}t - m) \\ &\quad + \sum_{l=-\infty}^{P} \sum_{m} \xi(l,m) 2^{-l/2} \psi(2^{-l}t - m), \end{aligned} \tag{8A.2}$$

where

$$\xi(l,m) = \int_{-\infty}^{+\infty} f(t) 2^{-l/2} \psi(2^{-l}t - m) dt$$

are wavelet (or *detail*) coefficients,

$$\zeta(m) = \int_{-\infty}^{+\infty} f(t) 2^{-P/2} \phi(2^{-P}t - m) dt$$

are scaling (or *approximation*) coefficients, and $\phi(t)$ is a lowpass scaling function [23, 24]. Particularly important is the interpretation of (8A.2) as a multiresolution analysis of $f(t)$: l indexes the *scale*, or *resolution* (the smaller the l, the higher the resolution), while m indexes the *spatial* location of analysis. If the mother wavelet is centered at time 0 and frequency f_c, $\xi(l,m)$ measures the content of $f(t)$ around time $2^l m$ and frequency $2^{-l} f_c$, while $\zeta(m)$ represents the local mean around time $2^P m$.

In this framework we can think of $f(t)$ as the finest scale ($P = 0$) representation of $f(t)$ itself. The function $\psi(t)$ has to satisfy some critical conditions to ensure that (8A.2) holds for any square integrable function $f(t)$. In particular, $\psi(t)$ has to satisfy the two scale equations

$$\phi(t) = \sqrt{2}\sum_n c_0(n)\phi(2t-n) \quad (8A.3)$$

$$\psi(t) = \sqrt{2}\sum_n c_1(n)\phi(2t-n), \quad (8A.4)$$

where the coefficients $c_i(n)$ can be nonzero[15] only over a finite number of consecutive values of n. Define $\psi_{l,m}(t) = 2^{-l/2}\psi(2^{-l}t - m)$ and $\phi_{l,m}(t) = 2^{-l/2}\phi(2^{-l}t - m)$. The three orthogonalities constraints imposed are **O1**: $\int \phi_{l,m}(t)\phi_{l,m'}(t)^* dt = \delta(m-m')$ **O2**: $\int \psi_{l,m}(t)\psi_{l',m'}(t)^* dt = \delta(m-m') = \delta(l-l')$ **O3**: $\int \psi_{l,m}(t)\phi_{l',m'}(t)^* dt = 0$. It is possible to show that **O1**, **O2**, and **O3** are satisfied by coefficients $c_i(n)$ that have the following properties:

$$\sum_n c_0(n)c_0(n+2k) = \delta(k)$$

$$c_1(n) = (-1)^n c_0(n).$$

A remarkable observation can be made: $c_i(n)$ are the coefficients of a perfect reconstruction two-band (or dyadic) filter bank with the quadrature mirror reconstruction property [27]. Therefore, the design of wavelets is equivalent to the design of filter banks. Observe that the conditions imposed up to now are not sufficient to construct in practice useful wavelets. In fact, they can lead to decompositions with not enough regularity. Regularity is imposed requiring a large number of vanishing moments [23, 24]. In practice, $\zeta(m)$ and $\xi(l,m)$ can be computed *recursively* from $\xi(l+1,m)$ using the efficient pyramid algorithm proposed by Mallat in [45]: It is not necessary to explicitly compute the shape of $\phi(t)$ and $\psi(t)$. The basic application of wavelets in this book stems from a multiresolution decomposition of a time-varying multipath response of the form $\tilde{h}(t,\tau)$ at any τ with respect

15. By doing so we restrict our description to compactly supported wavelets [27].

to t as

$$\tilde{h}(t,\tau) = \sum_m \zeta(m,\tau)2^{-P/2}\phi(2^{-P}t - m) + \sum_{l=-\infty}^{P} \sum_m \xi(l,m,\tau)2^{-l/2}\psi(2^{-l}t - m), \tag{8A.5}$$

where $\xi(l,m,\tau)$ are wavelet coefficients, $\zeta(m,\tau)$ are scaling coefficients. If the channel is modeled as deterministic, wavelet and scaling coefficients are considered deterministic, if the channel is modeled as random, wavelet and scaling coefficients are considered random parameters. The wavelet-based representation will exhibit *global* characteristics of the channel dynamics in the low resolution coefficients, while retaining *local* rapid transitions in just a few coefficients at higher resolutions. A practical advantage is that the decomposition decouples the variations in time and relegates them in $\psi(t)$ and $\phi(t)$, so that the wavelet coefficients are indeed time-invariant for any τ.

We would like to obtain a practical basis representation for the discrete-time channel $h(n,k)$ that is easily parameterized; it retains the essential features of the fading process and it characterizes these features using a small number of time-invariant coefficients. These requirements can be satisfied using a *discrete-time* wavelet series representation.

An important technical aspect arises. We do not need the coefficients of the continuous-time wavelet series expansion of $\tilde{h}(t,\tau)$ to obtain a representation of the form in (8A.5). We only need a discrete-time wavelet series representation of a sampled version of $\tilde{h}(t,\tau)$, that is, $h(n,k)$. It is possible to show that for a high resolution (that is, large P) and short T/R, the discrete-time series coefficients and continuous time series coefficients asymptotically converge to the same values. The orthonormal expansion can be regarded more properly as a multiscale subband decomposition using filter banks of the sampled response of a multipath channel. We direct the interested reader to [27] and in particular to [46] where a discrete-time approach to multiresolution representations is described. It is fundamental to observe that $h(n,k)$ can be represented at lower resolution applying a half band lowpass filter having impulse response $c_0(n)$, followed by a down-sampling-by-two operation

$$\zeta_{1,m}(k) = \sum_l c_0(l)h(2m-l,k). \tag{8A.6}$$

This equation represents a mapping from a vector space to itself. An added detail can be obtained by using a highpass filter with impulse response $c_1(n)$, then again down-sampling by two

$$\xi_{1,m}(k) = \sum_l c_1(l)h(2m-l,k). \tag{8A.7}$$

The filters $c_i(n)$ $i = 0,1$ satisfy some regularity constraints [23, 24] and form an orthonormal set. If $c_i(n)$ [in the z-domain $C_i(z)$] $i = 0,1$ is a dyadic perfect reconstruction filter bank [27], one can think of (8A.6) and (8A.7) as a *decomposition* of $h(n,k)$ onto a discrete-time orthonormal basis with the following perfect reconstruction rule

$$\begin{aligned} h(n,k) &= \sum_{m=0}^{NR/2-1} \zeta_{1,m}(k)c_0(2m-n) \\ &+ \sum_{m=0}^{NR/2-1} \xi_{1,m}(k)c_1(2m-n), \end{aligned} \tag{8A.8}$$

which is indeed a sum of orthogonal projections. Using the same filters $c_0(m), c_1(m)$, (8A.6) could be split again in lowpass and highpass down-sampled components, and then further the lowpass component split again to a desired depth or resolution, so that one can always explode the lowpass component resolution

$$\begin{aligned} \zeta_{l,n}(k) &= \sum_{m=0}^{NR/2-1} \zeta_{l+1,m}(k)c_0(2m-n) \\ &+ \sum_{m=0}^{NR/2-1} \xi_{l+1,m}(k)c_1(2m-n). \end{aligned} \tag{8A.9}$$

For example, consider (8A.9) for $l = 1$ and substitute in (8A.8) to get

$$h(n,k) = \sum_{m=0}^{NR/2-1} \left[\sum_{m'=0}^{NR/2-1} \zeta_{2,m'}(k)c_0(2m'-m) \right.$$

$$
\begin{aligned}
&+ \left.\sum_{m'=0}^{NR/2-1} \xi_{2,m'}(k)c_1(2m'-m)\right] c_0(2m-n) \\
&+ \sum_{m=0}^{NR/2-1} \xi_{1,m}(k)c_1(2m-n) \\
= &\sum_{m=0}^{NR/2^2-1} \zeta_{2,m}(k)c_0^{(2)}(2^2m-n) \\
&+\sum_{l=1}^{2}\sum_{m=0}^{NR/2^l-1} \xi_{l,m}(k)c_1^{(l)}(2^lm-n) \qquad (8A.10)
\end{aligned}
$$

where $c_0^{(2)}(n)$, $c_1^{(1)}(n)$, and $c_1^{(2)}(n)$ are filters obtained in the z-domain as $C_0^{(2)}(z) = C_0(z)C_0(z^2)$, $C_1^{(1)}(z) = C_1(z)$, and $C_1^{(2)}(z) = C_1(z^2)C_0(z)$. In the last equation we have used the Noble identity [27] which says that a downsampler by two, followed by a filter $F(z)$, is equivalent to the filter $F(z^2)$ followed by the downsampler by two. We can think of (8A.10) as a decomposition of $h(n,k)$ onto a discrete-time orthonormal basis at resolution level $P = 2$. Using recursively this method, one obtains at generic resolution depth P (with associated signal subspace $\mathcal{V}_P$) the following representation for the channel

$$
\begin{aligned}
h(n,k) = &\sum_{m=0}^{NR/2^P-1} \zeta_{P,m}(k)c_0^{(P)}(2^Pm-n) \\
&+\sum_{l=1}^{P}\sum_{m=0}^{NR/2^l-1} \xi_{l,m}(k)c_1^{(l)}(2^lm-n), \qquad (8A.11)
\end{aligned}
$$

where $\zeta_{P,m}(k)$ and $\xi_{l,m}(k)$ are coefficients of the wavelet-based decomposition (WBD), and the filters $c_0^{(P)}(n)$ and $c_1^{(l)}(n)$ (with real coefficients) are given in the z-domain by $C_0^{(P)}(z) = \prod_{i=1}^{P-1} C_0(z^{2^i})$ and $C_1^{(l)}(z) = C_1(z^{2^l-1})\prod_{i=1}^{l-2} C_0(z^{2^i})$.

9

Spatio-Temporal Radios for CDMA

9.1 Introduction

Second-generation spread spectrum multiple access systems have been successfully deployed in many cellular systems and will continue to play a dominant role in third-generation and fourth-generation systems development. In particular, direct-sequence code division multiple access (CDMA) relies on the inherent capability of spread spectrum modulation to handle both multipath distortions and cochannel interference. New standards and solutions are currently being proposed for increasing range and data services speeds. Wideband CDMA (WCDMA), for example, has been designed to facilitate the use of advanced receiver techniques, such as adaptive antennas and interference mitigation (in both the uplink and downlink) [1]. Motivated by important system level considerations, the use of multiuser/multiantenna detectors has attracted significant interest in recent years and a superficial review of available solutions would probably convey the impression that the problem is fully understood and resolved. However, a more careful analysis of the technical literature immediately reveals that there seems to be a lack of general agreement on a scheme that can reliably handle high mobility users: an environment that presents rapidly fading multipath channels.

The high data rate services provided by the system and small spreading factors contribute to make WCDMA extremely vulnerable to time-frequency selective multipath radio-frequency propagation. The crucial point of almost all of the proposed methods is based on the assumption that the multipath channels are quasi-static, that is, time-invariant over the length of the transmitted frame. Slow variations of the channels are then compensated by using adaptive algorithms that ultimately force the estimate to be constantly in search of a convergence point. As we have seen in Chapter 8, if the channel coefficients variations in time are fast with respect to the convergence time of the adaptive algorithm, significant degradation may result. In Chapter 8 we introduced the concept of localized basis expansions for the description of time-varying fading channels. The same idea is here applied to the problem of spatio-temporal detection of CDMA signals. We present in this chapter a new advanced CDMA receiver architecture that is able to overcome the problem of time-frequency selective multipath propagation.

9.2 Asynchronous CDMA

We now review the system model for the reverse link of a CDMA cellular system. No particular assumption is made on the geometry of the array at the base station [2, 3], and we intentionally neglect the propagation assumptions of the model. More details regarding these assumptions can be found in the previous chapters [4]. In a DS/CDMA communication system [5], the information signal relative to the kth user is given by

$$m_k(\tau) = \sum_{n=0}^{M-1} a_k(n) p_{T_s}(\tau - nT_s), \qquad k = 1, 2 \ldots K \tag{9.1}$$

where

$$p_{\tilde{\tau}}(\tau) = \begin{cases} 1 & \text{if } 0 < \tau < \tilde{\tau} \\ 0 & \text{otherwise} \end{cases}$$

and $\mathbf{a} = \{a_k(n)\}_{n=1}^{M}$ is the real or complex message (to include any type of digital modulation scheme) composed of $\tilde{m}$-ary symbols of duration T_s. The K users *asynchronously* share the common channel using the signature waveforms

$$a_{c,k}(\tau) = \sum_{n=1}^{J} a_{c,n}^{k} p_{T_c}(\tau - nT_c), \qquad k = 1, 2...K \tag{9.2}$$

where $T_c = \frac{T_s}{J}$ is the chip period and $\{a_{c,n}^k \in (-1;+1)\}_{n=1}^{J}$ is the pseudonoise spreading sequence of length J. The spreading signal waveform (9.2) of the kth user has duration T_s and is normalized to unit power. If an equivalent lowpass representation is employed, the kth transmitted spread spectrum signal can be written as $\sum_{i=0}^{M-1} a_k(i) p_{T_s}(\tau - iT_s)\ a_{c,k}(\tau - iT_s)$. This signal is distorted by multipath RF propagation and multiple access interference caused by other users. Each mobile transmitter has a single antenna while at the base station the multiplexed signal is received through an A-sensor antenna. The received signal at the base station receiver can be represented[1] as

$$r^{\kappa}(t) = \sum_{i=0}^{M-1} \sum_{k=1}^{K} a_k(i) \sqrt{E_k^s(i)} \tilde{f}_k^{\kappa}(t, t - iT_s) + n^{\kappa}(t)$$
$$\kappa = 1, 2, ..., A, \tag{9.3}$$

where $\tilde{f}_k^{\kappa}(t, \tau) = e^{j\theta_k}\ a_{c,k}(\tau) * \tilde{h}_k^{\kappa}(t, \tau)$ ($*$ denotes convolution) is the combined impulse response of each signal path of the kth spreading waveform and the channel from the kth user to the κth sensor, $E_k^s(i)$, θ_k are the received signal power and the carrier phase of the kth transmitted signal relative to the kth user, respectively. The channel $\tilde{h}_k^{\kappa}(t, \tau)$ between the kth transmitter and the κth sensor for $k = 1, 2, ..., K$ and $\kappa = 1, 2, ..., A$ is

$$\tilde{h}_k^{\kappa}(t, \tau) = \sum_{l=1}^{Q} f_{k,l}^{\kappa}(t) \delta(\tau - \tau_{k,l}^{\kappa}). \tag{9.4}$$

1. Equations (9.2) and (9.3) assume the code remains the same from bit to bit for a particular user. However, IS-95 and many other CDMA systems use different parts of a longer code for each bit. Besides substituting $a_{c,k}^{(i)}(\tau)$, where now the code utilized by the kth user depends on the bit index (i), the rest of the signal model is still valid for systems using long codes (defined aperiodic systems in [6]). For simplicity of exposition we limit our model description to periodic systems. Since the same multichannel model (9.6) can be derived for the aperiodic (long codes) case, the presentation of the algorithms that follow also apply to practical CDMA systems like third-generation (3G)-CDMA and IS-95.

$f^{\kappa}_{k,l}(t)$ are the normalized fading complex envelope processes so that $E^s_k(i)$ reflects the kth user received energy over the time period corresponding to iT_s, $\delta(n)$ is the delta function. The noise in (9.3) is white Gaussian, with two-sided power spectral density $\frac{N_0}{2}$; it can also represent the surrounding cell interference, plus noise. Multipath channels enhance interference among users, introducing intersymbol interference and additional correlation between the spreading waveforms. Received energies are assumed invariant over the duration of the message: $E^s_k(i) = E^s_k$ for $i = 0, ..., M-1$. We assume that the receiver has perfect knowledge of the powers, time delays $\tau^{\kappa}_{k,l}$ and phase lags of every received user signal. At the receiver, a bank of KQ filters for each sensor (see Figures 9.1 and 9.2), each filter matched to a delayed replica of the wanted spread spectrum signature, performs despreading with the signatures in (9.2) and is sampled at symbol rate to obtain the set of samples

$$y^{\kappa}_{k,l}(n) = \int_{nT_s}^{(n+1)T_s} r^{\kappa}(\tau)\, a_{c,k}(\tau - nT_s - \tau^{\kappa}_{k,l})d\tau. \tag{9.5}$$

The input/output relation for the multichannel discrete-time system can be written as

$$\begin{aligned} y^{\kappa}_{i,l}(n) &= \sum_{k=1}^{K}\sum_{q=1}^{Q}\sum_{m} \overline{h}^{\kappa}_{i,k,l,q}(n,m)\, a_k(n-m) + \eta^{\kappa}_{i,l}(n) \\ &= \sum_{k=1}^{K}\sum_{m} h^{\kappa}_{i,k,l}(n,m)\, a_k(n-m) + \eta^{\kappa}_{i,l}(n) \\ & i = 1,2,...,K \quad \kappa = 1,2,...,A \quad l = 1,2,...,Q \end{aligned} \tag{9.6}$$

where

$$\begin{aligned} \overline{h}^{\kappa}_{i,k,l,q}(n,m) &= \sqrt{E^s_k}\, e^{j\theta_k} f^{\kappa}_{k,q}(nT_s) \\ &\times \int_{-\infty}^{+\infty} a_{c,i}(u - \tau^{\kappa}_{i,l})a_{c,k}(u + mT_s - \tau^{\kappa}_{k,q})du, \\ h^{\kappa}_{i,k,l}(n,m) &= \sum_{q=1}^{Q} \overline{h}^{\kappa}_{i,k,l,q}(n,m), \end{aligned} \tag{9.7}$$

and $\eta^{\kappa}_{i,l}(n)$ is the noise component.

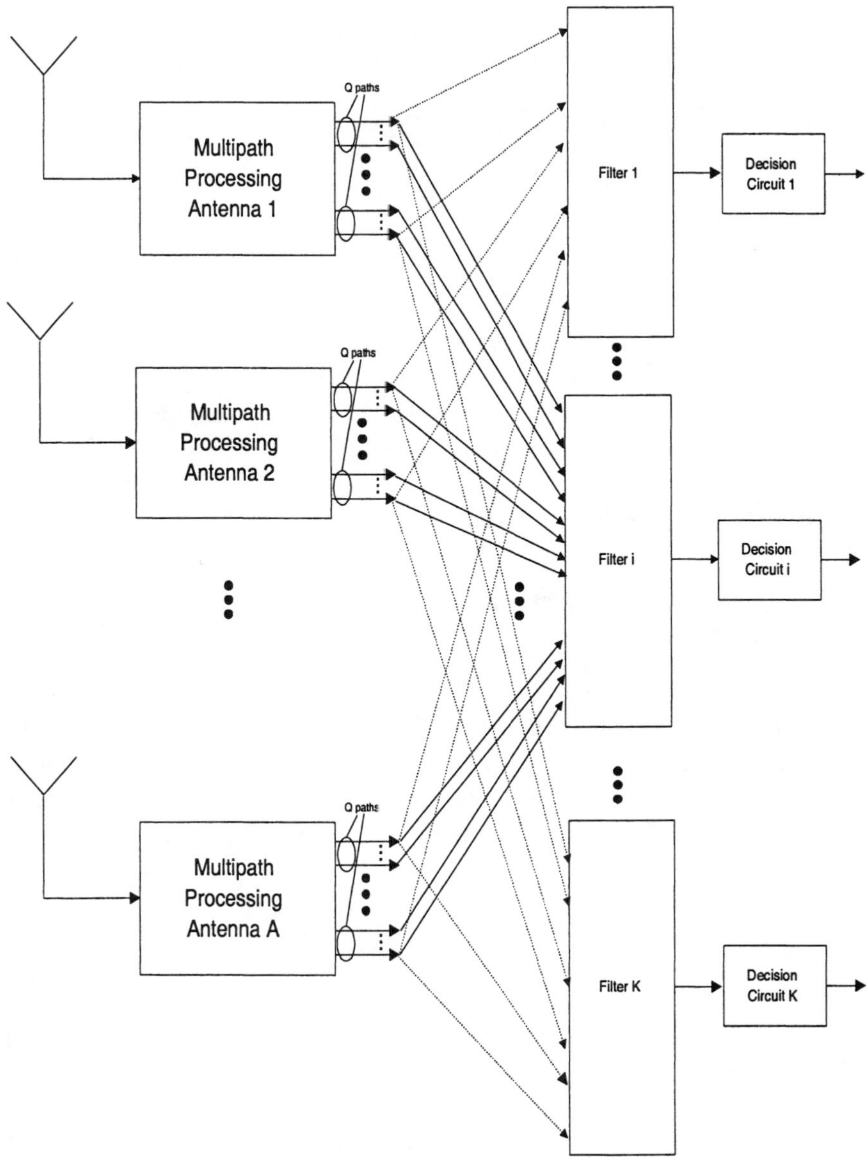

Figure 9.1 Architecture of the receiver.

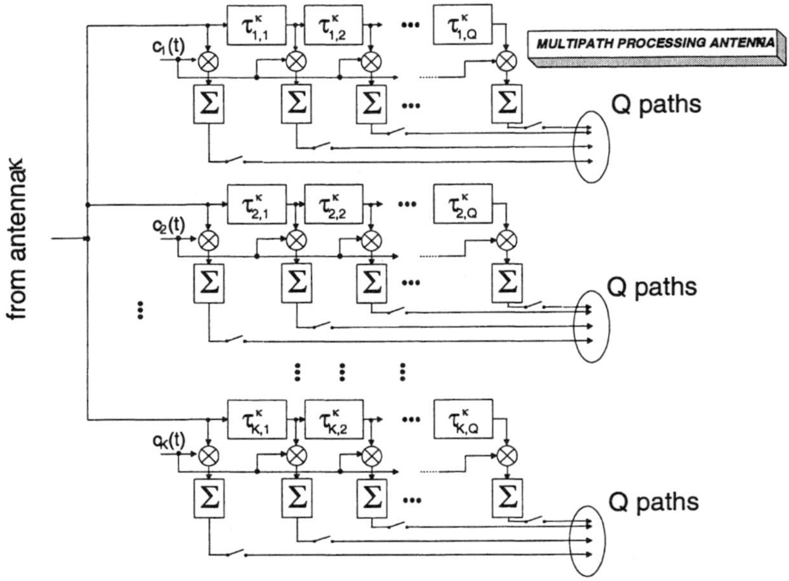

Figure 9.2 Signal processing subsection relative to antenna κ.

9.2.1 The Significance of Modeling Time-Varying Systems Using Wavelets

In this section we give a brief explanation of the concept of basis expansions by using an oversimplified continuous time *single-user* model. Consider a sequence of symbols collected in the vector $\mathbf{a}$, digitally modulated to form the signal $x(t, \mathbf{a})$. The signal $x(t, \mathbf{a})$ is input to $h(t, \tau)$, the time-varying impulse response of a multipath channel, as

$$y(t) = \int_{-\infty}^{t} h(t, t-\tau)x(\mathbf{a}, \tau)d\tau. \tag{9.8}$$

Assume that the kernel $h(t, \tau)$ can be represented by an expansion of the form

$$h(t, \tau) = \sum_{I \in \mathcal{I}_h} w_I(\tau)\psi_I(t) \tag{9.9}$$

where $\{\psi_I(t)\}_{I \in \mathcal{I}_h}$ is a set of functions (wavelets) and $\{w_I(\tau)\}_{I \in \mathcal{I}_h}$ are the wavelet coefficients of the expansion. Using this representation we can express (9.8) as

$$y(t) = \sum_{I \in \mathcal{I}_h} \psi_I(t) \int_{-\infty}^{t} w_I(t-\tau)x(\mathbf{a},\tau)d\tau$$
$$= \sum_{I \in \mathcal{I}_h} \psi_I(t)\tilde{x}_I(\mathbf{a},t) \qquad (9.10)$$

where $\tilde{x}_I(\mathbf{a},t) = \int_{-\infty}^{t} w_I(t-\tau)x(\mathbf{a},\tau)d\tau$. This representation can be economical if the set $\{\psi_I(t)\}_{I\in\mathcal{I}_h}$ contains a small number of functions. The multiplication $\psi_I(t)\tilde{x}_I(\mathbf{a},t)$ emphasizes the time selection of the input content and its frequency translations. Observe that $\tilde{x}_I(\mathbf{a},t)$ for any I in the set $\mathcal{I}_h$ is the output of a time-invariant linear filtering operation: It is simply obtained filtering $x(\mathbf{a},t)$ by $w_I(t)$. Another interesting property of this representation is that it gives us the ability to select for any given time interval only those wavelet coefficients responses $w_I(\tau)$ that affect the output during the time interval of interest. Assume now that white Gaussian noise $n(t)$ afflicts the channel output as $\tilde{y}(t) = y(t) + n(t)$. The maximization of the likelihood function with respect to the transmitted sequence of symbols $\mathbf{a}$ is easily seen to be equivalent to the maximization of the function [7]

$$L_{\mathcal{T}}(\mathbf{a}) = -\int_{t\in\mathcal{T}} \left| \tilde{y}(t) - \int_{\tau\in\mathcal{T}}^{t} h(t,t-\tau)x(\tau,\mathbf{a})d\tau \right|^2 dt$$
$$= -\int_{t\in\mathcal{T}} \left| \tilde{y}(t) - \sum_{I\in\mathcal{I}_h} \psi_I(t)\tilde{x}_I(t,\mathbf{a}) \right|^2 dt \qquad (9.11)$$

where $\mathcal{T}$ is the time interval over which we observe $\tilde{y}(t)$ and $\tilde{x}_I(t,\mathbf{a}) = \int_{\tau\in\mathcal{T}} w_I(t-\tau)x(\tau,\mathbf{a})d\tau$. The wavelet-based representation gives the possibility of an economical, but still accurate, formulation of the optimal detector based on the maximization of (9.11). In fact, it is possible to select a subset of the functions and coefficients in $\mathcal{I}_h$ such that $\int_{\tau\in\mathcal{T}}^{t} h(t,t-\tau)x(\tau,\mathbf{a})d\tau$ is well represented only over the time interval $\mathcal{T}$. This is quite valuable and somehow impossible using known nontime-localized models. Unfortunately we will find that the application of this idea to a multiuser system is very demanding in terms of computational complexity, so we will turn our attention to a simplified linear detector, one that is based on the design $g(t,\tau)$

that is expanded onto the same wavelet basis used in the channel model and whose coefficients minimize an averaged squared error

$$\begin{aligned} & M_{\mathcal{T}}\left[g(t,\tau)\right] \\ &= \int_{t\in\mathcal{T}} \left| x(t,\mathbf{a}) - \int_{\tau\in\mathcal{T}}^{t} g(t,t-\tau)\tilde{y}(\tau)d\tau \right|^2 dt \\ &= \int_{t\in\mathcal{T}} \left| x(t,\mathbf{a}) - \sum_{I\in\mathcal{I}_g} \psi_I(t)\tilde{y}_I(t) \right|^2 dt \end{aligned} \tag{9.12}$$

where $\tilde{y}_I(t) = \int_{-\infty}^{t} w_{g,I}(t-\tau)\tilde{y}(\tau)d\tau$. Again the wavelet-based representation gives the possibility of an economical, but still accurate, design of a time-varying filter by means of a selection of a subset of functions and coefficients in $\mathcal{I}_g$ such that $\int_{\tau\in\mathcal{T}}^{t} g(t,t-\tau)\tilde{y}(\tau)d\tau$ is well represented over the time interval $\mathcal{T}$.

The use of wavelet bases allows the selection of the subset in $\mathcal{I}_h$ (or $\mathcal{I}_g$) which will model the *local* and *global* characteristics of the channel dynamics. The idea that the wavelet-based representations can better model nonstationary as well as stationary processes is also the motivation for [8, 9].

9.3 Wavelet-Based Channel Basis Expansion

As in Chapter 8, we would like to obtain a practical basis representation for the discrete-time channel $h^{\kappa}_{i_1,i_2,i_3}(n,k)$ that is easily parameterized; it retains the essential features of the fading process and it characterizes these features using a small number of time-invariant coefficients. These requirements can be satisfied using a discrete-time wavelet series representation. As already pointed out in Appendix 8A, we do not need the coefficients of the continuous-time wavelet series expansion of

$$\begin{aligned} \tilde{h}^{\kappa}_{i,k,l,q}(t,\tau) &= \sum_{q=1}^{Q} \sqrt{E^s_k}\, e^{j\theta_k} f^{\kappa}_{k,q}(t) \\ &\times \int_{-\infty}^{+\infty} a_{c,i}(u-\tau^{\kappa}_{i,l})a_{c,k}(u+\tau-\tau^{\kappa}_{k,q})du. \end{aligned} \tag{9.13}$$

We only need a discrete-time wavelet series representation of a sampled version of $\tilde{h}^{\kappa}_{i,k,l,q}(t,\tau)$, that is, $h^{\kappa}_{i_1,i_2,i_3}(n,k) =$

$\tilde{h}^{\kappa}_{i,k,l,q}(t,\tau)\Big|_{t=nT_s,\tau=mT_s}$. It is possible to show that for a high resolution and short T_s, the discrete-time series coefficients and continuous time series coefficients asymptotically converge to the same values. The orthonormal expansion can be thought of more properly as a multiscale subband decomposition using filter banks of the sampled response of a multipath channel. We direct the interested reader to [10] and in particular to [11], where a discrete-time approach to multiresolution representations [12, 13] is described. Using the same exact approach described in Appendix 8A, we can express $h^{\kappa}_{i_1,i_2,i_3}(n,k)$ as

$$\begin{aligned} h^{\kappa}_{i_1,i_2,i_3}(n,k) &= \sum_{m=0}^{N/2^P-1} \zeta^{\kappa}_{P,m,i_1,i_2,i_3}(k) c_0^{(P)}(2^P m - n) \\ &\quad + \sum_{l=1}^{P} \sum_{m=0}^{N/2^l-1} \xi^{\kappa}_{l,m,i_1,i_2,i_3}(k) c_1^{(l)}(2^l m - n) \end{aligned} \tag{9.14}$$

where $\zeta^{\kappa}_{P,m,i_1,i_2,i_3}(k)$ and $\xi^{\kappa}_{l,m,i_1,i_2,i_3}(k)$ are discrete wavelet transform coefficients. Using vector notation, it is possible to express (9.14) simply as

$$h^{\kappa}_{i_1,i_2,i_3}(n,k) = \mathbf{c}(P,n)^T \mathbf{w}^{\kappa}_{i_1,i_2,i_3}(P,k), \tag{9.15}$$

where the organization of the wavelet coefficients in $\mathbf{w}^{\kappa}_{i_1,i_2,i_3}(P,k)$ is

$$\begin{aligned} &\mathbf{w}^{\kappa}_{i_1,i_2,i_3}(P,k) \\ &= \left[\boldsymbol{\zeta}^{\kappa}_{P,i_1,i_2,i_3}(k)^T \boldsymbol{\xi}^{\kappa}_{1,i_1,i_2,i_3}(k)^T \cdots \boldsymbol{\xi}^{\kappa}_{P,i_1,i_2,i_3}(k)^T\right]^T, \end{aligned}$$

with[2]

$$\left[\boldsymbol{\zeta}_{P,i_1,i_2,i_3}(k)\right]_m = \zeta^{\kappa}_{P,m,i_1,i_2,i_3}(k),$$

and

$$\left[\boldsymbol{\xi}^{\kappa}_{l,i_1,i_2,i_3}(k)\right]_m = \xi^{\kappa}_{l,m,i_1,i_2,i_3}(k).$$

Evidently $\mathbf{c}(P,n) = \left[\mathbf{c}_0^{(P)}(n)^T, \mathbf{c}_1^{(1)}(n)^T, ..., \mathbf{c}_1^{(P)}(n)^T\right]^T$ where $\left[\mathbf{c}_i^{(l)}(n)\right]_m = c_i^{(l)}(2^l m - n)$ for $i = 0, 1$.

2. The notation $[\mathbf{v}]_k$ is used for the kth element of vector $\mathbf{v}$.

The representation of the channel is shown at increasing resolutions. Observe that (9.14) is valid with equality because of the perfect reconstruction property of the dyadic filter bank. We will determine which components of the expansion (9.14) can be neglected (i.e., zeroed) without compromising the parsimony of the representation.

9.4 Validation of the DWT Channel

Statistical characteristics of the fading channel are typically available from experimental measurements in the form of the scattering function [14]. Assuming the channel to be WSSUS[3] [15], we can tune our channel representation: In other words, second-order statistics will provide general guidelines on how to select the low-resolution coefficients. Low-resolution coefficients are important for a global characterization of the fading process. It is, however, important to point out that retaining only the low-resolution coefficients will not give us the kind of parsimony that we expect from a wavelet-based representation. To achieve this goal a few high-resolution coefficients (in addition to the coefficients necessary for a global representation of the process) empirically selected will have to be maintained.

Having assumed the model (9.4), the knowledge of the second-order statistics of the channel variations is explicated in the knowledge of

$$E\left\{f_{k_1,l_1}^{\kappa_1^*}(t_1)f_{k_2,l_2}^{\kappa_2}(t_2)\right\}$$

which is then mapped to the knowledge of the discrete-time autocorrelation function of the channel

$$E\left\{h_{\mathbf{i}}^{\kappa}(n,k_1)^* h_{\mathbf{i}'}^{\kappa'}(n+l,k_2)\right\} = R_h(l,k_1,k_2,\mathbf{i},\kappa,\mathbf{i}',\kappa')$$

3. The WSSUS assumption implies

$$E\left\{h_{k_1}^{\kappa_1^*}(t_1,\tau_1)h_{k_2}^{\kappa_2}(t_2,\tau_2)\right\} = \mathcal{S}_{k_1,\kappa_1,k_2,\kappa_2}(\tau_1,t_2-t_1)\delta(\tau_2-\tau_1)$$

and the expectation is over the channel ensemble. The scattering function of the multipath channel is related to $\mathcal{S}_{k_1,\kappa_1,k_2,\kappa_2}(\tau,t)$ by the Fourier transform, first with respect to τ, and then with respect to t.

with $\mathbf{i}' = [i'_1, i'_2, i'_3]$, $\mathbf{i} = [i_1, i_2, i_3]$. [4] Using (9.14), we have

$$R_h(l, k_1, k_2, \mathbf{i}, \kappa, \mathbf{i}', \kappa') = \mathbf{c}(P, n)^T E\left\{\mathbf{w}_{\mathbf{i}}^{\kappa}(P, k_1)^* \mathbf{w}_{\mathbf{i}'}^{\kappa'}(P, k_2)^T\right\} \mathbf{c}(P, n+l). \quad (9.16)$$

Observe that due to the orthonormality of the DWT, defining $\mathbf{h}_{\mathbf{i}}^{\kappa}(k)$ as the vector with nth element

$$\left[\mathbf{h}_{\mathbf{i}}^{\kappa}(k)\right]_n = h_{\mathbf{i}}^{\kappa}(n, k),$$

we also have

$$E\left\{\mathbf{w}_{\mathbf{i}}^{\kappa}(P, k_1)^* \mathbf{w}_{\mathbf{i}'}^{\kappa'}(P, k_2)^T\right\} = E\left\{\mathbf{h}_{\mathbf{i}}^{\kappa}(k_1)^* \mathbf{h}_{\mathbf{i}'}^{\kappa'}(k_2)^T\right\}$$

which is a matrix constructed from $R_h(l, k_1, k_2, \mathbf{i}, \kappa, \mathbf{i}', \kappa')$. We can validate our model (9.14) using (9.16). In other words, we can determine which components of the expansion (9.14) can be neglected (i.e., zeroed) without compromising the parsimony of the wavelet-based representation. Define $\mathbf{w}_{\mathbf{i}}^{\kappa}(P, k)_{\mathcal{M}}$ as the vector obtained zeroing the last $\sum_{m=1}^{\mathcal{M}} N/2^m$ elements of $\mathbf{w}_{\mathbf{i}}^{\kappa}(P, k)$ and $\mathbf{c}(P, n)_{\mathcal{M}}$ as the vector obtained zeroing the last $\sum_{m=1}^{\mathcal{M}} N/2^m$ elements of $\mathbf{c}(P, n)$. A possible indication of the parsimony in the representation (9.14) retaining only $N/2^{\mathcal{M}}$ DWT coefficients, $\mathcal{M} = 0, 1, 2, ..., P$ is

$$\gamma(P, \mathcal{M}) = \sum_{\mathbf{i}, \mathbf{i}'} \sum_{\kappa, \kappa'} \sum_{k_1, k_2} ||\mathbf{r}_{\mathbf{i}, \mathbf{i}'}^{\kappa, \kappa'}(k_1, k_2) - \mathbf{v}_{\mathbf{i}, \mathbf{i}'}^{\kappa, \kappa'}(k_1, k_2)_{\mathcal{M}}||^2$$

where

$$\left[\mathbf{v}_{\mathbf{i}, \mathbf{i}'}^{\kappa, \kappa'}(k_1, k_2)_{\mathcal{M}}\right]_l = \frac{1}{N} \sum_{n=0}^{N-1} \mathbf{c}(P, n)_{\mathcal{M}}^T E\left\{\mathbf{h}_{\mathbf{i}}^{\kappa}(k_1)^* \mathbf{h}_{\mathbf{i}'}^{\kappa'}(k_2)^T\right\} \mathbf{c}(P, n+l)_{\mathcal{M}}$$

and $\left[\mathbf{r}_{\mathbf{i}, \mathbf{i}'}^{\kappa, \kappa'}(k_1, k_2)\right]_l = R_h(l, k_1, k_2, \mathbf{i}, \kappa, \mathbf{i}', \kappa')$. Once a particular $\mathcal{M}$ has been selected, we can write

$$h_{\mathbf{i}}^{\kappa}(n, k) \simeq \overline{\mathbf{c}}(n)^T \overline{\mathbf{w}}_{\mathbf{i}}^{\kappa}(k) \quad (9.17)$$

4. In fact, $R_h(l, k_1, k_2, \mathbf{i}, \kappa, \mathbf{i}', \kappa')$ can be computed from the knowledge of $\mathcal{S}_{k_1, \kappa_1, k_2, \kappa_2}(\tau_1, t_2 - t_1)$ and the spreading codes of each user.

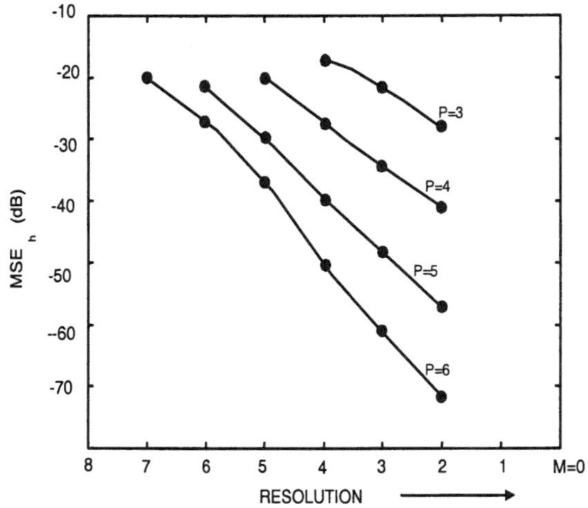

Figure 9.3 MSE (average over 100 experiments) of the multiresolution wavelet-based representation of a fading process (real part of the coefficient for a N=512 symbol spaced samples snapshot) for a time-varying multipath channel representing a mobile transmitter traveling at 100 km/hr. The maximum resolution depth is P and the mother wavelet used is generated from a Daubechies filter of order 6. The MSE is plotted versus M where N/2M is the number of wavelet coefficients maintained in the representation of the 512-point sample evolution.

where[5] $\overline{\mathbf{c}}(n) = \mathbf{I}_{[\mathcal{N},\overline{\mathcal{N}}]}\ \mathbf{c}(P,n)$, $\overline{\mathbf{w}}_{\mathbf{i}}^{\kappa}(k) = \mathbf{I}_{[\mathcal{N},\overline{\mathcal{N}}]}\ \mathbf{w}_{\mathbf{i}}^{\kappa}(P,k)$ with $\mathcal{N} = N2^{-\mathcal{M}}$, $\overline{\mathcal{N}} = N - \mathcal{N} = N\sum_{m=1}^{\mathcal{M}} 2^{-m}$. In Figure 9.3 we show the approximation error for different resolution depths P and for different values of $\mathcal{M}$ for $N = 512$. The approximation error (shown in dB) is defined as

$$\mathrm{MSE}_h = E\left\{\left|h_{\mathbf{i}}^{\kappa}(n,k) - \overline{\mathbf{c}}(n)^T \overline{\mathbf{w}}_{\mathbf{i}}^{\kappa}(k)\right|^2\right\},$$

where $\overline{\mathbf{c}}(n)$ and $\overline{\mathbf{w}}_{\mathbf{i}}^{\kappa}(k)$ contain, for any indicated P, only $512 \cdot 2^{-\mathcal{M}}$ components.

5. The notation $\mathbf{I}_A$ identifies the $A \times A$ identity matrix while the notation $\mathbf{I}_{[A,B]}$ identifies the $A \times (A+B)$ matrix whose first A columns are the columns of $\mathbf{I}_A$, and the last B columns have elements equal to zero.

9.4.1 Interpretations

The method we have outlined can be interpreted as a subspace selection procedure [16]. In fact the DWT of the N-long vector $\mathbf{h}_{\mathbf{i}}^{\kappa}(k)$ representing the dynamics of the channel at lag k can be expressed as

$$\mathbf{h}_{\mathbf{i}}^{\kappa}(k) = \mathbf{W} \;\; \mathbf{w}_{\mathbf{i}}^{\kappa}(P, k) \tag{9.18}$$

where $\mathbf{W}$ is an $N \times N$ orthonormal linear transformation expressing the operation of the wavelet transform. Consider $\overline{\mathbf{W}} = [\mathbf{W}_1, ..., \mathbf{W}_{\mathcal{N}}]^T$, the $N \times \mathcal{N}$ matrix formed by the first $\mathcal{N} = N2^{-\mathcal{M}}$ columns of $\mathbf{W}$ corresponding to the *shrinkage* (that is, the zeroing) that yields tolerably small values of the metric $\gamma(P, \mathcal{N})$. Applying the transformation

$$\boldsymbol{\theta}_{\mathbf{i}}^{\kappa}(k) = \overline{\mathbf{W}}^H \mathbf{h}_{\mathbf{i}}^{\kappa}(k) \tag{9.19}$$

we essentially define a subspace parameter $\boldsymbol{\theta}_{\mathbf{i}}^{\kappa}(k)$. The $\mathcal{N}$ orthonormal columns of $\overline{\mathbf{W}}$ span an $\mathcal{N}$-dimensional subspace defined $\mathcal{W}_{\mathcal{N}}$ such that $\mathcal{W}_{\mathcal{N}} \subseteq \mathcal{C}^N$ (the space of any complex vector of length N). Observe that when the transformation $\overline{\mathbf{W}}$ is selected with $\mathcal{N} < N$, any estimate (or representation) of $\boldsymbol{\theta}_{\mathbf{i}}^{\kappa}(k)$ is insensitive to disturbance components of $\mathbf{h}_{\mathbf{i}}^{\kappa}(k)$ for $m = \mathcal{N}+1, \mathcal{N}+2, ..., N$ and the resulting estimator (or representation) has a smaller variance than the full space estimator. Also, any traditional adaptive scheme for equalization (or channel tracking) designed to follow the variations of $h_{i_1,i_2,i_3}^{\kappa}(n, k)$ attempts a full-space estimation of the channel. In other words traditional schemes will generate point-by-point estimates spanning $\mathcal{C}^N$. The proposed channel representation method when incorporated in practical detectors will achieve a decreased variance of the channel estimate error. It is, however, important to emphasize that this advantage is obtained at the expense of estimation bias because in practice

$$\mathbf{h}_{\mathbf{i}}^{\kappa}(k) = \overline{\mathbf{W}}\boldsymbol{\theta}_{\mathbf{i}}^{\kappa}(k) = \mathbf{P}_{\mathcal{W}}\mathbf{h}_{\mathbf{i}}^{\kappa}(k) \tag{9.20}$$

with $\mathbf{P}_{\mathcal{W}} = \overline{\mathbf{W}} \;\; \overline{\mathbf{W}}^H$ holds only approximately true and while the variance increases with the dimensions of the subspace (that is $\mathcal{N}$) the bias decreases [16]. Evidently the goal of selecting a transformation $\overline{\mathbf{W}}$ is to optimize the trade-off between bias, variance, and computational load.

9.5 Detectors

Once we have achieved a parsimonious representation of the form (9.17), we can write (9.6) as

$$\begin{aligned} y_{i,l}^{\kappa}(n) &= \sum_{k=1}^{K}\sum_{m}\overline{\mathbf{c}}(n)^T\overline{\mathbf{w}}_{i,k,l}^{\kappa}(m)a_k(n-m)+\eta_{i,l}^{\kappa}(n) \\ &= \mathbf{a}(n)^T\mathbf{C}(n)\mathbf{w}_{i,l}^{\kappa}+\eta_{i,l}^{\kappa}(n) \\ &= \tilde{\mathbf{a}}(n)^T\mathbf{w}_{i,l}^{\kappa}+\eta_{i,l}^{\kappa}(n) \\ & \quad i=1,2,...,K \quad \kappa=1,2,...,A \quad l=1,2,...,Q \end{aligned} \quad (9.21)$$

where $\tilde{\mathbf{a}}(n) = \mathbf{C}(n)^T\mathbf{a}(n)$, and the organization of the filterbank coefficients into $\mathbf{C}(n)$ and the wavelet coefficients into $\mathbf{w}_{i,l}^{\kappa}$ is evident from the context. We can compact in matrix notation (9.21) organizing the samples $y_{i,l}^{\kappa}(n)$ for $n = 0,1,2,...,M-1$, $i = 1,2,...,K$, $\kappa = 1,2,...,A$, $l = 1,2,...,Q$ in a vector $\mathbf{y}_M$ to obtain the familiar model

$$\mathbf{y}_M = \mathbf{A}_M(\mathbf{a})\mathbf{w} + \mathbf{n}, \quad (9.22)$$

where $\mathbf{w}$ is a vector that contains the wavelets coefficients and $\mathbf{A}_M(\mathbf{a})$ depends uniquely on the MK transmitted symbols $\mathbf{a} = [a_1(0),...a_K(M-1)]^T$ and the coefficients of the filterbank (ultimately the selected orthonormal basis). Using a generalized likelihood [7] argument we can obtain the (blind) maximum likelihood rule [17]

$$\hat{\mathbf{a}} = \arg\min_{\boldsymbol{\alpha}} \mathbf{y}_M^H\mathbf{Q}_\alpha\mathbf{y}_M \quad (9.23)$$

$$= \arg\max_{\boldsymbol{\alpha}} \mathbf{y}_M^H\mathbf{P}_\alpha\mathbf{y}_M \quad (9.24)$$

where $\boldsymbol{\alpha}$ is the hypothesized sequence, $\mathbf{Q}_\alpha = \mathbf{I} - \mathbf{P}_\alpha$, and

$$\mathbf{P}_\alpha = \mathbf{A}_M(\boldsymbol{\alpha})[\mathbf{A}_M(\boldsymbol{\alpha})^H\mathbf{R}_n^{-1}\mathbf{A}_M(\boldsymbol{\alpha})]^{-1}\mathbf{A}_M(\boldsymbol{\alpha})^H\mathbf{R}_n^{-1}$$

is the novel separating kernel with $\mathbf{R}_n = E\left\{\mathbf{n}\mathbf{n}^H\right\}$. The generalized maximum likelihood rule in (9.24) presents significant complexity, although it is quite attractive conceptually. Indeed, it is rarely practical. A more feasible approach is to design a time-variant minimum average squared error filter that attempts restoration of the desired user.

9.5.1 A Time-Variant Filter

A possible solution to detection is to implement a time-variant minimum average square error (MASE) filter for the kth user indicated at time n as $g_{i,l}^{\kappa,k}(n,m)$, $m = L_1, L_1+1, ..., L_2$, $i = 1,2,...,K$, $l = 1,2,...,Q$, $\kappa = 1,2,...,A$ as indicated in Figure 9.4. Denote $\hat{a}_k(n)$ as the output of the filter for the detection of the nth symbol of the kth user

$$\sum_{i=1}^{K}\sum_{l=1}^{Q}\sum_{\kappa=1}^{A}\sum_{m=L_1}^{L_2} g_{i,l}^{\kappa,k}(n,m) y_{i,l}^{\kappa}(n-m) = \hat{a}_k(n). \tag{9.25}$$

The filter coefficients can be obtained from the minimization of the averaged squared error

$$\left\langle \left| \sum_{i=1}^{K}\sum_{l=1}^{Q}\sum_{\kappa=1}^{A}\sum_{m=L_1}^{L_2} g_{i,l}^{\kappa,k}(n,m) y_{i,l}^{\kappa}(n-m) - a_k(n) \right|^2 \right\rangle_N , \tag{9.26}$$

where we have used the notation $\langle x_n \rangle_N = \frac{1}{N}\sum_{n=1}^{N} x_n$ for an N-sample time average. The orthogonality principle dictates that

$$\left\langle \left(\sum_{i=1}^{K}\sum_{l=1}^{Q}\sum_{\kappa_1=1}^{A}\sum_{m_1=L_1}^{L_2} g_{i,l}^{\kappa_1,k}(n,m) y_{i,l}^{\kappa}(n-m_1) \right.\right.$$
$$\left.\left. - a_k(n) \right) y_{p,q}^{\kappa_2^*}(n-m_2) \right\rangle_N = 0 \tag{9.27}$$

for $\kappa_2 = 1, ..., A$, $p = 1, ..., K$, $q = 1, ..., Q$, $m_2 = L_1, ..., L_2$.

Since we have established that a reasonably accurate representation of the fading channel variations is given by (9.17), it is also possible to show that the optimum time-variant filter can be expressed as

$$g_{i_1,i_2}^{\kappa,k}(n,m) \simeq \overline{\mathbf{c}}(n)^T \overline{\mathbf{w}}_{[\mathbf{g}]}{}_{i_1,i_2}^{\kappa,k}(m).$$

If the time variations of the multipath responses are well represented by an expansion of the form

$$h_{i_1,i_2,i_3}^{\kappa}(n,m) = \mathbf{c}(P,n)^T \mathbf{w}_{i_1,i_2,i_3}^{\kappa}(P,m), \tag{9.28}$$

where $\mathbf{c}(P,n)^T$ are coefficients of the discrete-time basis and $\mathbf{w}_{i_1,i_2,i_3}^{\kappa}(P,k)$ are coefficients of the expansion, then it is very

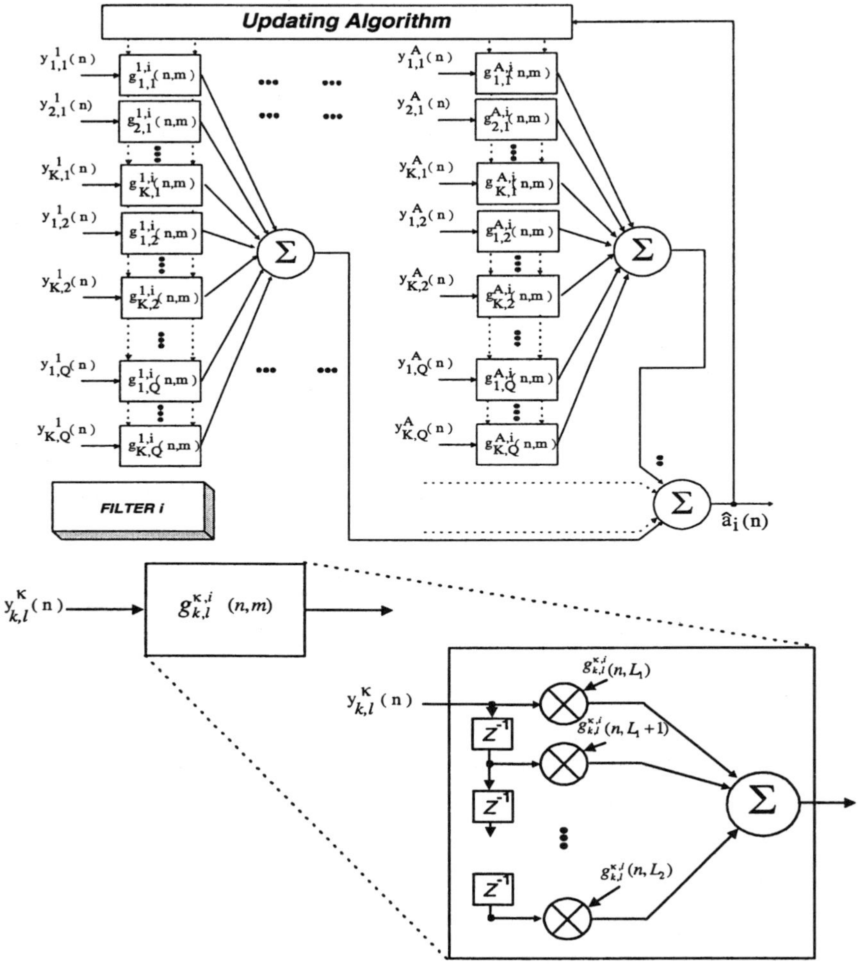

Figure 9.4 The time-varying filter relative to the ith user.

reasonable to expect that the time-varying filter that minimizes the MSE between $a_k(n)$ and

$$\sum_{i=1}^{K}\sum_{l=1}^{Q}\sum_{\kappa=1}^{A}\sum_{m=L_1}^{L_2} g_{i,l}^{\kappa,k}(n,m)y_{i,l}^{\kappa}(n-m)$$

also can be represented as

$$g_{i_1,i_2}^{\kappa,k}(n,m) \simeq \mathbf{c}(n)^T \mathbf{w}_{[\mathbf{g}]}{}_{i_1,i_2}^{\kappa,k}(m),$$

where $\mathbf{w}_{[\mathbf{g}]}{}_{i_1,i_2}^{\kappa,k}(m)$ are coefficients of an expansion equivalent (that is, using the same basis) to the expansion valid for $h_{i_1,i_2,i_3}^{\kappa}(n,m)$.

Define

$$\mathbf{h}_{i,k,l}^{\kappa}(n,m) = \Big[h_{i,k,l}^{\kappa}(n+L_1,m+L_1),..., h_{i,k,l}^{\kappa}(n+L_2,m+L_2)\Big]^T$$

and

$$\mathbf{y}_{i,l}^{\kappa}(n) = \Big[y_{i,l}^{\kappa}(n+L_1),..., y_{i,l}^{\kappa}(n+L_2)\Big]^T,$$

$$\mathbf{n}_{i,l}^{\kappa}(n) = \Big[\eta_{i,l}^{\kappa}(n+L_1),..., \eta_{i,l}^{\kappa}(n+L_2)\Big]^T$$

so that (9.6) can be written

$$\mathbf{y}_{i,l}^{\kappa}(n) = \sum_{k=1}^{K}\sum_{m}\mathbf{h}_{i,k,l}^{\kappa}(n,m)\, a_k(n-m) + \mathbf{n}_{i,l}^{\kappa}(n) \tag{9.29}$$

or

$$\mathbf{y}(n) = \sum_{k=1}^{K}\sum_{m}\mathbf{h}_k(n,m)\, a_k(n-m) + \mathbf{n}(n) \tag{9.30}$$

where

$$\mathbf{y}(n) = \Big[\mathbf{y}_{1,1}^{1}(n)^T,..., \mathbf{y}_{K,Q}^{A}(n)^T\Big]^T,$$

$$\mathbf{n}(n) = \Big[\mathbf{n}_{1,1}^{1}(n)^T,..., \mathbf{n}_{K,Q}^{A}(n)^T\Big]^T,$$

$$\mathbf{h}_k(n,m) = \Big[\mathbf{h}_{1,k,1}^{1}(n)^T,..., \mathbf{h}_{K,k,Q}^{A}(n)^T\Big]^T.$$

Then we can express (9.25) as

$$\hat{a}_k(n) = \mathbf{g}_k^H(n)\left[\sum_{k=1}^{K}\sum_{m}\mathbf{h}_k(n,m)\,a_k(n-m) + \mathbf{n}(n)\right] \tag{9.31}$$

where $\mathbf{g}_k^H(n)$ contains the weights $g_{i,l}^{\kappa,k}(n,m)$ properly organized. We can also write

$$\begin{aligned}\hat{a}_k(n) &= \mathbf{g}_k^H(n)\Big[\mathbf{h}_k(n,0)\,a_k(n) \\ &\quad + \sum_{l=1,l\neq k}^{K}\sum_{m\neq 0}\mathbf{h}_k(n,m)\,a_l(n-m) + \mathbf{n}(n)\Big] \\ &= \mathbf{g}_k^H(n)\left[\mathbf{h}_k(n,0)\,a_k(n) + \mathbf{b}_k(n) + \mathbf{n}(n)\right].\end{aligned} \tag{9.32}$$

The MMSE solution for $\mathbf{g}_k(n)$ [18, 19] is well approximated by

$$\begin{aligned}\mathbf{g}_k(n) &= \left[\mathbf{h}_k(n,0)\mathbf{h}_k(n,0)^H + \mathbf{S}_k\right]^{-1}\mathbf{h}_k(n,0) \\ &= \frac{1}{1+\mathbf{h}_k(n,0)^H\mathbf{S}_k^{-1}\mathbf{h}_k(n,0)}\mathbf{S}_k^{-1}\mathbf{h}_k(n,0) \\ &\simeq \mathbf{S}_k^{-1}\mathbf{h}_k(n,0)\end{aligned} \tag{9.33}$$

where $\mathbf{S}_k = E\{\mathbf{b}_k(n)\mathbf{b}_k(n)^H\} + E\{\mathbf{n}(n)\mathbf{n}(n)^H\}$ and where we have used the fact that the scaling factor $1+\mathbf{h}_k(n,0)^H\mathbf{S}_k^{-1}\mathbf{h}_k(n,0)$ changes more slowly than $\mathbf{h}_k(n,0)$. The expression in (9.33) is justified in [18] and is valid also for decision feedback filters with some modifications. The approximation was proved asymptotically valid by experimental results and in fact used also in [19] to analyze the performance of multichannel decision feedback equalizers.

$\mathbf{S}_k$ in our CDMA system model is very close to a diagonal matrix in practice if $a_k(n)$ are i.i.d. symbols, the fading processes $h_{i_1,i_2,i_3}^{\kappa}(n,m)$ are assumed independent (across antennas and paths), and the PN codes used are assumed ideally uncorrelated with shifted replicas of themselves and different codes. Equation (9.33) makes evident that the dynamics of the linear time-varying filter $\mathbf{g}_k(n)$ are asymptotically equivalent (up to scale factors) to the variations of $\mathbf{h}_k(n,0)$.

Renaming more conveniently the components of $\overline{\mathbf{c}}(n)$ and $\overline{\mathbf{w}}[\mathrm{g}]_{i_1,i_2}^{\kappa}(k)$ as

$$\overline{\mathbf{c}}(n) = [c_1(n), c_2(n), ..., c_{\mathcal{N}}(n)]^T$$

$$\overline{\mathbf{w}}_{[\mathbf{g}]_{i_1,i_2}^{\kappa}}(k) = \left[\overline{w}_{[g]_{i_1,i_2}^{\kappa}}(k)_1, \overline{w}_{[g]_{i_1,i_2}^{\kappa}}(k)_2, ..., \overline{w}_{[g]_{i_1,i_2}^{\kappa}}(k)_{\mathcal{N}}\right]^T,$$

we can use the following expression

$$g_{i_1,i_2}^{\kappa,k}(n,m) \simeq \overline{\mathbf{c}}(n)^T \overline{\mathbf{w}}_{[\mathbf{g}]_{i_1,i_2}^{\kappa,k}}(m) = \sum_{p=1}^{\mathcal{N}} \overline{w}_{[g]_{i_1,i_2}^{\kappa,k}}(m)_p c_p(n)$$

and obtain

$$\left\langle \left| \sum_{i=1}^{K} \sum_{l=1}^{Q} \sum_{\kappa=1}^{A} \sum_{m=L_1}^{L_2} g_{i,l}^{\kappa,k}(n,m) y_{i,l}^{\kappa}(n-m) - a_k(n) \right|^2 \right\rangle_N$$
$$= \left\langle \left| \sum_{i=1}^{K} \sum_{l=1}^{Q} \sum_{\kappa=1}^{A} \sum_{m=L_1}^{L_2} \sum_{p=1}^{\mathcal{N}} \overline{w}_{[g]_{i,l}^{\kappa,k}}(m)_p c_p(n) y_{i,l}^{\kappa}(n-m) \right.\right.$$
$$\left.\left. - a_k(n) \right|^2 \right\rangle_N.$$

If we denote $\tilde{y}_{i,l,p}^{\kappa}(n,m) = c_p(n) y_{i,l}^{\kappa}(n-m)$, the orthogonality principle can be expressed as

$$\left\langle \left(\sum_{i_1=1}^{K} \sum_{l_1=1}^{Q} \sum_{\kappa_1=1}^{A} \sum_{m_1=L_1}^{L_2} \sum_{p_1=1}^{\mathcal{N}} \overline{w}_{[g]_{i_1,l_1}^{\kappa_1,k}}(m_1)_{p_1} \right.\right.$$
$$\left.\left. \cdot \tilde{y}_{i_1,l_1,p_1}^{\kappa}(n,m_1) - a_k(n) \right) \tilde{y}_{i_2,l_2,p_2}^{\kappa_2}(n,m_2) \right\rangle_N = 0 \tag{9.34}$$

for $\kappa_2 = 1,...,A$, $i_2 = 1,...,K$, $l_2 = 1,...,Q$, $m_2 = L_1,...,L_2$, and $p_2 = 1,...,\mathcal{N}$. It is easily shown that the solution to this estimation problem is

$$\mathbf{w}_{[\mathbf{g}]_k} = \mathbf{R}^{-1}\mathbf{d}_k \tag{9.35}$$

where

$$\begin{aligned}
\mathbf{w}_{[\mathbf{g}]_k} &= \left[\mathbf{w}_{[\mathbf{g}]_k^{1^T}}, \mathbf{w}_{[\mathbf{g}]_k^{2^T}}, ..., \mathbf{w}_{[\mathbf{g}]_k^{A^T}}\right]^T \\
\mathbf{w}_{[\mathbf{g}]_k^{\kappa}} &= \left[\mathbf{w}_{[\mathbf{g}]_{1,1}^{\kappa,k^T}}, ..., \mathbf{w}_{[\mathbf{g}]_{K,1}^{\kappa,k^T}}, \cdots \mathbf{w}_{[\mathbf{g}]_{K,Q}^{\kappa,k^T}}\right]^T \\
\mathbf{w}_{[\mathbf{g}]_{j,l}^{\kappa,k}} &= \left[\overline{w}_{[g]_{j,l}^{\kappa,k}}(L_1)_1, ..., \overline{w}_{[g]_{j,l}^{\kappa,k}}(L_2)_1, ..., \overline{w}_{[g]_{j,l}^{\kappa,k}}(L_2)_{\mathcal{N}}\right]^T,
\end{aligned}$$

$$\mathbf{R} = \left\langle \tilde{\mathbf{y}}_{\mathbf{c}}^{*}(n)\, \tilde{\mathbf{y}}_{\mathbf{c}}(n)^{T} \right\rangle_{N} \tag{9.36}$$

$$\mathbf{d}_{k} = \langle \tilde{\mathbf{y}}_{\mathbf{c}}^{*}(n) a_{k}(n) \rangle_{N} \tag{9.37}$$

with

$$\begin{aligned}
\tilde{\mathbf{y}}_{\mathbf{c}}(n) &= \left[\check{\mathbf{y}}_{\mathbf{c}}^{1}(n)^{T}, \check{\mathbf{y}}_{\mathbf{c}}^{2}(n)^{T}, ..., \check{\mathbf{y}}_{\mathbf{c}}^{A}(n)^{T}\right]^{T} \\
\check{\mathbf{y}}_{\mathbf{c}}^{\kappa}(n) &= \left[\mathbf{y}\mathbf{c}_{1,1}^{\kappa}(n)^{T}, ..., \mathbf{y}\mathbf{c}_{K,2}^{\kappa}(n)^{T}, ..., \mathbf{y}\mathbf{c}_{K,Q}^{\kappa}(n)^{T}\right]^{T} \\
\mathbf{y}\mathbf{c}_{i,l}^{\kappa}(n) &= \left[\tilde{y}_{i,l,1}^{\kappa}(n, L_{1}), \tilde{y}_{i,l,1}^{\kappa}(n, L_{2}), ..., \tilde{y}_{i,l,\mathcal{N}}^{\kappa}(n, L_{2})\right]^{T}.
\end{aligned} \tag{9.38}$$

Typical approaches to achieve real-time implementation are gradient-based methods and RLS methods. The well-known recursion in both cases is

$$\mathbf{w}_{[\mathbf{g}]_k}(n+1) = \mathbf{w}_{[\mathbf{g}]_k}(n) + \mathbf{p}_{n}\left[\tilde{\mathbf{y}}_{\mathbf{c}}(n), \mathbf{w}_{[\mathbf{g}]_k}(n)\right] \tag{9.39}$$

where

$$\mathbf{p}_{n}\left[\tilde{\mathbf{y}}_{\mathbf{c}}(n), \mathbf{w}_{[\mathbf{g}]_k}(n)\right] = \mu\left(a_{k}(n) - \tilde{\mathbf{y}}_{\mathbf{c}}(n)^{T}\mathbf{w}_{[\mathbf{g}]_k}(n)\right)\tilde{\mathbf{y}}_{\mathbf{c}}^{*}(n)$$

if the algorithm of choice is the LMS and μ is a step-size parameter that controls the rate of adjustment. Using an RLS approach, we have

$$\mathbf{p}_{n}\left[\tilde{\mathbf{y}}_{\mathbf{c}}(n), \mathbf{w}_{[\mathbf{g}]_k}(n)\right] = \lambda \mathbf{K}_{n}^{*}\epsilon_{n}$$

where

$$\epsilon_{n} = a_{k}(n) - \tilde{\mathbf{y}}_{\mathbf{c}}(n)^{T}\mathbf{w}_{[\mathbf{g}]_k}(n),$$

$$\mathbf{K}_{n} = \frac{\mathbf{P}_{n-1}\mathbf{a}_{n}}{\lambda + \tilde{\mathbf{y}}_{\mathbf{c}}(n)^{H}\mathbf{P}_{n-1}\tilde{\mathbf{y}}_{\mathbf{c}}(n)}$$

and

$$\mathbf{P}_{n} = \lambda^{-1}(\mathbf{P}_{n-1} - \tilde{\mathbf{y}}_{\mathbf{c}}(n)^{H}\mathbf{P}_{n-1}\mathbf{K}_{n}).$$

λ is a forgetting factor that can be set approximately equal to 1 because in reality the wavelet coefficients are time-invariant and need only to be identified, not tracked.

It is important to observe that the true vector of wavelet coefficients is time-invariant, so the task of the adaptive algorithm in the proposed approach is to converge to the channel parameters as opposed to *tracking* them. This makes the LMS algorithm perfectly adequate for the task. If the WBD retains $N2^{-\mathcal{M}}$ wavelet coefficients, then the complexity of the proposed approach is $\mathcal{O}\left[LKAN2^{-\mathcal{M}}\right]$ for an L long filter, K users and A antennas. In the traditional MMSE approach one is forced to use an RLS approach because directly tracking the taps of the filter using a gradient approach is ineffective in fast fading. Evidently one ends up with a complexity $\mathcal{O}\left[(LKA)^2\right]$ which in most cases is much higher than the complexity presented by the WBD multiscale filter approach for the same length of the filter, number of antennas, and number of users. This advantage in fast fading environments is extremely valuable and, in fact, results in significant performance advantages.

Note also that there is an apparent similarity between (9.35) and the MMSE linear multiuser detector of [5]. In particular the two detectors converge asymptotically to the same detector in the case of time-invariant channels. It is then appropriate to state that the detector described by (9.35) in a rapidly varying environment is a practical approximation of the instantaneous exact Wiener solution. The instantaneous Wiener filter can not be implemented in a time-varying environment using average sample-statistics estimators because the processes involved are nonstationary. This is exactly why the detector (9.35) outperforms the traditional multiuser MMSE detector which uses sample-statistics averages for the estimation of second-order statistics.

9.6 Performance Analysis

Consider a mobile radio system when the uplink applies DS/CDMA with a gross bit rate equal to 48.5 Kbps. The users' codes are known at the receiver as the base station has allocated one code for each user. We use Gold codes of 15 chips, seven equipower users, and $Q = 2$ (two-path Rayleigh fading channel) in each channel impulse response between each user and each antenna array element. The antenna is a three-element uniform linear array with 3/4 of a wavelength spacing (this spacing guarantees uncorrelated scattering at the different antennas). The impulse responses of the multipath channels are generated so that the delays are constrained to be an integer number of chip periods according to Table 9.1. Users

Table 9.1
Channel Propagation Environments for Performance Evaluation Results: Delay Spreads of the Mobiles

User	$\tau_{k,1}^1, \tau_{k,2}^1$	$\tau_{k,1}^2, \tau_{k,2}^2$	$\tau_{k,1}^3, \tau_{k,2}^3$	Speed
1	$0T_c, 2T_c$	$1T_c, 2T_c$	$1T_c, 4T_c$	100 km/hr
2	$1T_c, 3T_c$	$0T_c, 2T_c$	$3T_c, 4T_c$	80 km/hr
3	$0T_c, 4T_c$	$1T_c, 5T_c$	$3T_c, 4T_c$	50 km/hr
4	$1T_c, 2T_c$	$2T_c, 3T_c$	$3T_c, 4T_c$	10 km/hr
5	$1T_c, 1T_c$	$2T_c, 4T_c$	$3T_c, 4T_c$	20 km/hr
6	$2T_c, 4T_c$	$1T_c, 3T_c$	$3T_c, 4T_c$	80 km/hr
7	$1T_c, 5T_c$	$2T_c, 3T_c$	$3T_c, 4T_c$	100 km/hr

modulate data using QPSK. A synchronization unit is assumed to estimate the delays exactly. We assume the first user to be the reference user. SNR in the figures is equal to E_1^s/N_0. A training sequence of 14 symbols is transmitted every 256 symbols, to train the time-variant filter.

The slots are generated of dyadic length and $C_0(z)$ $C_1(z)$ are Daubechies filters of order 3. We assume [20] that

$$R_h(l, k_1, k_2, \mathbf{i}, \kappa, \mathbf{i}', \kappa') = \begin{cases} \sigma_k^2 J_0(2\pi l f_D T/R) & for\ k_1 = k_2 = k\ and\ \mathbf{i} = \mathbf{i}' \\ 0 & \text{elsewhere} \end{cases}$$

where $J_0(x)$ is the Bessel function of order zero, f_D is Doppler spread which depends on the velocity of the mobile transmitter, and σ_k^2 is the variance of the fading process. The Doppler frequency describes the second order statistics of channel variations. Doppler frequency is related through wavelength $\tilde{\lambda}$ to the ith mobile transmitter velocity V_i expressed in km/hr. The complex weights are generated as filtered Gaussian processes fully specified by the scattering function. In particular, each process has a frequency response equal to the square-root of the Doppler power density spectrum. It is then straightforward to verify (using the described validation method) that for speeds of the mobile up to 300 km/hr and $P = 4$ or

$P = 5$, it is possible to retain only $N/2^{\mathcal{M}} = 32$ or $N/2^{\mathcal{M}} = 16$ wavelet coefficients in $\mathbf{w}^{\kappa}_{i_1,i_2,i_3}(P,k)$ without significant penalty in the matching metric $\gamma(P,\mathcal{M})$. In other words the excellent "compacting" properties (see [21]) of the wavelet transform are able to compress most of the energy of the time variations of the channel in the low resolution representation of the fading process which makes the approach very attractive in practice. The output of the filter at time step n is $\tilde{\mathbf{y}}_{\mathbf{c}}(n)^T\mathbf{w}_{[\mathbf{g}]_k}(n)$ where the vector $\mathbf{w}_{[\mathbf{g}]_k}(n)$ is obtained from $\mathbf{w}_{[\mathbf{g}]_k}(n-1)$, $\hat{a}_k(n-1)$, $\tilde{\mathbf{y}}_{\mathbf{c}}(n-1)$ as in (9.39).[6] The vector $\tilde{\mathbf{y}}_{\mathbf{c}}(n)$ is computed from the baseband samples and the coefficients of the Daubechies filters $c_0(n)$ and $c_1(n)$ at each antenna as detailed in (9.38). A decision for the nth symbol for the kth user is obtained from a QPSK slicer whose input is the scalar $\tilde{\mathbf{y}}_{\mathbf{c}}(n)^T\mathbf{w}_{[\mathbf{g}]_k}(n)$. BER analysis results are shown in Figures 9.5 and 9.6.

We compare with the multiuser MMSE detector updated using the QR-RLS algorithm (probably the best method in terms of tracking performance). The MMSE-QR-RLS at every iteration solves the following minimization problem

$$\min_{\mathbf{g}_k} \left\| \begin{bmatrix} \lambda\mathbf{Y}(n) \\ \tilde{\mathbf{y}}(n+1)^T \end{bmatrix} \mathbf{g}_k - \begin{bmatrix} \lambda\mathbf{a}_k(n) \\ a_k(n+1) \end{bmatrix} \right\|^2 \tag{9.40}$$

where $\mathbf{Y}(n)$ and $\mathbf{a}(n)$ are recursively defined as

$$\mathbf{Y}(n) = \begin{bmatrix} \lambda\mathbf{Y}(n-1) \\ \tilde{\mathbf{y}}(n)^T \end{bmatrix}, \quad \mathbf{a}_k(n) = \begin{bmatrix} \lambda\mathbf{a}_k(n-1) \\ a_k(n) \end{bmatrix},$$

with

$$\begin{aligned}
\tilde{\mathbf{y}}(n) &= \left[\check{\mathbf{y}}^1(n)^T, \check{\mathbf{y}}^2(n)^T, \ldots, \check{\mathbf{y}}^A(n)^T\right]^T \\
\check{\mathbf{y}}^{\kappa}(n) &= \left[\mathbf{y}^{\kappa}_{1,1}(n)^T, \ldots, \mathbf{y}^{\kappa}_{1,Q}(n)^T, \ldots, \mathbf{y}^{\kappa}_{K,Q}(n)^T\right]^T \\
\mathbf{y}^{\kappa}_{i,l}(n) &= \left[y^{\kappa}_{i,l}(n-L_1), \ldots, y^{\kappa}_{i,l}(n-L_2)\right]^T, \\
\mathbf{g}_k &= \left[\mathbf{g}_k^{1^T}, \mathbf{g}_k^{2^T}, \ldots, \mathbf{g}_k^{A^T}\right]^T
\end{aligned}$$

6. $\hat{a}_k(n-1)$ is the decision on the complex QPSK nth symbol, kth user. Evidently the filter runs in decision-directed mode. While in training the exact knowledge of the first 14 symbols allow the estimation of $\mathbf{w}_{[\mathbf{g}]_k}(n)$ using (9.35).

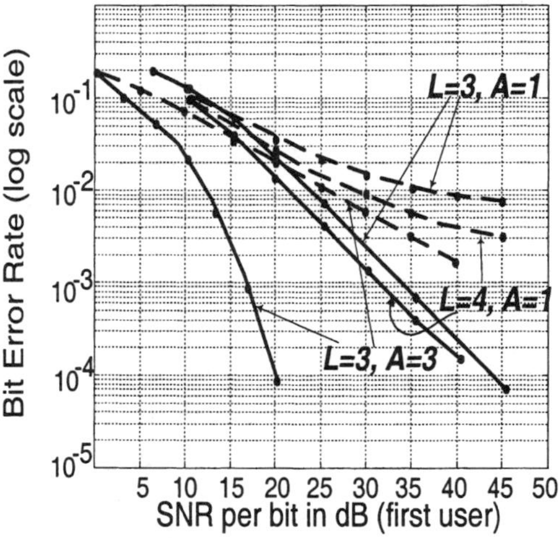

Figure 9.5 BER for seven users, 15-length Gold codes in fast frequency selective fading channels. For propagation parameters, see Table 9.1. Solid curves are for the time-varying separating filter based on the DWT representation. Dashed curves are for the MMSE time-invariant filter based on the traditional adaptive QR-RLS scheme.

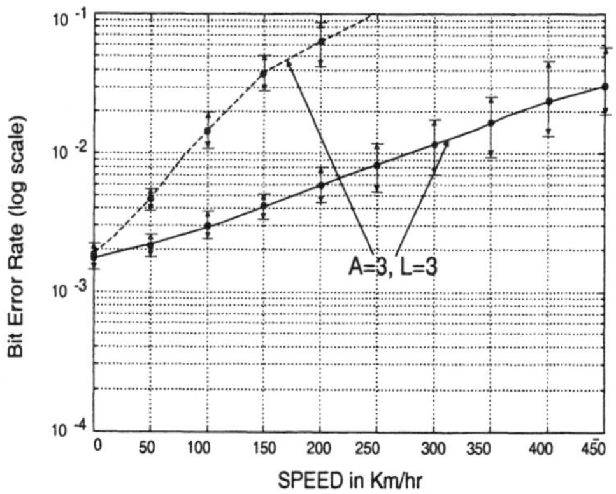

Figure 9.6 BER for three users, 15-length Gold codes in fast frequency selective fading channels with increasing speed of user 1, 2, 3. For propagation parameters, see Table 9.1, first three users. Solid curves are for the time-variant separating filter based on the DWT representation. Dashed curves are for the MMSE time-invariant filter based on the traditional adaptive QR-RLS scheme. The 90% confidence intervals for the measured BER are shown.

$$\mathbf{g}_k^{\kappa} = \left[\mathbf{g}_{1,1}^{\kappa,k^T}, ..., \mathbf{g}_{K,2}^{\kappa,k^T} \cdots \mathbf{g}_{1,Q}^{\kappa,k^T}, ..., \mathbf{g}_{K,Q}^{\kappa,k^T}\right]^T$$

$$\mathbf{g}_{j,l}^{\kappa,k} = \left[\overline{g}_{j,l}^{\kappa,k}(L_1), \overline{g}_{j,l}^{\kappa,k}(L_1+1), ..., \overline{g}_{j,l}^{\kappa,k}(L_2)\right]^T,$$

The solution obtained applying a recursive QR decomposition to the data matrix $\mathbf{Y}(n)$ defines the QR-RLS algorithm which attempts convergence to the MMSE solution for $\mathbf{g}_k$. We use in the DWT representation of the time-variant filter $P = 4$ and retain $N/2^{\mathcal{M}} = 32$ (for $N = 256$). BER is relative to the first mobile transmitter. Ideal frame and symbol synchronization is assumed and the LMS is used to update the wavelet coefficients. A sample size of 10^{k+4} was used to estimate an error probability of 10^{-k}. Figure 9.5 shows result for the propagation environment reported in Table 9.1 which evidently represents a fast fading environment, such as those practically found in high-mobility cellular systems. $L = L_2 - L_1 + 1$ is the length of the filter, and A is the number of sensors at the receiving antenna array. It is clear that the MMSE approach (dashed curves in Figure 9.5) is inadequate. Figure 9.6 shows the results of experiments with only the first three users active simultaneously. The speed of the three mobile transmitters is increased up to 400 km/hr, with the SNR per bit equal to 32 dB. The DWT-based filter is significantly less sensitive to Doppler spread increases.

9.6.1 Hardware Implementation Results

A simpler scenario ($K = 2$, $A = 2$, $Q = 2$) was studied using baseband data collected from the DSP receiver section of $Base_2$, the dual-mode wideband base station implemented at WJ Communications. The dispersive channel is created using two hardware fading emulators (NoiseCom MP1600) which model two-ray Rayleigh fading channels. The signal generator simulates transmission of CDMA frames from two different mobiles with the frame format described in the previous subsection. The DSP modem receives two complex I,Q pairs at the rate of 727.5 kHz after RF conversion and digital quadrature downconversion. The word length used is 24 bits, and the algorithm has been implemented using simulated fixed-point arithmetic. While perfect synchronization was assumed in the previous simulations, in the hardware experiments there is an open-loop synchronizer (both PN code and frame synchronization). We use a resolution depth equal to $P = 5$ and

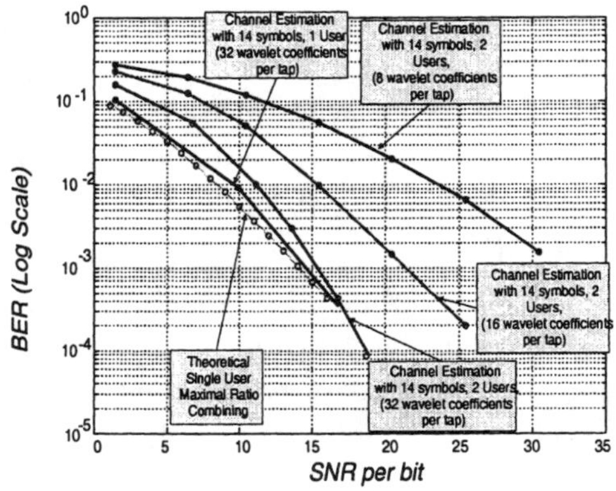

Figure 9.7 BER for hardware experiments with one or two users, two antennas, and 15-length Gold codes in fast frequency selective fading channels. The length of the filter is L=4. For propagation parameters, see Table 9.1, first two users, and first two antennas.

$\mathcal{M} = 3, 4, 5$ (that is, we keep $8, 16, 32$ wavelet coefficients). The results of extensive BER measurements are summarized in Figure 9.7. The propagation environment is described in Table 9.1, considering only the first two users and the first two antennas. The complexity of the algorithm is for $L = 4, A = 2, K = 2$ equivalent to 78 MIPS which is well within current processing capability of most DSP processors today on the market.

9.7 Time-Varying MIMO Deconvolution

The super-exponential equations presented and derived in Chapter 6 also apply to the CDMA case. Since the super-exponential algorithm is iterative in nature, the transformation of the technique into a real-time adaptive scheme places a significant challenge whenever the multipath fading channel cannot be assumed static, which is the case in most wireless macrocell systems. The super-exponential equations are obviously time-varying in the presence of a time-varying channel, which is the case of interest here.

Previously we have seen that the input/output relation for the multichannel discrete-time system can be written as in (9.6). The recovery of the input signals can be achieved by means of a linear KAQ-input K-output time-variant filter that can be represented at time step n in the z-domain as $\mathcal{W}(n,z) = \sum_{k=L_1}^{L_2} \mathbf{W}(n,k)z^{-k}$ with length $L = L_2 - L_1 + 1$. The main objective for $\mathcal{W}(n,z)$ is to achieve *distortionless reception* [2, 4] at any time step n. If we define

$$\begin{aligned}
\mathcal{W}(n,z) &= \left[\begin{array}{cccc} \tilde{\mathcal{W}}^1(n,z) & \tilde{\mathcal{W}}^2(n,z) & \cdots & \tilde{\mathcal{W}}^A(n,z) \end{array}\right], \\
\tilde{\mathcal{W}}^\kappa(n,z) &= \left[\begin{array}{cccc} \breve{\mathcal{W}}_1^\kappa(n,z) & \breve{\mathcal{W}}_2^\kappa(n,z) & \cdots & \breve{\mathcal{W}}_Q^\kappa(n,z) \end{array}\right]
\end{aligned}$$

with

$$\breve{\mathcal{W}}_l^\kappa(n,z) = \left[\begin{array}{ccc} \mathcal{W}_{1,1,l}^\kappa(n,z) & \cdots & \mathcal{W}_{1,K,l}^\kappa(n,z) \\ \mathcal{W}_{2,1,l}^\kappa(n,z) & \cdots & \mathcal{W}_{2,K,l}^\kappa(n,z) \\ \vdots & \cdots & \vdots \\ \mathcal{W}_{K,1,l}^\kappa(n,z) & \cdots & \mathcal{W}_{K,K,l}^\kappa(n,z) \end{array}\right],$$

distortionless reception means that

$$\mathcal{W}(n,z)\,\mathcal{H}(n,z) = \mathbf{P}\mathcal{D}(n,z) \tag{9.41}$$

is verified at any n where $\mathbf{P}$ is a permutation matrix,

$$\mathcal{D}(n,z) = \mathrm{diag}\{e^{j\phi_1}z^{-n_1}, e^{j\phi_2}z^{-n_2}, ..., e^{j\phi_K}z^{-n_K}\},$$

$\phi_i \in [-\pi,\pi]$ n_i integer for $i = 1,2...,K$,

$$\mathcal{H}(n,z) = \sum_k \mathbf{H}(n,k)z^{-k} \tag{9.42}$$

and the organization of the $\overline{\mathcal{H}}_{i,j,l,n}^\kappa(n,z)$ polynomials (z-transforms of $\overline{h}_{i,j,l,n}^\kappa(n,k)$) in $\mathcal{H}(n,z)$ is given by

$$\mathcal{H}(n,z) = \left[\begin{array}{c} \tilde{\mathcal{H}}^1(n,z) \\ \tilde{\mathcal{H}}^2(n,z) \\ \vdots \\ \tilde{\mathcal{H}}^A(n,z) \end{array}\right], \quad \tilde{\mathcal{H}}^\kappa(n,z) = \left[\begin{array}{c} \breve{\mathcal{H}}_1^\kappa(n,z) \\ \breve{\mathcal{H}}_2^\kappa(n,z) \\ \cdots \\ \breve{\mathcal{H}}_Q^\kappa(n,z) \end{array}\right],$$

$$\breve{\mathcal{H}}_l^\kappa(n,z) = \left[\begin{array}{ccc} \mathcal{H}_{1,1,l}^\kappa(n,z) & \cdots & \mathcal{H}_{1,K,l}^\kappa(n,z) \\ \mathcal{H}_{2,1,l}^\kappa(n,z) & \cdots & \mathcal{H}_{2,K,l}^\kappa(n,z) \\ \vdots & \cdots & \vdots \\ \mathcal{H}_{K,1,l}^\kappa(n,z) & \cdots & \mathcal{H}_{K,K,l}^\kappa(n,z) \end{array}\right],$$

$$\mathcal{H}^{\kappa}_{i,j,l}(n,z) = \sum_{m=1}^{Q} \overline{\mathcal{H}}^{\kappa}_{i,j,l,m}(n,z).$$

Conditions for the existence of a filter $\mathcal{W}(n,z)$ were detailed in Chapter 4. The output of the filter relative to the nth symbol relative to the kth user is denoted $z_k(n)$ and can be expressed as

$$\sum_{i=1}^{K}\sum_{l=1}^{Q}\sum_{\kappa=1}^{A}\sum_{m=L_1}^{L_2} w^{\kappa}_{k,i,l}(n,m)y^{\kappa}_{i,l}(n-m) = z_k(n). \tag{9.43}$$

9.7.1 The Algorithmic Approach

Minor modifications of the derivations reported in Chapter 6 (in fact, time-varying cumulants have to be used due to the variations of the fading coefficients–see Appendix 6A) give the super-exponential time-varying equations[7]

$$\tilde{\mathbf{w}}_k^{[1]}(n) = \tilde{\mathbf{R}}(n)^{\dagger}\ \mathbf{D}_k(n) \tag{9.44}$$

$$\tilde{\mathbf{w}}_k^{[2]}(n) = \frac{\tilde{\mathbf{w}}_k^{[1]}(n)}{\sqrt{\tilde{\mathbf{w}}_k^{[1]^H}(n)\tilde{\mathbf{R}}(n)\tilde{\mathbf{w}}_k^{[1]}(n)}} \tag{9.45}$$

where $\tilde{\mathbf{w}}_k^{[1]}(n)$ and $\tilde{\mathbf{w}}_k^{[2]}(n)$ are the weights of the filter $\tilde{\mathbf{w}}_k(n)$

$$\begin{aligned}
\tilde{\mathbf{w}}_k(n) &= \left[\mathbf{w}_k^{1^T}(n), \mathbf{w}_k^{2^T}(n), ..., \mathbf{w}_k^{A^T}(n)\right]^T \\
\mathbf{w}_k^{\kappa}(n) &= \left[\mathbf{w}_{1,1}^{\kappa,k^T}(n), \cdots, \mathbf{w}_{K,Q}^{\kappa,k^T}(n)\right]^T \\
\mathbf{w}_{j,l}^{\kappa,k}(n) &= \left[w^{\kappa}_{k,j,l}(n,L_1), \cdots, w^{\kappa}_{k,j,l}(n,L_2)\right]^T,
\end{aligned}$$

at time n obtained in the first and second step, respectively, of the iterative procedure, the matrix $\tilde{\mathbf{R}}(n)$ with dimensions $AQKL \times AQKL$ and the vector $\mathbf{D}_i(n)$ of fourth-order cumulants with dimensions $AQKL \times 1$ are given by

$$\tilde{\mathbf{R}}(n) = E\left\{\tilde{\mathbf{y}}(n)\tilde{\mathbf{y}}(n)^H\right\} \tag{9.46}$$

7. The notation $\mathbf{A}^{\dagger}$ is used for the pseudoinverse of the matrix $\mathbf{A}$.

and

$$\tilde{\mathbf{D}}_i(n) = E\{\tilde{\mathbf{y}}^*(n)\tilde{z}_i(n)\}$$

where

$$\begin{aligned}
\tilde{\mathbf{y}}(n) &= \left[\check{\mathbf{y}}^1(n)^T, ..., \check{\mathbf{y}}^A(n)^T\right]^T \\
\check{\mathbf{y}}^\kappa(n) &= \left[\mathbf{y}^\kappa_{1,1}(n)^T, ..., \mathbf{y}^\kappa_{K,Q}(n)^T\right]^T \\
\mathbf{y}^\kappa_{i,l}(n) &= \left[y^\kappa_{i,l}(n-L_1), ..., y^\kappa_{i,l}(n-L_2)\right]^T, \\
\tilde{z}_i(n) &= \frac{\sigma_a^2}{\gamma_{4a}}\left(|z_i(n)|^2 - 2\,\sigma_a^2\right)\, z_i(n).
\end{aligned}$$

Observe at this point that we cannot proceed as in Chapter 6, because time variations of the fading coefficients and as a consequence of the optimum distortionless reception filter do not allow the super-exponential iterations to achieve convergence. Ideally one should iterate several times to obtain convergence the super-exponential procedure at every time instant n: In other words, one would need to "freeze" the evolution of the fading coefficients until convergence. This naive method prevents the application of a real-time algorithm where recursive operations are also iterations of the super-exponential algorithm, the strategy of the methods reported in [2, 22]. We will attack the problem exploiting the ideas on basis expansions already introduced in the first part of this chapter.

9.7.2 A Basis Expansion for the Time-Variant Filter

It was shown in Chapter 6 that the filter coefficients obtained from Step 1 of the procedure (9.44)–(9.45) are equivalently obtained minimizing the following cost function

$$E\left\{\left|\sum_{i=1}^{K}\sum_{l=1}^{Q}\sum_{\kappa=1}^{A}\sum_{m=L_1}^{L_2} w^\kappa_{k,i,l}(n,m)y^\kappa_{i,l}(n-m) - \tilde{z}_k(n)\right|^2\right\}, \qquad (9.47)$$

or alternatively by solving an orthogonality-principle–like equation

$$E\left\{\left(\sum_{i=1}^{K}\sum_{l=1}^{Q}\sum_{\kappa_1=1}^{A}\sum_{m_1=L_1}^{L_2} w_{k,i,l}^{\kappa_1}(n,m)y_{i,l}^{\kappa}(n-m_1)\right.\right.$$
$$\left.\left. -\tilde{z}_k(n)\right)\cdot y_{p,q}^{\kappa_2^*}(n-m_2)\right\}=0 \qquad (9.48)$$

for $\kappa_2 = 1,...,A$, $p = 1,...,K$, $q = 1,...,Q$, $m_2 = L_1,...,L_2$. These two expressions will be useful in the development that follows.

Consider the fading coefficients $f_{k,q}^{\kappa}(nT_s)$ and assume that their variations can be represented for a large enough $\mathcal{N}$ as

$$f_{k,q}^{\kappa}(nT_s) \simeq \sum_{p=1}^{\mathcal{N}} \overline{w}_{[f]}{}_{k,q,p}^{\kappa} c_p(n),$$

where $c_p(n)$ is a known basis; in [23] the basis is an exponential $c_p(n) = e^{j\omega_p n}$, and $\overline{w}_{[f]}{}_{k,q,p}^{\kappa}$ are the coefficients of the expansion. This representation is accurate under very general assumptions detailed in [23]. A number of interesting practical techniques can be developed because the method decouples the unknown time variations of the channel from the set of time-invariant coefficients. This method reveals its value in rapidly fading environments where algorithms based on the typical quasi-static assumption of the fading channel usually fail.

We follow here a similar approach with the important difference that the basis expansion is performed on the time-variant distortionless reception filter $\mathcal{W}(n,z)$. Moreover, we depart from the simple exponential basis to improve the representation accuracy.

The Wavelet Basis

We proceed to obtain a practical basis representation for the time variant filter $w_{i,k,l}^{\kappa}(n,m)$ applying a DWT (discrete wavelet transform) with respect to n over N available samples of the processes in (9.6) following an approach first proposed in [17, 23].

Define $g_i(n)$ $i = 0,1$ as a dyadic perfect reconstruction filter bank [10] [in the z-domain $G_i(z)$]. The wavelet coefficients at resolution $P+1=2$ for $w_{i_1,i_2,i_3}^{\kappa}(n,k)$ are given by [10]

$$\zeta_{1,n,i_1,i_2,i_3}(k) \;=\; \sum_l g_0(l)w_{i_1,i_2,i_3}^{\kappa}(2m-l,k) \qquad (9.49)$$

$$\xi^{\kappa}_{1,n,i_1,i_2,i_3}(k) = \sum_{l} g_1(l) w^{\kappa}_{i_1,i_2,i_3}(2m-l,k). \tag{9.50}$$

The inverse wavelet transform is obtained from

$$\begin{aligned} w^{\kappa}_{i_1,i_2,i_3}(n,k) &= \sum_{m=0}^{N/2-1} \zeta^{\kappa}_{1,m,i_1,i_2,i_3}(k) g_0^{(1)}(2^P m - n) \\ &\quad + \sum_{m=0}^{N/2-1} \xi^{\kappa}_{1,m,i_1,i_2,i_3}(k) g_1^{(1)}(2m-n). \end{aligned} \tag{9.51}$$

This method can be applied recursively to obtain at generic resolution depth P

$$\begin{aligned} w^{\kappa}_{i,i_1,i_2}(n,k) &= \sum_{m=0}^{N/2^P-1} \zeta^{\kappa}_{P,m,i,i_1,i_2}(k) g_0^{(P)}(2^P m - n) \\ &\quad + \sum_{l=1}^{P} \sum_{m=0}^{N/2^l-1} \xi^{\kappa}_{l,m,i,i_1,i_2}(k) g_1^{(l)}(2^l m - n) \end{aligned} \tag{9.52}$$

where $\zeta^{\kappa}_{P,m,i,i_1,i_2}(k)$ and $\xi^{\kappa}_{l,m,i,i_1,i_2}(k)$ are discrete wavelet transform coefficients and the filters $g_0^{(P)}(n)$ and $g_1^{(l)}(n)$ (with real coefficients) are given in the z-domain by $G_0^{(P)}(z) = \prod_{i=1}^{P-1} G_0(z^{2^i})$ and $G_1^{(l)}(z) = G_1(z^{2^l-1}) \prod_{i=1}^{l-2} G_0(z^{2^i})$. Using vector notation, it is possible to express (9.52) simply as

$$w^{\kappa}_{i,i_1,i_2}(n,k) = \mathbf{g}(P,n)^T \mathbf{w}_{[\mathbf{g}]}{}^{\kappa}_{i,i_1,i_2}(P,k), \tag{9.53}$$

where the organization of the wavelet coefficients in $\mathbf{w}_{[\mathbf{g}]}{}^{\kappa}_{i,i_1,i_2}(P,k)$ is

$$\begin{aligned} &\mathbf{w}_{[\mathbf{g}]}{}^{\kappa}_{i,i_1,i_2}(P,k) \\ &= \left[\boldsymbol{\zeta}^{\kappa}_{P,i,i_1,i_2}(k)^T \boldsymbol{\xi}^{\kappa}_{1,i,i_1,i_2}(k)^T \cdots \boldsymbol{\xi}^{\kappa}_{P,i,i_1,i_2}(k)^T \right]^T, \end{aligned}$$

with[8]

$$\left[\boldsymbol{\zeta}_{P,i,i_1,i_2}(k) \right]_m = \zeta^{\kappa}_{P,m,i,i_1,i_2}(k),$$

8. The notation $[\mathbf{v}]_k$ is used for the kth element of vector $\mathbf{v}$.

$$\left[\boldsymbol{\xi}^{\kappa}_{l,i,i_1,i_2}(k)\right]_m = \xi^{\kappa}_{l,m,i,i_1,i_2}(k).$$

Evidently $\mathbf{g}(P,n) = \left[\mathbf{g}_0^{(P)}(n)^T, \mathbf{g}_1^{(1)}(n)^T, ..., \mathbf{g}_1^{(P)}(n)^T\right]^T$ where $\left[\mathbf{g}_i^{(l)}(n)\right]_m = g_i^{(l)}(2^l m - n)$ for $i = 0, 1$.

As it is, (9.52) is of little practical interest; however, one can determine which components of the expansion (9.52) can be neglected (i.e., zeroed) without compromising the parsimony of the wavelet-based representation, and show in fact that it is possible to represent the time variations of the filter coefficients in a reduced order dimensional space. Indeed, this is one of the celebrated merits of the wavelet transform [21]. Defining $\overline{\mathbf{w}}_{[\mathbf{g}]}{}^{i,\kappa}_{i_1,i_2}(k)$ as the vector obtained zeroing the last $N - \mathcal{N} = \sum_{m=1}^{\mathcal{M}} N2^{-m}$ elements of $\mathbf{w}_{[\mathbf{g}]}{}^{\kappa}_{i,i_1,i_2}(k)$ and $\overline{\mathbf{c}}(n)$ as the vector obtained zeroing the last $N - \mathcal{N} = \sum_{m=1}^{\mathcal{M}} N2^{-m}$ elements of $\mathbf{g}(P,n)$, we can then write

$$w^{\kappa}_{k,i_1,i_2}(n,m) \simeq \overline{\mathbf{c}}(n)^T \overline{\mathbf{w}}_{[\mathbf{g}]}{}^{\kappa,k}_{i_1,i_2}(m). \tag{9.54}$$

It is possible to verify, for example, that for $N = 256$, speeds of the mobiles up to 300 km/hr and $P = 3$ or $P = 4$, it is possible to impose $\xi^{\kappa}_{l,m,i,i_1,i_2}(k) = 0$ in (9.52) for any l, m, k without significant loss of representation accuracy. This means that we can have only $N/2^{\mathcal{M}} = N/2^P = 32$ or $N/2^{\mathcal{M}} = N/2^P = 16$ wavelet coefficients in $\overline{\mathbf{w}}_{[\mathbf{g}]}{}^{\kappa,i}_{i_1,i_2}(k)$.

Solution of the Time-Varying Super-Exponential Equations

We have shown that an accurate representation of the filter variations is given by (9.54), or renaming the components of the vectors

$$\overline{\mathbf{c}}(n) = [c_1(n), c_2(n), ..., c_{\mathcal{N}}(n)]^T$$

$$\overline{\mathbf{w}}_{[\mathbf{g}]}{}^{\kappa,k}_{k,i_1,i_2}(m) = \left[\overline{w}_{[g]}{}^{\kappa,k}_{i_1,i_2}(m)_1, ..., \overline{w}_{[g]}{}^{\kappa,k}_{i_1,i_2}(m)_{\mathcal{N}}\right]^T,$$

and by

$$w^{\kappa}_{k,i_1,i_2}(n,m) \simeq \sum_{p=1}^{\mathcal{N}} \overline{w}_{[g]}{}^{\kappa,k}_{i_1,i_2}(m)_p c_p(n).$$

Then we can evidently write (9.47) as

$$E\left\{\left|\sum_{i=1}^{K}\sum_{l=1}^{Q}\sum_{\kappa=1}^{A}\sum_{m=L_1}^{L_2}\sum_{p=1}^{\mathcal{N}}\overline{w}_{[g]}{}_{i,l}^{\kappa,k}(m)_p c_p(n) y_{i,l}^{\kappa}(n-m)\right.\right.$$
$$\left.\left. -\tilde{z}_k(n)\right|^2\right\}.$$

If we denote $\tilde{y}_{i,l,p}^{\kappa}(n,m) = c_p(n)y_{i,l}^{\kappa}(n-m)$, then (9.48) can be expressed as

$$E\left\{\left(\sum_{i_1=1}^{K}\sum_{l_1=1}^{Q}\sum_{\kappa_1=1}^{A}\sum_{m_1=L_1}^{L_2}\sum_{p_1=1}^{\mathcal{N}}\overline{w}_{[g]}{}_{i_1,l_1}^{\kappa_1,k}(m_1)_{p_1}\right.\right.$$
$$\left.\left.\cdot\tilde{y}_{i_1,l_1,p_1}^{\kappa}(n,m_1) - \tilde{z}_k(n)\right)\tilde{y}_{i_2,l_2,p_2}^{\kappa_2}(n,m_2)\right\} = 0 \qquad (9.55)$$

for $\kappa_2 = 1, ..., A$, $i_2 = 1, ..., K$, $l_2 = 1, ..., Q$, $m_2 = L_1, ..., L_2$, and $p_2 = 1, ..., \mathcal{N}$. It is easily shown that the solution to the problem of obtaining $\overline{w}_{[g]}{}_{i_1,l_1}^{\kappa_1,k}(m_1)_{p_1}$, $p_1 = 1, ..., \mathcal{N}$ from (9.55) is given by

$$\overline{\mathbf{w}}_{[\mathbf{g}]_k} = \mathbf{R}^{\dagger}\mathbf{d}_k \qquad (9.56)$$

where

$$\begin{aligned}
\overline{\mathbf{w}}_{[\mathbf{g}]_k} &= \left[\mathbf{w}_{[\mathbf{g}]_k^{1}}^{T}, ..., \mathbf{w}_{[\mathbf{g}]_k^{A}}^{T}\right]^T \\
\mathbf{w}_{[\mathbf{g}]_k^{\kappa}} &= \left[\mathbf{w}_{[\mathbf{g}]_{1,1}^{\kappa,k}}^{T}, ..., \mathbf{w}_{[\mathbf{g}]_{K,Q}^{\kappa,k}}^{T}\right]^T \\
\mathbf{w}_{[\mathbf{g}]_{j,l}^{\kappa,k}} &= \left[\overline{w}_{[g]}{}_{j,l}^{\kappa,k}(L_1)_1, ..., \overline{w}_{[g]}{}_{j,l}^{\kappa,k}(L_2)_{\mathcal{N}}\right]^T,
\end{aligned}$$

$$\mathbf{R} = \left\langle\tilde{\mathbf{y}}_{\mathbf{c}}^{*}(n)\,\tilde{\mathbf{y}}_{\mathbf{c}}(n)^T\right\rangle_N \qquad (9.57)$$
$$\mathbf{d}_k = \langle\tilde{\mathbf{y}}_{\mathbf{c}}^{*}(n)\tilde{z}_k(n)\rangle_N \qquad (9.58)$$

and we have used the notation $\langle x_n\rangle_N = \frac{1}{N}\sum_{n=1}^{N} x_n$ for an N-sample time average with

$$\begin{aligned}
\tilde{\mathbf{y}}_{\mathbf{c}}(n) &= \left[\breve{\mathbf{y}}_{\mathbf{c}}^{1}(n)^T, ..., \breve{\mathbf{y}}_{\mathbf{c}}^{A}(n)^T\right]^T \\
\breve{\mathbf{y}}_{\mathbf{c}}^{\kappa}(n) &= \left[\mathbf{y}_{\mathbf{c}}{}_{1,1}^{\kappa}(n)^T, ..., \mathbf{y}_{\mathbf{c}}{}_{K,Q}^{\kappa}(n)^T\right]^T \\
\mathbf{y}_{\mathbf{c}}{}_{i,l}^{\kappa}(n) &= \left[\tilde{y}_{i,l,1}^{\kappa}(n,L_1), ..., \tilde{y}_{i,l,\mathcal{N}}^{\kappa}(n,L_2)\right]^T.
\end{aligned}$$

Observe that we used a technique similar to [24] where time-varying moments of the form

$$E\{\tilde{y}^{\kappa_1}_{i_1,l_1,p_1}(n,m_1)\tilde{y}^{\kappa_2}_{i_2,l_2,p_2}(n,m_2)^*\}$$

and

$$E\{\tilde{y}^{\kappa_1}_{i_1,l_1,p_1}(n,m_1)^*\tilde{z}_k(n)\}$$

are estimated by means of instantaneous products

$$\tilde{y}^{\kappa_1}_{i_1,l_1,p_1}(n,m_1)\tilde{y}^{\kappa_2}_{i_2,l_2,p_2}(n,m_2)^*$$

and

$$\tilde{y}^{\kappa_1}_{i_1,l_1,p_1}(n,m_1)^*\tilde{z}_k(n),$$

respectively, which are then collected for $n = 1, 2, ..., N$, and arranged to form an overdetermined system of linear equations in $\overline{\mathbf{w}}_{[\mathbf{g}]_k}$. Observe also that the first step of the super-exponential algorithm becomes

$$\tilde{\mathbf{w}}_k^{[1]}(n) = \overline{\mathbf{c}}(n)^T\, \mathbf{R}^\dagger\, \mathbf{d}_k, \tag{9.59}$$

while the second step can be in practice performed by scaling $\tilde{\mathbf{w}}_k^{[1]}(n)$ [2, 22].

9.8 Key Points of This Chapter

A new multipath fading channel representation based on wavelets allowed a new formulation of the CDMA detection problem in frequency-selective rapidly fading channels. The channel time variations were decomposed using optimal unconditional bases such as orthonormal wavelet bases. The use of these expansions is well motivated by the excellent time-frequency localization properties of the wavelet functions. Particularly the time-localization allows a parsimonious, that is, economical while still accurate, representation of the channel dynamics over the time of observation. This can not possibly be achieved using traditional time-only or frequency-only processing techniques. Using the new model, we have formulated the generalized likelihood detection statistics which, however, present an unmanageable complexity. As a practical solution we focused on a multiscale time-variant adaptive filter characterized by the same dynamic features exhibited by the channel and we proved the approach extremely effective with respect to the traditional MMSE multiuser receiver. Simulation results and hardware experiments have been shown to demonstrate that the approach can be more effective than traditional algorithms in time-varying channels characterized by large Doppler spread. We have also shown that the CDMA system can be reconnected to a multichannel time-varying deconvolution problem when the channel is possibly non-minimum-phase and the input distribution is non-Gaussian. The fundamental contributions of the chapter are summarized as follows:

- A novel practical approach to MIMO blind time-variant deconvolution based on wavelet basis expansion was introduced.
- The application of the new method to the DS/CDMA detection problem was detailed.
- An important observation was made: Multiscale decomposition imposes a subspace constraint to the dynamics of any deconvolution filter and appears to accelerate dramatically the convergence and tracking capability of any blind or non-blind adaptive algorithm.

References

[1] E. Dahlamn et al., "WCDMA-The Radio Interface for Future Mobile Multimedia Communications," *IEEE Trans. Veh. Tech.* Vol. 47, pp. 1105–1118, Nov. 1998.

[2] M. Martone, "An Adaptive Algorithm for Adaptive Antenna Array Low-Rank Processing in Cellular TDMA Base-Stations," *IEEE Trans. Comm.*, Vol. 46, No. 5, May 1998.

[3] M. Martone, "Cumulant-Based Adaptive Multichannel Filtering for Wireless Communication Systems with Multipath RF Propagation Using Antenna Arrays," *IEEE Trans. Veh. Tech.*, Vol. 47, No. 2, pp. 377–391, May 1998.

[4] M. Martone, "On MMSE Real-Time Antenna Array Processing Using Fourth-Order Statistics in the US Cellular TDMA System," *IEEE Journal Sel. Areas in Comm.*, Vol. 16, No. 8, pp. 1396–1410, Oct. 1998.

[5] S. Verdu, *Multiuser Detection*, New York: Cambridge University Press, 1998.

[6] H. Liu and M. D. Zoltowski "Blind Equalization in Antenna Array CDMA Systems," *IEEE Trans. Sig. Proc.*, Vol. 45, pp. 161–172, Jan. 1997.

[7] L. Van Trees, *Detection, Estimation, and Modulation Theory,* Part I, New York: Wiley, 1968.

[8] J. Zhang and G. Walter, "A Wavelet-Based KL-Like Expansion for Wide-Sense Sstationary Random Processes," *IEEE Trans. Sig. Proc.*, Vol. 42, No. 7, pp. 1737–1745, July 1994.

[9] R. W. Dijkerman and R. R. Mazumdar, "Wavelet Representations of Stochastic Processes and Multiresolution Stochastic Models," *IEEE Trans. Sig. Proc.*, Vol. 42, No. 7, pp. 1640–1652, July 1994.

[10] P. P. Vaidyanathan, *Multirate Systems and Filter Banks*, Upper Saddle River, NJ: Prentice Hall, 1993.

[11] O. Rioul, "A Discrete-Time Multiresolution Theory," *IEEE Trans. Sig. Proc.*, Vol. 41, No. 8, pp. 2591–2606, Aug. 1993.

[12] I. Daubechies, *Ten Lectures in Wavelets*, New York: SIAM, 1992.

[13] I. Daubechies, "Orthogonal Bases of Compactly Supported Wavelets" *Commun. Pure Appl. Math.*, Vol. 41, pp. 909–996, 1988.

[14] J. G. Proakis, *Digital Communications*, New York: McGraw-Hill, 1989.

[15] P. A. Bello, "Characterization of Randomly Time Variant Linear Channels," *IEEE Trans. Comm. Syst. Tech.*, Vol. 11, pp. 360–393, Dec. 1963.

[16] L. L. Scharf, *Statistical Signal Processing*, Reading, MA: Addison Wesley, 1991.

[17] M. Martone, "Wavelet-Based Separating Kernels for Sequence Estimation with Unknown Rapidly Time-Varying Channels," *IEEE Comm. Lett.*, March 1999.

[18] P. Monsen, "Theoretical and Measured Performance of a DFE Modem on a Fading Multipath Channel," *IEEE Trans. Comm.*, Vol. 25, pp. 1144–1153, Oct. 1977.

[19] M. Stojanovic, J. G. Proakis, and J. A. Catipovic, "Analysis of the Impact of Channel Estimation Errors on the Performance of a Decision-Feedback Equalizer in Fading Multipath Channels," *IEEE Trans. Comm.* Vol. 43, pp. 877–886, Feb./March/April 1995.

[20] W. C. Jakes, Jr., *Microwave Mobile Communications,* New York: Wiley, 1974.

[21] D. L. Donoho, "Unconditional Bases Are Optimal Bases for Data Compression and for Statistical Estimation," *App. Comp. Harmonic Anal.*, pp. 100–115, Dec. 1993.

[22] M. Martone, "On-board Regeneration of Uplink Signals Using a Blind Adaptive Multichannel Estimator," *IEEE Trans. Aerospace Electronic Systems*, Vol. 34, No. 1, pp. 49–62, Jan. 1998.

[23] G. B. Giannakis and C. Tepedelenlioglu, "Basis-Expansion Models and Diversity Techniques for Blind Identification and Equalization of Time-Varying Channels," *Proc. of the IEEE*, Vol. 86, No. 10, pp. 1969–1986, Oct. 1998.

[24] M. K. Tsatsanis and G. B. Giannakis, "Equalization of Rapidly Fading Channels: Self-Recovering Methods," *IEEE Trans. Comm.*, Vol. 44, pp. 619–630, May 1996.

About the Author

Massimiliano "Max" Martone was born in Rome, Italy, and received a doctor in electronic engineering degree in 1990 from the University of Rome, La Sapienza. In Europe he worked for the Italian Air Force and the On Board Equipment Division of Alenia Spazio, where he was involved in the design of satellite receivers and spread spectrum transponders for the European Space Agency. When he moved to the United States, he was appointed visiting scientist at the Rensselaer Polytechnic Institute in Troy, New York. He has worked for many years at WJ Communications (formerly the Watkins-Johnson Company), where he is the head of advanced wireless development for the coordination of digital hardware and digital signal processing design for cellular digital base station transceivers, acting as a leader in the development of several proprietary signal processing algorithms and digital modems.

Dr. Martone has consulted for several communications companies in the United States and Europe and has published more than 100 papers in journals and proceedings of international conferences. He holds several patents and continues to dedicate most of his professional life to the development of systematic methodologies for the design and implementation of advanced communication systems.

Index

Recent Titles in the Artech House Mobile Communications Series

John Walker, Series Editor

Advances in Mobile Information Systems, John Walker, editor

CDMA for Wireless Personal Communications, Ramjee Prasad

CDMA Mobile Radio Design, John B. Groe and Lawrence E. Larson

CDMA RF System Engineering, Samuel C. Yang

CDMA Systems Engineering Handbook, Jhong S. Lee and Leonard E. Miller

Cell Planning for Wireless Communications, Manuel F. Cátedra and Jesús Pérez-Arriaga

Cellular Communications: Worldwide Market Development, Garry A. Garrard

Cellular Mobile Systems Engineering, Saleh Faruque

The Complete Wireless Communications Professional: A Guide for Engineers and Managers, William Webb

Emerging Public Safety Wireless Communication Systems, Robert I. Desourdis, Jr., et al.

The Future of Wireless Communications, William Webb

GSM and Personal Communications Handbook, Siegmund M. Redl, Matthias K. Weber, and Malcolm W. Oliphant

GSM Networks: Protocols, Terminology, and Implementation, Gunnar Heine

GSM System Engineering, Asha Mehrotra

Handbook of Land-Mobile Radio System Coverage, Garry C. Hess

Handbook of Mobile Radio Networks, Sami Tabbane

High-Speed Wireless ATM and LANs, Benny Bing

Introduction to 3G Wireless Communications, Juha Korhonen

An Introduction to GSM, Siegmund M. Redl, Matthias K. Weber, and Malcolm W. Oliphant

Introduction to Mobile Communications Engineering, José M. Hernando and F. Pérez-Fontán

Introduction to Radio Propagation for Fixed and Mobile Communications, John Doble

Introduction to Wireless Local Loop, Second Edition: Broadband and Narrowband Systems, William Webb

IS-136 TDMA Technology, Economics, and Services, Lawrence Harte, Adrian Smith, and Charles A. Jacobs

Mobile Data Communications Systems, Peter Wong and David Britland

Mobile IP Technology for M-Business, Mark Norris

Mobile Satellite Communications, Shingo Ohmori, Hiromitsu Wakana, and Seiichiro Kawase

Mobile Telecommunications Standards: GSM, UMTS, TETRA, and ERMES, Rudi Bekkers

Mobile Telecommunications: Standards, Regulation, and Applications, Rudi Bekkers and Jan Smits

Multiantenna Digital Radio Transmission, Massimiliano "Max" Martone

Personal Wireless Communication with DECT and PWT, John Phillips and Gerard Mac Namee

Practical Wireless Data Modem Design, Jonathon Y. C. Cheah

Radio Propagation in Cellular Networks, Nathan Blaunstein

Radio Resource Management for Wireless Networks, Jens Zander and Seong-Lyun Kim

RDS: The Radio Data System, Dietmar Kopitz and Bev Marks

Resource Allocation in Hierarchical Cellular Systems, Lauro Ortigoza-Guerrero and A. Hamid Aghvami

RF and Microwave Circuit Design for Wireless Communications, Lawrence E. Larson, editor

Signal Processing Applications in CDMA Communications, Hui Liu

Spread Spectrum CDMA Systems for Wireless Communications, Savo G. Glisic and Branka Vucetic

Understanding Cellular Radio, William Webb

Understanding Digital PCS: The TDMA Standard, Cameron Kelly Coursey

Understanding GPS: Principles and Applications, Elliott D. Kaplan, editor

Understanding WAP: Wireless Applications, Devices, and Services, Marcel van der Heijden and Marcus Taylor, editors

Universal Wireless Personal Communications, Ramjee Prasad

WCDMA: Towards IP Mobility and Mobile Internet, Tero Ojanperä and Ramjee Prasad, editors

Wireless Communications in Developing Countries: Cellular and Satellite Systems, Rachael E. Schwartz

Wireless Intelligent Networking, Gerry Christensen, Paul G. Florack, and Robert Duncan

Wireless LAN Standards and Applications, Asunción Santamaría and Francisco J. López-Hernández, editors

Wireless Technician's Handbook, Andrew Miceli

For further information on these and other Artech House titles, including previously considered out-of-print books now available through our In-Print-Forever® (IPF®) program, contact:

Artech House
685 Canton Street
Norwood, MA 02062
Phone: 781-769-9750
Fax: 781-769-6334
e-mail: artech@artechhouse.com

Artech House
46 Gillingham Street
London SW1V 1AH UK
Phone: +44 (0)20 7596-8750
Fax: +44 (0)20 7630-0166
e-mail: artech-uk@artechhouse.com

Find us on the World Wide Web at:
www.artechhouse.com